AF849995

VOLUME ONE HUNDRED AND EIGHTY SEVEN

Progress in MOLECULAR BIOLOGY AND TRANSLATIONAL SCIENCE

Micro/Nanofluidics and Lab-on-Chip Based Emerging Technologies for Biomedical and Translational Research Applications - Part B

VOLUME ONE HUNDRED AND EIGHTY SEVEN

PROGRESS IN MOLECULAR BIOLOGY AND TRANSLATIONAL SCIENCE

Micro/Nanofluidics and Lab-on-Chip Based Emerging Technologies for Biomedical and Translational Research Applications - Part B

Edited by

ALOK PANDYA

Department of Engineering and Physical Sciences, Institute of Advanced Research, Gandhinagar, Gujarat, India

VIJAI SINGH

Department of Biosciences, School of Science, Indrashil University, Rajpur, Mehsana, Gujarat, India

ELSEVIER

ACADEMIC PRESS

An imprint of Elsevier

Academic Press is an imprint of Elsevier
50 Hampshire Street, 5th Floor, Cambridge, MA 02139, United States
525 B Street, Suite 1650, San Diego, CA 92101, United States
The Boulevard, Langford Lane, Kidlington, Oxford OX5 1GB, United Kingdom
125 London Wall, London EC2Y 5AS, United Kingdom

First edition 2022

Copyright © 2022 Elsevier Inc. All rights reserved.

No part of this publication may be reproduced or transmitted in any form or by any means, electronic or mechanical, including photocopying, recording, or any information storage and retrieval system, without permission in writing from the publisher. Details on how to seek permission, further information about the Publisher's permissions policies and our arrangements with organizations such as the Copyright Clearance Center and the Copyright Licensing Agency, can be found at our website: www.elsevier.com/permissions.

This book and the individual contributions contained in it are protected under copyright by the Publisher (other than as may be noted herein).

Notices
Knowledge and best practice in this field are constantly changing. As new research and experience broaden our understanding, changes in research methods, professional practices, or medical treatment may become necessary.

Practitioners and researchers must always rely on their own experience and knowledge in evaluating and using any information, methods, compounds, or experiments described herein. In using such information or methods they should be mindful of their own safety and the safety of others, including parties for whom they have a professional responsibility.

To the fullest extent of the law, neither the Publisher nor the authors, contributors, or editors, assume any liability for any injury and/or damage to persons or property as a matter of products liability, negligence or otherwise, or from any use or operation of any methods, products, instructions, or ideas contained in the material herein.

ISBN: 978-0-323-85303-3
ISSN: 1877-1173

For information on all Academic Press publications
visit our website at https://www.elsevier.com/books-and-journals

Publisher: Zoe Kruze
Acquisitions Editor: Ashlie M. Jackman
Developmental Editor: Jhon Michael Peñano
Production Project Manager: James Selvam
Cover Designer: Matthew Limbert

Typeset by STRAIVE, India

Working together to grow libraries in developing countries

www.elsevier.com • www.bookaid.org

Contents

Contributors

Khalid J. Alzahrani
Department of Clinical Laboratories Sciences, College of Applied Medical Sciences, Taif University, Taif, Saudi Arabia

T.G. Satheesh Babu
Amrita Biosensor Research Lab; Department of Sciences, Amrita School of Engineering, Amrita Vishwa Vidyapeetham, Coimbatore, India

Stuti Bhagat
Nanomaterials and Toxicology Laboratory, Biological and Life Sciences, School of Arts and Sciences, Ahmedabad University, Ahmedabad, Gujarat; DBT-National Institute of Animal Biotechnology (DBT-NIAB), Hyderabad, Telangana, India

Gargi Bhattacharjee
Department of Biosciences, School of Science, Indrashil University, Rajpur, Mehsana, Gujarat, India

Cesare M. Cejas
Microfluidics, MEMS, Nanostructures (MMN), CNRS UMR 8231, Institut Pierre Gilles de Gennes (IPGG) ESPCI Paris, PSL Research University, Paris, France

Gianaurelio Cuniberti
Institute for Materials Science and Max Bergmann Center for Biomaterials; Center for Advancing Electronics Dresden (CFAED), Technische Universität Dresden, Dresden, Germany

Nisarg Gohil
Department of Biosciences, School of Science, Indrashil University, Rajpur, Mehsana, Gujarat, India

Bergoi Ibarlucea
Institute for Materials Science and Max Bergmann Center for Biomaterials; Center for Advancing Electronics Dresden (CFAED), Technische Universität Dresden, Dresden, Germany

Andrea Idili
Department of Chemistry and Biochemistry, University of California, Santa Barbara, CA, United States; Department of Chemical Science and Technologies, University of Rome, Tor Vergata, Rome, Italy

Itthipon Jeerapan
Division of Physical Science, Faculty of Science; Center of Excellence for Trace Analysis and Biosensor, Prince of Songkla University, Hat Yai, Songkhla, Thailand

Proespichaya Kanatharana
Division of Physical Science, Faculty of Science; Center of Excellence for Trace Analysis and Biosensor, Prince of Songkla University, Hat Yai, Songkhla, Thailand

Navid Kashaninejad
Queensland Micro- and Nanotechnology Centre, Nathan Campus, Griffith University, Brisbane, QLD, Australia

Rutuparna Kulkarni
Department of Biological Sciences and Biotechnology, Institute of Advanced Research, Gandhinagar, Gujarat, India

Jerome Lacombe
Center for Applied Nanobioscience and Medicine; Department of Basic Medical Sciences, College of Medicine-Phoenix, The University of Arizona, Phoenix, AZ, United States

Navya L. Lam
The J. David Gladstone Institutes, San Francisco, CA, United States

Bonolo S.P. Mathekga
Drizzle Health LLC, Baltimore, MD, United States

Rupesh Maurya
Department of Biosciences, School of Science, Indrashil University, Rajpur, Mehsana, Gujarat, India

Mariana Medina-Sánchez
Institute for Integrative Nanosciences, Leibniz IFW Dresden, Dresden, Germany

Hajar Moghadas
Department of Mechanical Engineering, Gas and Petroleum Faculty, Yasouj University, Yasuj, Iran

Helena Montón
Department of Chemistry and Biochemistry, University of California, Santa Barbara, CA, United States

Chochanon Moonla
School of Chemistry, Institute of Science, Suranaree University of Technology, Nakhon Ratchasima, Thailand

Ehsanollah Moradi
Department of Biomedical Engineering, Amirkabir University of Technology, Tehran, Iran

Zandile Nxumalo
Synthetic Nanobiotechnology and Biomachines Group, Synthetic Biology and Precision Medicine Centre, CSIR, Pretoria, South Africa

Alok Pandya
Department of Engineering and Physical Sciences, Institute of Advanced Research, Gandhinagar, Gujarat, India

Claudio Parolo
Department of Chemistry and Biochemistry, University of California, Santa Barbara, CA, United States; Barcelona Institute for Global Health (ISGlobal) Hospital Clínic, Barcelona, Spain

Gabriele Pitingolo
Bioassays, Microsystems and Optical Engineering Unit, BIOASTER, Paris, France

Kevin W. Plaxco
Department of Chemistry and Biochemistry; Interdepartmental Program in Biomolecular Science and Engineering University of California, Santa Barbara, CA, United States

Aarathi Pradeep
Amrita Biosensor Research Lab; Department of Sciences, Amrita School of Engineering, Amrita Vishwa Vidyapeetham, Coimbatore, India

Suresh Ramakrishna
Graduate School of Biomedical Science and Engineering; College of Medicine, Hanyang University, Seoul, South Korea

Jeethu Raveendran
Amrita Biosensor Research Lab, Amrita School of Engineering, Amrita Vishwa Vidyapeetham, Coimbatore, India

Maria Russo
Microfluidics, MEMS, Nanostructures (MMN), CNRS UMR 8231, Institut Pierre Gilles de Gennes (IPGG) ESPCI Paris, PSL Research University, Paris, France

Oliver G. Schmidt
Institute for Integrative Nanosciences, Leibniz IFW Dresden; School of Science, TU Dresden, Dresden; Research Center for Materials, Architectures and Integration of Nanomembranes (MAIN), Chemnitz, Germany

Sanjay Singh
Nanomaterials and Toxicology Laboratory, Biological and Life Sciences, School of Arts and Sciences, Ahmedabad University, Ahmedabad, Gujarat; DBT-National Institute of Animal Biotechnology (DBT-NIAB), Hyderabad, Telangana, India

Vijai Singh
Department of Biosciences, School of Science, Indrashil University, Rajpur, Mehsana, Gujarat, India

Maria Soldevila
Center for Applied Nanobioscience and Medicine, Phoenix, AZ, United States

Panote Thavarungkul
Division of Physical Science, Faculty of Science; Center of Excellence for Trace Analysis and Biosensor, Prince of Songkla University, Hat Yai, Songkhla, Thailand

Deepak B. Thimiri Govinda Raj
Synthetic Nanobiotechnology and Biomachines Group, Synthetic Biology and Precision Medicine Centre, CSIR, Pretoria, South Africa

Nidhi Verma
Department of Engineering and Physical Sciences, Institute of Advanced Research, Gandhinagar, Gujarat, India

Frederic Zenhausern
Center for Applied Nanobioscience and Medicine; Department of Basic Medical Sciences, College of Medicine-Phoenix, The University of Arizona, Phoenix, AZ, United States

Preface

Microfluidics is a rapidly growing area of science and is being used in a wide range of applications from biology to medicine and advanced biotechnology. Microfluidic systems are prevalent in many areas of science given their advantage in miniaturization, development of unique tools for disease diagnosis, and biomolecule separation, as well as detection followed by screening. Lately, the relevance of microfluidics has been quite significant, considering its range of features, which leads to system miniaturization. The microfluidic toolset has influenced many areas of science by developing lab-on-a-chip, organ-on-a-chip, and also advanced drug delivery systems. Microfluidics and lab-on-a-chip technologies have significantly impacted the research in clinical diagnostics outside of laboratories, in particular, at the point of care (POC).

This has led to the development of a series of new components and adoption of innovative techniques, such as pumping, mixing, and introducing, as well as storing fluids in the channels of microfluidic devices. Despite rapid progress in this area, there remain several unresolved issues, such as suitable system integration, portability, and versatility. In this volume, we provide a comprehensive overview of recent advances in nano/microfluidic technologies in terms of design, fabrication, system integration and development, and application for personalized health care.

A pressing need has arisen globally to devise more advanced and next-generation micro/nanofluidic devices and systems for accelerating biomedical, biotechnological, and translational research. This volume covers various advanced applications of microfluidics in health care, high-throughput drug screening, 3-D cell culture, organ-on-a-chip, and real-time monitoring of physiological parameters for noninvasive diagnostics. This book focuses on the emerging demands/research of micro/nanofluidic-based devices in different fields. It also highlights a number of aspects of micro/nanofluidic-based miniaturized lab-on-a-chip devices in a manner that can help future investigators, researchers, students, and stakeholders to perform their research with greater ease. The book will be an excellent basis from which scientific knowledge would grow and widen in the field of micro/nanofluidics and lab-on-a-chip.

Alok Pandya
Vijai Singh

CHAPTER ONE

Design and fabrication of microfluidics devices for molecular biology applications

Rupesh Maurya[a], Gargi Bhattacharjee[a], Nisarg Gohil[a], Navya L. Lam[b], Khalid J. Alzahrani[c], and Vijai Singh[a,*]
[a]Department of Biosciences, School of Science, Indrashil University, Rajpur, Mehsana, Gujarat, India
[b]The J. David Gladstone Institutes, San Francisco, CA, United States
[c]Department of Clinical Laboratories Sciences, College of Applied Medical Sciences, Taif University, Taif, Saudi Arabia
*Corresponding author: e-mail address: vijaisingh15@gmail.com

Contents

Abstract

In the past decade, microfluidics has emerged as a rapidly growing area with potential to reduce cost and reagent consumption. It has been used for detection of nucleic acids and high-throughput screening of cells and metabolites. It is extensively used for extraction of DNA, RNA, proteins, biomolecules, as well as for cloning and transformation of plasmid into cells. Microfluidics is made up of polydimethylsiloxane (PDMS) polymer which is transparent and is used for preparation of a wide range of devices and systems. In this chapter, we discuss advances and challenges of using microfluidics in molecular biology and its biomedical applications.

1. Introduction

Microfluidics is a rapidly growing area and one of the powerful tools with a wide range of applications in basic sciences to medical as well as translational research. It has been applied to a number of uses such as for PCR,

Progress in Molecular Biology and Translational Science, Volume 187
ISSN 1877-1173
https://doi.org/10.1016/bs.pmbts.2021.07.017
Copyright © 2022 Elsevier Inc.
All rights reserved.

cloning, diagnosis, drug screening, *in vitro* toxicity, monitoring of single cell dynamics and many more. Microfluidics chips are made up of polydimethylsiloxane (PDMS) or dimethicone. PDMS is the most widely used silicon-based polymer. It is optically clear, inert, non-flammable and non-toxic in nature.[1,2] Microfluidics is widely used for extraction of DNA and its detection, RNA, proteins, biomolecules, cloning and transformation of plasmid into cells, etc.[3]

Microfluidics has the potential to reduce the process cost and reagent consumption. Microfluidics-based miniaturized lab-on-chip (LOC) devices are an emerging technology aiding the improvement of human health. Microfluidics-based devices offer a powerful platform for automation and integration of complex operations onto a single chip with high resolution and sensitivity. With these excellent properties, in recent years, micro/nanofluidics-based devices have been explored for various applications in healthcare, smart contact lens sensors, real-time monitoring of physiological parameters from tear fluid for non-invasive diagnosis, tattoo-based sensors and space sciences.[4] This chapter highlights the emerging demands and research of microfluidics-based devices in different fields.

2. Design and fabrication of microfluidics chip

An AutoCAD is a tool for basic 2D sketching and applied for the design of microfluidics chips. This design is printed on silicon wafer with at least 1200 dpi. Chips are fabricated using a mold created by soft photolithography and printed directly onto photomask (transparent film of silicon wafer). Wafer is used for the development of chips using PDMS and curing agent that is properly mixed together and centrifuged at 4000 rpm for 10 min so as to remove air bubbles if any. Wafers can be placed on a big sized petri plate. Subsequently, the mixture of PDMS and curing agent is poured on top of the wafer and kept at 80 °C for 1 h to allow polymerization.[1] Following that, the solid transparent polymer of PDMS is removed from the wafer using a sharp knife to cut properly and detached as a whole unit, which is subsequently used for loading culture and fresh media for growing cells on the chip. Solid polymer and glass slide is kept under plasma activation for 30 s, then both are sticked together and kept overnight to have proper bonding between PDMS and glass slide. Once the chip is ready, it can be used for loading the microbial culture and visualize under a microscope to obtain phase contrast and fluorescent images for monitoring cell behavior. Fig. 1 shows (A) a designed and fabricated chip. (B) single channel microfluidics chip designed using PDMS.[5]

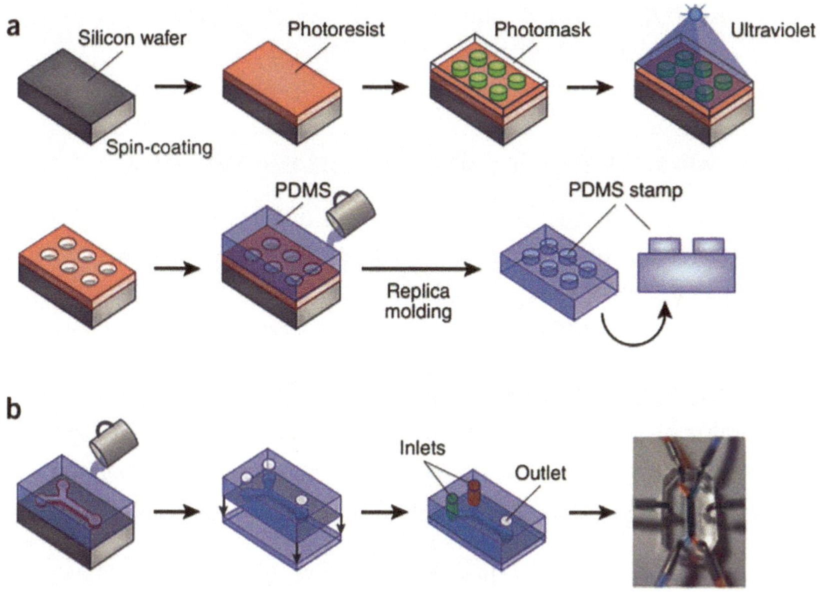

Fig. 1 Microfluidics chip fabrication methods. (A) Chip was designed and fabricated. (B) Single channel microfluidics chip was designed using PDMS. *Figure reproduced with permission from Bhatia SN, Ingber DE. Microfluidic organs-on-chips.* Nat Biotechnol. *2014;32:760–772 © Springer Nature.*

3. Microfluidics for extraction of nucleic acids

Extraction of nucleic acids is one of the tedious steps in molecular biology. The time and cost of DNA or RNA extraction can be reduced by using high-throughput microfluidics chips. Chip can be designed with different chambers, and with different buffers and solutions, used for moving samples for the extraction of DNA. Brassard et al.[6] designed a chip for extraction of nucleic acid from blood samples based on rotation of liquid samples. It reduced the time and expense of chemicals in extraction of DNA. In a study, Geissler et al.[7] designed automated devices for isolation of bacteria, lysis and PCR amplification of genes for identification of micro-organism. It was found that this device was rapid and efficient for automatic detection of organisms. Very recently, Sullivan et al.[8] developed microfluidics devices for purification of nucleic acids from blood samples using isotachoresis (ITP) that can be used for point-of-care (POC) diagnostics. Nucleic acid was extracted and purified from 33 μL of blood sample within 30 min. Further amplification with recombinase polymerase amplification (RPA) reactions amplified 3×10^3 copies of nucleic acid per mL input blood. This tool has the potential of a single step robustness, low-cost design and can directly process a large number of blood samples.

4. Microfluidics for detection of nucleic acids

In the past decade, nucleic acid-based diagnostics have widely been used for detection of genes for identification of organisms. It has an advantage over immunoassay-based diagnostic assays with regard to specificity and sensitivity. Microfluidics chips have been used for the detection of nucleic acids from blood, clinical specimens, food, water and environmental samples. Such chips have been coupled with solid-phase extraction, isothermal amplification including RPA, LAMP (Loop-mediated AMPlification), NASBA (Nucleic Acid Sequence Based Amplification), and real-time optical detection of DNA or RNA.[9] Kim et al.[3] fabricated a continuous flow (CF) microchip using PDMS for PCR. It has three different temperature zones similar to thermal cyclers that make use of external PID temperature controllers. When PCR reagents were introduced using pump in to the device, the amplified product was obtained within 8–30 min. This study is useful for quick amplification of PCR products for rapid diagnosis. Fig. 2 depicts the design and fabrication of self-propelled continuous-flow PCR (SP-CF-PCR) microfluidics chip. It contains two heat blocks

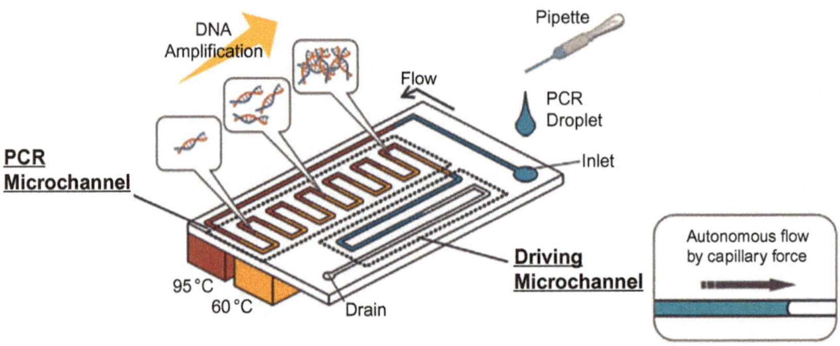

Fig. 2 Design and fabrication of self-propelled continuous-flow PCR (SP-CF-PCR) microfluidics chip with two heat blocks (95 and 60 °C) for maintaining temperature. Sample loaded onto that inlet can move forward through capillary action, reach PCR thermal temperature and amplify the target DNA. *Figure reproduced with permission from Tachibana H, Saito M, Tsuji K, Yamanaka K, Hoa LQ, Tamiya E. Self-propelled continuous-flow PCR in capillary-driven microfluidic device: microfluidic behavior and DNA amplification.* Sens Actuators B. *2015a;206:303–310.*

(95 and 60 °C) for maintaining temperature. When the sample is loaded onto the inlet, it moves forward through capillary action, reaches PCR thermal temperature and amplifies the target DNA sequence.[10]

In another work, Kopp et al.[11] developed a micromachined chemical amplifier for PCR in a continuous flow at high speed. It is used for amplification that is independent of input concentration. A small fragment of DNA gyrase gene (176 bp) of *Neisseria gonorrhoeae* was amplified for 20 cycles with various flow rates within 90 s to 18.7 min. Similarly, Hashimoto et al.[12] fabricated continuous-flow polymerase chain reaction (CF-PCR) devices for rapid DNA amplification. In this device, PCR reagent reaches quickly to different temperature zones with high flow velocities in a spiral 20 cycles device. A 500 bp DNA fragment was observed in 1.7 min (5.2 s per cycle) and 997 bp fragment has been detected within 3.2 min (9.7 s per cycle). Longer time was required due to enzymes Taq DNA polymerase kinetics for incorporating dNTPs. Tachibana et al.[13] developed CF-PCR in microfluidic devices which had the potential to detect several pathogens from various sources rapidly and with great sensitivity. A syringe pump is an essential component in microfluidics. However, in this device, they developed self-propelled CF-PCR, so no external pump was required to control flow. When PCR reagent was added onto the inlet it moved automatically by capillary action. This way, Tachibana's group amplified β-actin (295 bp) from human, *AH1pdm* gene (95 bp) from Influenza virus and 16S rRNA of *E. coli* in less than 14 min of time. Therefore, this study was instrumental in demonstrating rapid amplification of genes for detection of pathogens. Similarly, Li et al.[14] developed CF-PCR microfluidic chips wherein, they integrated pumps for electrophoresis and hybridization all-in-one. It has been used for rapid and sensitive detection of *Porphyromonas gingivalis*, *Treponema denticola* (amplification time was 2 min and 31 s while PCR product detection time was 3 min 43 s) and *Tannerela forsythia*. Therefore, minimum 125 cfu/mL bacteria could be detected. This device has the potential to be applied in POC nucleic acid detection for disease diagnosis.

Schabmueller et al.[15] developed a silicon/Pyrex-based chip for small quantities of DNA detection. Sometimes, it may happen that the amount of DNA available is meager. Therefore, with the help of this pyrex-based chip 1 μL of sample could be used for detection. This has allowed to incorporate heating and sensing elements that enables rapid heating and control of the chip. The study demonstrated that a small amount of sample is enough for DNA detection. In a similar study, Qiu et al.[16] designed a fully disposable heat capillary tube without electric supply for DNA amplification. PCR reagents repeatedly moved through different temperature zones. A single step nucleic acid dipstick assay was adopted to visualize amplification by naked eye that has been used for POC detection of H1N1 virus within 35 min. This device can be a good resource for diagnosis of pathogens, especially in areas with poor diagnostic settings. Li et al.[17] developed a CF-PCR based on microfluidics chips. It was fabricated in a way that automatic injection of flow could be made possible. They used this device to amplify *T. denticola* DNA within 8 min. This study offers rapid and specific amplification of organisms in a short period of time and can be extended in other organisms for quick detection in order to begin with accurate treatment in case of a disease.

5. Microfluidics devices for detection of metabolites and toxins

A metabolite is an intermediate product of metabolic pathways which is catalyzed by several enzymes or proteins. It is naturally produced within cells. Metabolite sensing is equally important for detection of availability of molecules and detecting them. In order to perform high-throughput screening methods, a microfluidics platform plays a key role in detecting metabolites that may be labeled or unlabeled. Kim et al.[18] developed microfluidics for multiplexed sensing of metabolites concentration from blood and non-invasive fluid samples. It has been combined with sampling and transport systems and can be used for routine monitoring of metabolites. Trace amount of primary fatty acid amides present in biological systems that are difficult to quantify and analyze require a highly sensitive system. Now microfluidics have been designed that enable the detection of even trace amounts (fmol) of fluorescently labeled primary fatty acid amines.[19]

Herr et al.[20] developed a microfluidics device for detecting tetanus toxin C-fragment in less than 3 min. Similarly, Meagher et al.[21] developed portable toxins (biotoxins ricin, Shiga toxin I, and Staphylococcal enterotoxin B (SEB)) detecting devices with sample volume less than 10 μL. These devices were able to detect toxins at pmol concentration within 10 min. These devices have the potential to simultaneously detect and quantify toxins. In addition to that Yang et al.[22] designed eight channels LOC carbon nanotube for detecting SEB. Using that, total 8 samples of SEB were assayed simultaneously and observed at 0.1 ng/mL detection limit.

6. Conclusion and future remarks

Microfluidics chips are rapidly fluorishing areas and have the potential to be used for several applications. Rapid high-throughput screening of molecules, metabolites sensing in cells, detection of DNA, RNA, toxin and proteins have been discussed. This technology is further finding applications in fermentation technology for rapid screening and optimization of biomolecules production using a wide range of substrates. Several groups are using chip for culturing and growing organs, cells (for testing toxicity effects), for drug testing, and as regenerative medicine. In the near future, chips are bound to exhibit huge potential and used for a number of biotechnological, diagnostics, and sensing applications.

Acknowledgments

The financial assistance from Gujarat State Biotechnology Mission (GSBTM Project ID: 5LY45F), Gujarat, India to G.B. and V.S. is duly acknowledged. N.G. acknowledges the Indian Council of Medical Research, Government of India for financial support as Senior Research Fellowship (File No. 5/3/8/63/ITR-F/2020).

References

1. Carrera J, Rodrigo G, Singh V, Kirov B, Jaramillo A. Empirical model and *in vivo* characterization of the bacterial response to synthetic gene expression show that ribosome allocation limits growth rate. *Biotechnol J*. 2011;6:773–783.
2. Wolf MP, Salieb-Beugelaar GB, Hunziker P. PDMS with designer functionalities—properties, modifications strategies, and applications. *Prog Polym Sci*. 2018;83:97–134.
3. Kim JA, Lee JY, Seong S, et al. Fabrication and characterization of a PDMS–glass hybrid continuous-flow PCR chip. *Biochem Eng J*. 2006;29:91–97.
4. Cui P, Wang S. Application of microfluidic chip technology in pharmaceutical analysis: a review. *J Pharm Anal*. 2019;9:238–247.
5. Bhatia SN, Ingber DE. Microfluidic organs-on-chips. *Nat Biotechnol*. 2014;32:760–772.
6. Brassard D, Geissler M, Descarreaux M, et al. Extraction of nucleic acids from blood: unveiling the potential of active pneumatic pumping in centrifugal microfluidics for integration and automation of sample preparation processes. *Lab Chip*. 2019;19: 1941–1952.
7. Geissler M, Brassard D, Clime L, et al. Centrifugal microfluidic lab-on-a-chip system with automated sample lysis, DNA amplification and microarray hybridization for identification of enterohemorrhagic *Escherichia coli* culture isolates. *Analyst*. 2020; 145:6831–6845.
8. Sullivan BP, Bender AT, Ngyuen DN, Zhang JY, Posner JD. Nucleic acid sample preparation from whole blood in a paper microfluidic device using isotachophoresis. *J Chromatogr B*. 2021;1163:122494.
9. Mauk MG, Liu C, Sadik M, Bau HH. Microfluidic devices for nucleic acid (NA) isolation, isothermal NA amplification, and real-time detection. *Methods Mol Biol*. 2015;1256:15–40.

10. Tachibana H, Saito M, Tsuji K, Yamanaka K, Hoa LQ, Tamiya E. Self-propelled continuous-flow PCR in capillary-driven microfluidic device: Microfluidic behavior and DNA amplification. *Sens Actuators B.* 2015;206:303–310.
11. Kopp MU, Mello AJ, Manz A. Chemical amplification: continuous-flow PCR on a chip. *Science.* 1998;280:1046–1048.
12. Hashimoto M, Chen PC, Mitchell MW, Nikitopoulos DE, Soper SA, Murphy MC. Rapid PCR in a continuous flow device. *Lab Chip.* 2004;4:638–645.
13. Tachibana H, Saito M, Shibuya S, et al. On-chip quantitative detection of pathogen genes by autonomous microfluidic PCR platform. *Biosens Bioelectron.* 2015;74:725–730.
14. Li Z, Ju R, Sekine S, Zhang D, Zhuang S, Yamaguchi Y. All-in-one microfluidic device for on-site diagnosis of pathogens based on an integrated continuous flow PCR and electrophoresis biochip. *Lab Chip.* 2019;19:2663–2668.
15. Schabmueller CGJ, Lee MA, Evans AGR, Brunnschweiler A, Ensell GJ, Leslie DL. Closed chamber PCR chips for DNA amplification. *Eng Sci Educ J.* 2000;9:259–264.
16. Qiu X, Zhang S, Xiang F, et al. Instrument-free point-of-care molecular diagnosis of H1N1 based on microfluidic convective PCR. *Sens Actuators B.* 2017;243:738–744.
17. Li Z, Li Y, Sekine S, et al. Design and fabrication of portable continuous flow PCR microfluidic chip for DNA replication. *Biomed Microdevices.* 2020;22:5.
18. Kim J, Campbell AS, de Ávila BE, Wang J. Wearable biosensors for healthcare monitoring. *Nat Biotechnol.* 2019;37:389–406.
19. Davic A, Cascio M. Development of a microfluidic platform for trace lipid analysis. *Metabolites.* 2021;11:130.
20. Herr AE, Throckmorton DJ, Davenport AA, Singh AK. On-chip native gel electrophoresis-based immunoassays for tetanus antibody and toxin. *Anal Chem.* 2005;77:585–590.
21. Meagher RJ, Hatch AV, Renzi RF, Singh AK. An integrated microfluidic platform for sensitive and rapid detection of biological toxins. *Lab Chip.* 2008;8:2046–2053.
22. Yang M, Sun S, Kostov Y, Rasooly A. Lab-on-a-chip for carbon nanotubes based immunoassay detection of staphylococcal enterotoxin B (SEB). *Lab Chip.* 2010;10:1011–1017.

CHAPTER TWO

Micro/nanofluidic devices for drug delivery

Navid Kashaninejad[a,*], Ehsanollah Moradi[b], and Hajar Moghadas[c]
[a]Queensland Micro- and Nanotechnology Centre, Nathan Campus, Griffith University, Brisbane, QLD, Australia
[b]Department of Biomedical Engineering, Amirkabir University of Technology, Tehran, Iran
[c]Department of Mechanical Engineering, Gas and Petroleum Faculty, Yasouj University, Yasuj, Iran
[*]Corresponding author: e-mail address: n.kashaninejad@griffith.edu.au

Contents

Abstract

Micro/nanofluidic drug delivery systems have attracted significant attention as they offer unique advantages in targeted and controlled drug delivery. Based on the desired application, these systems can be categorized into three different groups: *in vitro, in situ* and *in vivo* microfluidic drug delivery platforms. *In vitro* microfluidic drug delivery platforms are closely linked with the emerging concept of lab-on-a-chip for cell culture studies. These systems can be used to administer drugs or therapeutic agents, mostly at the cellular or tissue level, to find the therapeutic index and can potentially be used for personalized medicine. *In situ* and *in vivo* microfluidic drug delivery platforms are still at the developmental stage and can be used for drug delivery at tissue or organ levels. A famous example of these systems are microneedles that can be used for painless

Progress in Molecular Biology and Translational Science, Volume 187
ISSN 1877-1173
https://doi.org/10.1016/bs.pmbts.2021.07.018
Copyright © 2022 Elsevier Inc.
All rights reserved.

and controllable delivery of drugs or vaccines through human skin. This chapter presents the cutting edge advances in the design and fabrication of *in vitro* microfluidic drug delivery systems that can be used for both cellular and tissue drug delivery. It also briefly discusses the *in situ* drug delivery platforms using microneedles.

1. Microfluidic *in vitro* drug delivery at the cellular level

Drug delivery has different stages in the body, and dissolvable drug molecules go through a process that finally ends up with delivering the drug molecules to the site of interest and taking up the drug by the targeted cells; therefore, drug delivery experiments at the cellular level are of great importance and commonly applied for the study of cytotoxicity testing, drug screening and drug discovery.[1] In the last decade, various *in vitro* cell culture platforms have been developed for drug screening and development practices. Although these conventional methods are cost-efficient, robust, and easy to manipulate, they lack a physiologically-relevant microenvironment, sustained cell viability, and functional stability. Hence, there is a dire need in the drug development pipeline for developing predictive *in vitro* models with high preclinical efficacy that faithfully recapitulate cell-cell interactions and cellular phenotype.[2–6]

Microfluidic systems offer a high surface-to-volume ratio that gives rise to unique surface-dominating phenomena such as slip flow.[7] Surface-dominated forces in microfluidic systems offer significant advantages for drug delivery application. For instance, the laminar and diffusion-based flow in microchannels can precisely control the mechanism of drug release.

Recently, microtechnologies and microfluidic cell culture systems have paved the way for examining complex biologic processes, and with controlling spatiotemporal parameters, they can mimic the intricate cellular microenvironment properties like cell morphology, cell-drug interactions, and chemical concentration gradients.[6,8] Microfluidic devices can offer chemical concentration gradients with great accuracy and sensitivity, ideal for producing controllable drug profiles and evaluating concentration-dependent cellular responses to different drug doses.[1,9–11]

1.1 Microfluidic concentration gradient generators (MCGGs)

Microfluidic concentration gradient generators, unlike traditional platforms, possess extraordinary characteristics like cost-efficient operation, the ability to generate accurate and flexible drug profiles, allow real-time cellular response monitoring, and offer high-throughputs.[1,12] Microfluidic concentration gradient generators are generally divided into diffusion-based and flow-based platforms and have become an essential tool for studying biological phenomena like cell migration, proliferation, chemotaxis, drug screening, and personalized medicine.[1,13,14] The following section overview different cell-based MCGGs for drug testing experiments.

1.1.1 Diffusion-based gradient generators

The diffusion base method's gradient generators mostly consist of a membrane or hydrogel to create an environment suitable for diffusive transport.[12,15] There are two compartments with high and low reagent concentrations in most cases, and the targeted cells are cultured in the bridging channel between the source and the sink.[1] In one study, a shear stress-free MCGG was developed for studying cytoskeleton dynamics molecular pathways and the effect of the Cytochalasin D gradient on MG-63 cellular response.[16] In this study, a profile of Cytochalasin D concentration ranging between 0 and 5 μg/mL was produced in five independent cell chambers, and the concentration-dependent cell responses were examined by measuring cell area and cell eccentricity.[16]

Different biological phenomena like chemotaxis and morphogenesis are driven by essential cell signaling cues that are often delivered by biochemical concentration gradients.[17,18] A microfluidic biochip was developed in one study for generating diffusion-based orthogonal linear concentration gradients in a 3D hydrogel-based scaffold in two different designs: Simultaneous and Sequential (Fig. 1A).[18] It was demonstrated that the two orthogonal gradients of retinoic acid and smoothened agonist emulated the localized differentiation of stem cells, and the neural tube was preferentially developed in the high morphogens concentration regions. Moreover, HT1080, as a highly metastatic fibrosarcoma cell line, was exposed to a rotating gradient of Fetal Bovine Serum (FBS) for probing chemotactic response time and adaptation in the tumor microenvironment. The direction of cell migration and cell's migratory behavior was evaluated by comparing average cell displacements, distribution, and migration speed.[18]

In work done by Kilinc et al., a diffusion-based microchip was constructed to establish multiple compounds linear concentration gradient over different cells for cell-based drug combination assays.[19] The microfluidic tool consisted of a cell culture chamber and parallel flow channels for creating a linear concentration gradient across the chamber. Producing orthogonally-aligned gradients of two compounds, flowing concentrated

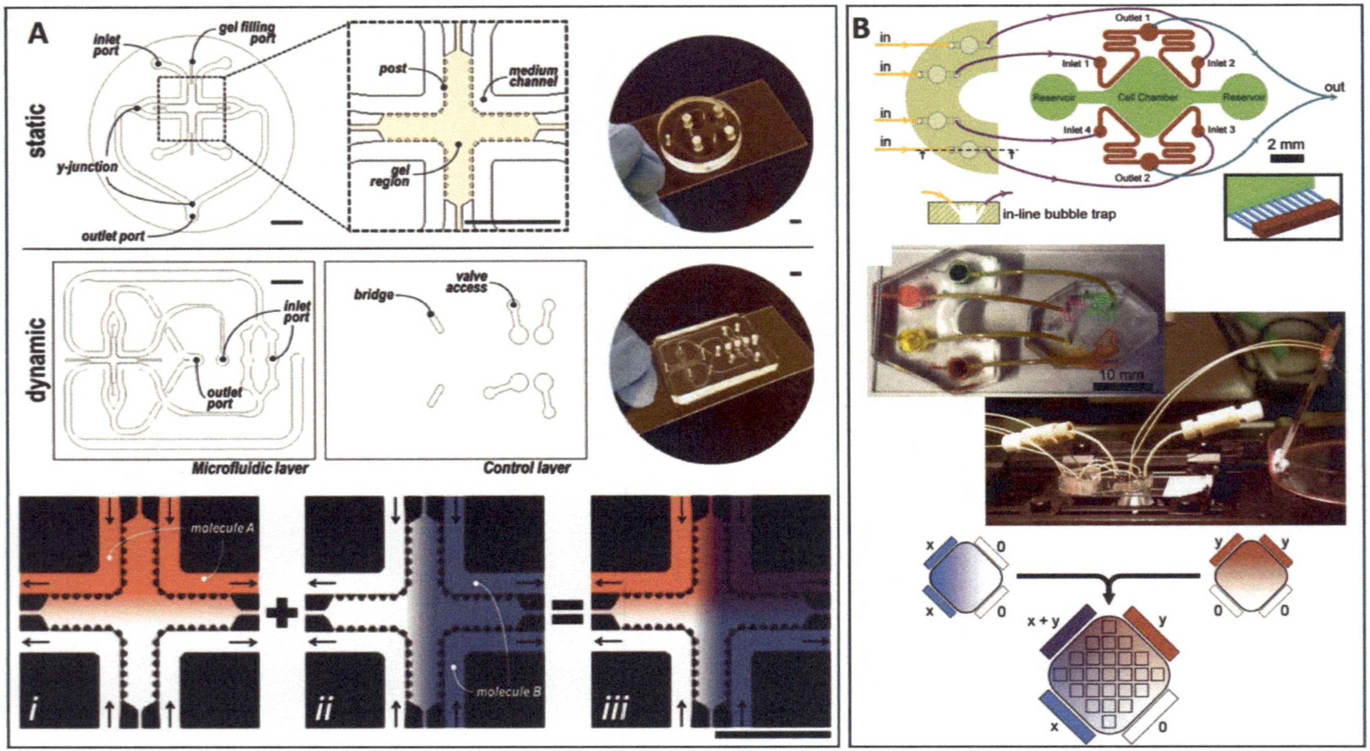

Fig. 1 See figure legend on opposite page.

solutions in two perpendicular directions is needed (Fig. 1B). A431 epidermoid carcinoma cell line responds dissimilarly to EGF concentrations, and while low concentration of EGF induces cell proliferation, higher concentrations result in cell apoptosis. This device generates orthogonal linear gradients of EGFR and MEK inhibitors, and the resultant cell apoptosis gradient in the MCGG demonstrated the ability of the device to determine the combinatorial cell response to specific signaling pathways.[19]

1.1.2 Flow-based gradient generators

In the last decade, microfluidic flow-based gradient generators have revolutionized a wide range of chemical and biological practices and produced biochemical concentration gradients for cellular monitoring, and drug screening applications have been one of them.[1,10,20] MCGGs based on laminar flow can provide desired ranges of biochemical concentration gradients with high precision and controllability by manipulating convective flow.[12,21] In one study, a microfluidic gradient device was utilized for drug screening and disease modeling with evaluating iPSC-derived motoneurons responses to the riluzole concentration gradient.[22] The Microfluidic biochip was constructed with micropillar arrays for trapping uniform-sized neurospheres and promoting neurite network formation. The peripheral channels were connected to the central flow channel through microgrooves that generated a stable concentration gradient with a combination of convection and diffusion. To examine the effect of drug concentration gradient on neurospheres, cells were treated with thapsigargin, a neurotoxin that promotes neuronal death and is a model reagent for neurodegenerative diseases. A concentration gradient of riluzole as a drug to alleviate thapsigargin's neurotoxicity was produced by injecting drugged and drug-free culture

Fig. 1 Design and operation principle of diffusion-based microfluidic orthogonal gradient generators. (A) Static and dynamic versions of the microfluidic gradient generator for producing a combinatorial concentration gradient of morphogens to simulate motor neurons localized differentiation in the process of neural tube development and probing chemotactic response time and adaptation in the tumor microenvironment. (B) A microfluidic device for generating linear orthogonally-aligned gradients of MEK and EGFR inhibitors and analyzing the A431 epidermoid carcinoma cells apoptosis pathway. *Panel (A) reproduced with permission from Uzel SG, Amadi OC, Pearl TM, Lee RT, So PT, Kamm RD. Simultaneous or sequential orthogonal gradient formation in a 3D cell culture microfluidic platform. Small (Weinheim an der Bergstrasse, Germany). 2016;12(5):612–622. Panel (B) reproduced with permission from Kilinc D, Schwab J, Rampini S, et al. A microfluidic dual gradient generator for conducting cell-based drug combination assays.* Integr. Biol. *2015;8(1):39–49.*

mediums into different inlets. It was shown that riluzole promoted neuro-rescue or neuroprotection in the regions with higher drug concentration, while thapsigargin induced considerable damage to neurospheres in the dead zone.[22]

Christmas tree-like gradient generators have been among the most commonly used tools for creating biochemical concentration gradients. In these flow-based microdevices, two reagents with different concentrations combine at repeated T-junctions, and the desired concentration gradient is generated in a direction perpendicular to the fluid flow by diffusion across the laminar flow interface.[10,23,24] In one study by Tang et al., a bone-on-a-chip microdevice was developed for recapitulating bone tissue microenvironment.[25] A hydroxyapatite-PDMS microfluidic concentration gradient chip was fabricated by the stereolithographic printing process, and it contained a ceramic layer and a PDMS cover with Christmas tree-like structure. In high media flow rates, the reagents had inadequate time to diffuse and mix uniformly and could cause excessive shear stress damage to cells, while in low flow rates, there was a lack of sufficient drive to push the fluid forward. To determine the ideal flow rate to generate the appropriate concentration range, a COMSOL flow analysis was applied (Fig. 2A).[25] To examine the device's functionality for bone-related drug studies, mouse osteosarcoma cells (UMR-106) were cultured in the device, and a concentration gradient of doxorubicin as a primary antitumor drug for the treatment of osteosarcoma was produced by the Christmas tree gradient generator. The confocal microscopy examinations showed that the number of UMR-106 cells in the region with high drug concentration decreased significantly. Moreover, the DOX concentration in the zone with half of the control group cell population was measured to determine the IC50 of DOX, and the results were validated with the conventional culture plate model.[25]

Shimizu et al. created an ECM-based MCGG from transglutaminase enzymatically cross-linked gelatin with a porous nature that permitted interstitial flow and generated a culture surface with a controlled biochemical gradient on the surface.[26] To evaluate the applicability of the device for drug screening purposes, the response of HUVECs to histamine concentration gradient induced by interstitial flow was examined (Fig. 2B). Endothelial cells have a contracting response to histamine introduction to increase the barrier permeability of the vessel. In this study, the cellular area of the endothelial cells before and after the histamine treatment was measured as the indication of cellular shrinkage, and HUVECs cultured in the high histamine concentration region shrunk significantly greater than the histamine-free region.[26]

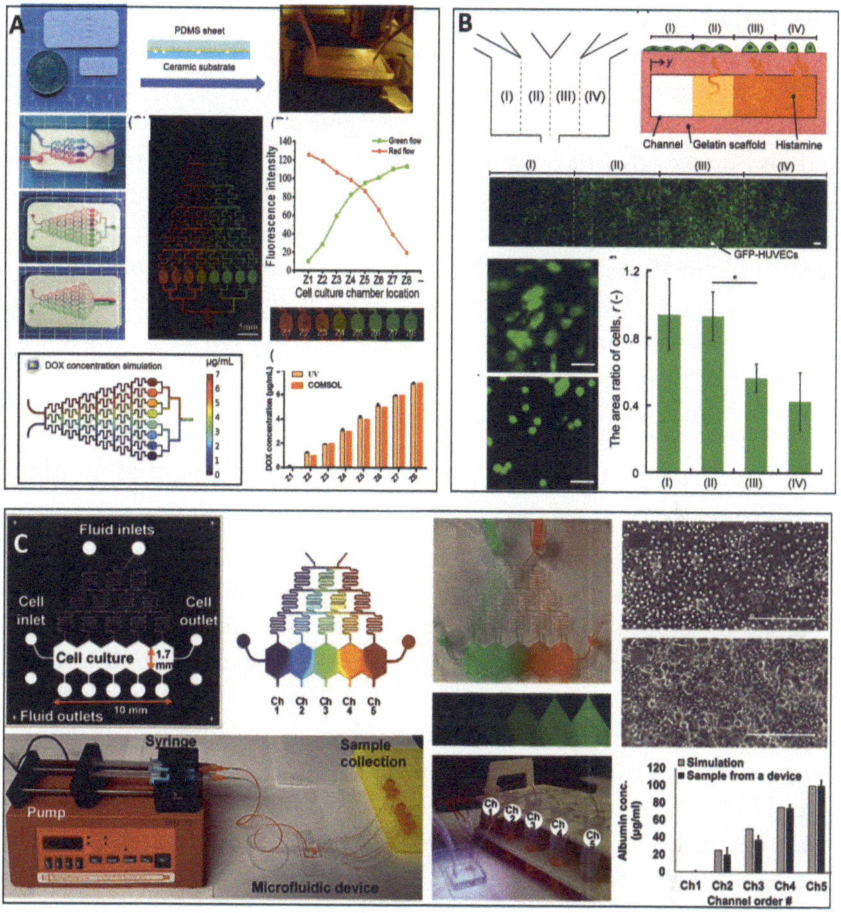

Fig. 2 Design and applications of flow-based microfluidic drug concentration gradient generators. (A) A christmas tree-like HA-PDMS-based MCGG for creating a concentration gradient of doxorubicin. (B) An ECM-based MCGG for evaluating HUVECs response to histamine concentration. (C) The Metabolic Patterning on a Chip platform for recapitulating liver zonation through concentration gradients of glucagon and insulin to induce the carbohydrate/nitrogen metabolism pattern and examine the drug-induced hepatotoxicity *via* Cytochrome P450 activity-dependent response to acetaminophen. *Panel (A) reproduced from Tang Q, Li X, Lai C, et al. Fabrication of a hydroxyapatite-PDMS microfluidic chip for bone-related cell culture and drug screening.* Bioact. Mater. *2021;6 (1):169–178, under a Creative Commons License. Panel (B) reproduced with permission from Shimizu A, Goh WH, Itai S, Karyappa R, Hashimoto M, Onoe H. ECM-based microfluidic gradient generator for tunable surface environment by interstitial flow.* Biomicrofluidics *2020;14(4):044106. Panel (C) reproduced from Kang YB, Eo J, Mert S, Yarmush ML, Usta OB. Metabolic patterning on a chip: towards in vitro liver zonation of primary rat and human hepatocytes.* Sci. Rep. *2018;8(1):8951, under a Creative Common Attribution 4.0 International License.*

In another organ-on-chip, a metabolic patterning on a chip (MPOC) was constructed to mimic the liver zonation.[27] One key physiological feature of the liver sinusoid is hepatic zonation so that cells display various activities and characteristics based on their location in the liver sinusoid. In this MPOC device, tree-like shape concentration gradients of hormones and enzymatic inducers were produced to create a dynamic metabolic pattern across the cell microchamber and mimic *in vivo*-like liver zonation and zonal toxic response (Fig. 2C).[27] The hepatocytes' cell viability and mitochondrial activity gradually decreased across the cell chamber's width from zone 1-like to zone 3-like region due to the higher CYP1A2 activity of the hepatocytes in zone 3.[27]

1.2 Gradient generators for high-throughput drug screening

In the last years, fast-paced progress in developing microfluidic tools for cell studies and integrating robotics, automated analyzing devices, and biosensors in biomedical microdevices have made the microfluidic biochips promising tools for high-throughput drug screening; therefore, microfluidic platforms like MCGGs and microarrays have been the gold standard in drug testing experiments.[28–32] A microfluidic device was established to cultivate human high-grade glioma cells (UVW) spheroids obtained from human prostate biopsies and test the anticancer drugs by an MCGG tool.[33] In this device, drug solution and culture media were injected into two separate reservoirs on two sides of the cell culture channel to create a symmetrical hydrostatic pressure-driven flow through the cell microarray that concurrently transported drug and media from each side of the channel to the central channels reservoirs (Fig. 3A). This structure produced a self-generated concentration gradient tool for cancer drug screening. UVW spheroids were treated with a concentration gradient of a chemotherapeutic agent called cisplatin, and the cell viability was assessed with PI and FDA double staining. The results showed that the drug concentration was proportional to the spheroid's shape factor parameter, which is a sign of spheroids disaggregation and outline roughness and is a readout for spheroids' health.[33]

A disk-like microfluidic gradient generator with compact double spiral mixers was developed in one research for producing three stable, accurate, and controllable drug gradients.[34] Two cancer cell lines (MCF-7 and HepG2) were cultured in the device and treated with three different concentration profiles of anticancer drugs (doxorubicin and cisplatin). Doxorubicin, cisplatin, and drug-free medium were injected into the device

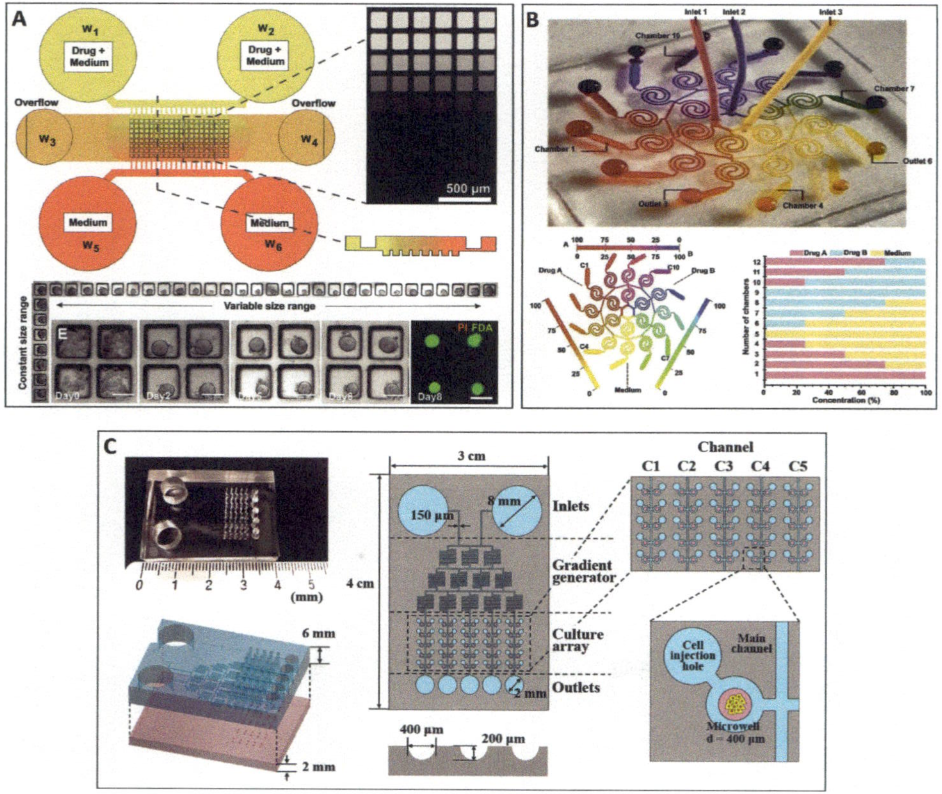

Fig. 3 See figure legend on next page.

via inlet 1,2 and 3, respectively. Consequently, three concentration gradients of doxorubicin, cisplatin, and the combination of the drugs were generated in 12 separate chambers (Fig. 3B). Cells that were cultured in the chambers with high drug concentration for a prolonged time had significantly lower live cell populations, and the most effective concentration profile with the highest cell apoptosis induction was the combination of doxorubicin (75%) and cisplatin (25%) in Chamber 12. The results demonstrated that this device is a powerful tool to generate multidrug concentration gradients for evaluating dose- and time-dependent cell-based studies and high-throughput drug screenings.[34]

A microfluidic spheroid culture device with a concentration gradient generator was developed for examining cancer drug efficacy (Fig. 3C).[35] The colon cancer cell line (HCT116) and glioma cell line (U87) were cultured in the microdevice with concave microwells to form the cancer spheroids with uniform size and distribution. As an anticancer drug, Irinotecan was injected into the chip to produce a concentration gradient, and the functionality of the device was evaluated by monitoring the shape and viability of the cells. Spheroids in the microwells with high irinotecan concentration lost their roundness, collapsed severely, and live/dead staining images indicated that cell viability declined continuously with elevated drug concentration.[35]

A paper-based microfluidic biochip was constructed to generate a drug concentration gradient and assess cell response to drugs based on high-throughput screening.[28] For creating the drug concentration gradient, two solutions containing doxorubicin and buffer were injected in two

Fig. 3 Layout and functions of drug concentration gradient generators for high-throughput drug screening. (A) A microfluidic device for culturing human high-grade glioma cells (UVW) spheroids obtained from human prostate biopsies and producing a self-generated concentration gradient for cancer drug screening. (B) Disc-like microfluidic three concentration gradient generator for analyzing different combinations of doxorubicin and cisplatin on MCF-7 and HepG2 tumor cell lines. (C) A microchip with concave microwells for culturing colon cancer cell line (HCT116) and glioma cell line (U87) spheroids and generating Irinotecan concentration gradient. *Panel (A) reproduced from Mulholland T, McAllister M, Patek S, et al. Drug screening of biopsy-derived spheroids using a self-generated microfluidic concentration gradient.* Sci. Rep. *2018;8(1):14672, under a Creative Common Attribution 4.0 International License. Panel (B) reproduced with permission from Shen S, Zhang X, Zhang F, Wang D, Long D, Niu Y. Three-gradient constructions in a flow-rate insensitive microfluidic system for drug screening towards personalized treatment.* Talanta *2020;208:120477. Panel (C) reproduce from Lim W, Park S. A microfluidic spheroid culture device with a concentration gradient generator for high-throughput screening of drug efficacy.* Molecules *2018;23 (12):3355, under an open access Creative Common CC BY license.*

separate inlets, and HeLa cells encapsulated in the collagen type I matrix were cultured in a microwells array, and a contrary trend of cell viability compared with drug concentration was shown. In paper-based microdevices, fluid flow is mainly derived by capillary force in porous paper channels, and there is no need for additional equipment like pumps and valves. Moreover, in flow-based MCGGs, operating the device for a prolonged time would be quite costly due to the excessive reagent requirement, while in paper-based tools, the solution consumption is significantly reduced.[28]

1.3 Challenges and possible solutions for microfluidic drug delivery at cellular level

Although various microfluidic platforms have been developed so far for recapitulating drug–cell interactions and mimic *in vivo* cellular microenvironment of different tissues, there are still some limitations in the microphysiological complexity and drug-induced cellular pathways resemblance.[11] Microfluidic systems for dynamic fluid manipulation and cell culture have revolutionized the research in concentration gradient generators by introducing state-of-the-art designs and configurations with extraordinary precision in flow control geometric constraints to form gradients with high stability and accuracy. These systems, based on the type of the microdevice, could rely on only molecular diffusion or diffusion-convection phenomena for generating gradients.[1,12]

Diffusion-based MCGGs compared to flow-based platforms, have a more cost-effective setup, are needless of expensive pumping systems, and minimize the cell damage due to a shear stress-free environment. However, these flow-free gradient generators are commonly much slower in producing the concentration gradient and require improvements in gradient stability and significant reduction of source-sink replenishment frequency.[1,12,18] Convection-based tools are able to sustain a drug concentration gradient within a fluid-flow environment and are suitable for the maintenance of a biochemical gradient within a specific tissue.[21] Moreover, these flow-based systems can create concentration gradients rapidly in a small area because of their short relaxation time and large Péclet (*Pe*) number, making them appropriate for applications where quick response is required. Though the significant hurdle in these devices is that, the gradient at a fast flow rate can damage the cells and ultimately trigger adverse efficiency in the cell signaling system.[19,20] To alleviate the problem, one recommended solution has been applying interconnected microgrooves and separate chambers for minimizing the effect of shear stress and generating long-term stable concentration gradients.[22]

In traditional tree-like microdevices, cells cultivated on the surface of the microchambers are exposed to direct biochemical factors from the fluid flow. In contrast, in the physiological microenvironment, cells are exposed to a complex and controlled concentration gradient of soluble factors through an interstitial flow.[26] Biomimetic hydrogels like collagen and agarose can increase the system's physiological relevance, generate a 3D scaffold environment for embedded cells, and recapitulate *in vivo*-like concentration gradients. These 3D ECM-like constructs could mimic tissue physiological functions such as microvasculature formation and axon guidance by chemotaxis due to their porous nature, albeit hydrogel-based systems usually are limited in throughput and spatiotemporal gradient conditions due to slow diffusion rate and also add experimental complexity and variability to the system.[15,16,18,32] The gradient in most common MCGGs is limited to one or multiple fixed perpendicular directions that restrict the positioning and orientation of drug gradients for cell re-orientation and migration assays. An omnidirectional radially symmetric MCGG was developed to overcome this concern for orienting laminar flow in 360° freely and manipulating chemical profiles in time and space.[36]

There is currently an urgent need to execute accurate and high-throughput predictive *in vitro* platforms in the pharmaceutical industry that resembles human biology like biochemical concentration gradients at a higher level and cut the time-consuming and costly drug procedures discovery pipeline.[11,28,30] Microfluidic systems have displayed remarkable benefits for drug screening assays due to their capacity for miniaturization and parallelization. Repeatable and robust results could be generated for an extended period with a high level of organotypicalness and *in vivo* microenvironment resemblance by employing scaled-down biosensors and automated features. The balance between the operational simplicity and biological complexity may end up with the commercialization of MCGGs and improve the user-friendliness of these microdevices.[11,29,32]

2. Microfluidic *in vitro* drug delivery at tissue level

There is an urgent necessity in the drug screening development pipeline for emerging predictive laboratory tissue models that truly orchestrate *in vivo* microenvironment and drug–cell interactions.[37,38] Conventional 2D or 3D *in vitro* platforms lack physiologically-relevant tissue microstructure, sustained cell viability, functional stability, and unchanged cellular morphology and phenotype.[39,40] In the past years, microfluidic tissue on chips have transformed the study of complex biologic practices like drug metabolism/toxicity and cancer research, and with monitoring spatiotemporal factors, these microdevices are capable of recapitulating the *in vivo*-like microenvironment properties like cell–cell and cell–ECM interactions, cellular composition, and tissue 3D microarchitecture.[2–4,41]

Novel patient-specific drug-testing tools are needed in cancer research and personalized medicine due to the current oncology drug development pipeline.[41,42] In one research, microdissected tissues (MDT) were obtained from live tumor tissue and were segmented reproducibly to submillimeter sections and cultured on an MDT-on-chip (Fig. 4A). The objective was to obtain tissue sections large enough to resemble biologically relevant gradients of nutrients, waste, and signaling molecules while also being small enough to preserve high viability all over the tissue without oxygen deficiency in the center. The microfluidic platform was able to trap the microtissues in a low shear stress environment, and the chemosensitivity of patient-derived ovarian cancer MDTs to carboplatin was evaluated by loading the drug into the microchannels and measuring the chemoresponse of treated and nontreated cell populations.[41]

In a study by Yin et al., a Microfluidic Kidney Chip was constructed for coculturing renal proximal tubular epithelial cells (RPTECs) and peritubular capillary endothelial cells (PCECs) and nephrotoxicity assessment.[43] The kidney-on-chip consisted of three parts: drug concentration gradient generator, three-layer microfluidic organ chip, and flow-temperature control device (Fig. 4B). The two groups of drug concentrations were produced on the gradient chip, then RPTECs and PCECs were seeded on opposite sides of membranes, and after the drug treatment, RPTECs' viability was evaluated to determine the influence of that drug. Finally, a flow-temperature control device was activated to stabilize the environments for prolonged cell culture. The results demonstrated a significant reduction in cisplatin-induced nephrotoxicity caused by the intervention of cimetidine in the microchip. This controllable biomimetic kidney model replicated critical aspects of kidney biological features, drug response, resistance, and toxicity could speed up the drug discovery and screening practices.[43]

In the drug delivery at the cellular level section, we described the MCGGs for drug delivery and toxicity testing; however, these platforms can generate concentration gradients in a micron-sized level (e.g., cells).[20,28,44] Therefore, there is a dire need to introduce an MCGG that can produce continuous concentration gradients of multi mixtures (e.g., drugs) in a millimeter-sized sample (e.g., tissue, spheroids). Rismanian et al. proposed a new class of MCGG

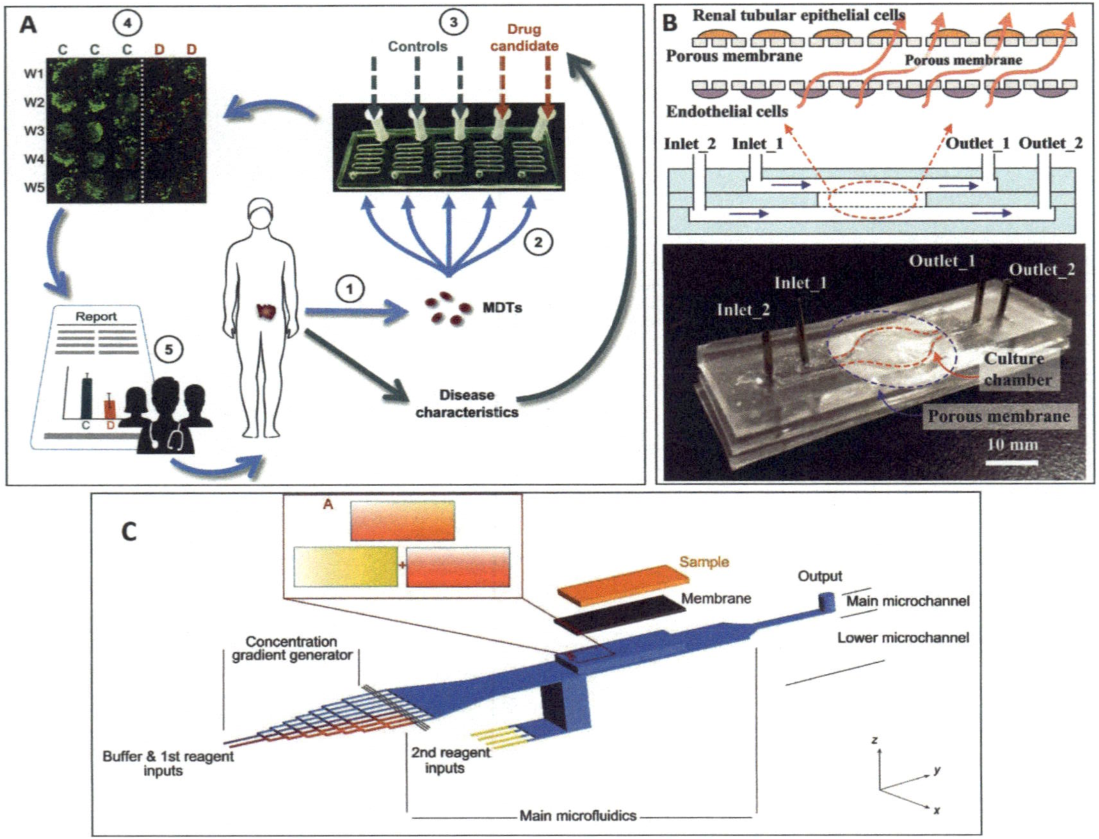

Fig. 4 See figure legend on opposite page.

by linking a modified tree-like CGG with a micromixer to develop a so-called Millimeter-sized sample /Multireagents/ Continuous CG with a diffusion-convection-based approach (Fig. 4C).[20] This microdevice introduced two reagents concurrently with a continuous concentration gradient of two inlets to deliver the reagents to a millimeter-sized sample (e.g., microtissue). This microchip with measurable concentration relaxation time and low effect of variability in the sample's porosity and permeability and diffusivity of various drugs could be an ideal tool for *ex vivo* drug chemosensitivity testing in personalized medicine.[20]

The high complexity and heterogeneity of the tumor microenvironment play a substantial part in drug resistance and chemoresponse.[45–47] Moreover, the endothelium is essential in systemic drug delivery, provides nutrients and oxygen to the tumor, and plays a crucial role in promoting new blood vessel formation *via* angiogenesis. Thus, a simple capillary force-based microchip for 3D tumor tissue-2D endothelium interfaces and drug testing was developed in a study. In this work, breast tumor cells were cultured in 3D microwells, and the endothelial barrier was modeled on top of the 3D culture by seeding HUVEC cells as an endothelium.[47] The endothelium was efficient and displayed physiological features, like the preferential proliferation of cells with higher access to nutrients and oxygen. Next, the drug penetration and toxicity assays were conducted in the tumor-endothelium model, and death ligand TRAIL (TNF-related apoptosis-inducing ligand) conjugates with a sizeable unilamellar vesicle (LUV) was tested. The results demonstrated that LUV-TRAIL prompted cell death in a higher number and was more effective than TRAIL.[47]

Fig. 4 Microfluidic systems for drug delivery and assessment at the tissue level. (A) Microdissected Tumor on-chip for *ex vivo* drug testing and personalized medicine. (B) A Microfluidic Kidney Chip for coculturing kidney cells and efficient drug screening and cisplatin-induced nephrotoxicity assessment. (C) A Millimeter sized sample/multireagents/continuous concentration generator with functionality in drug chemosensitivity testing for personalized medicine. *Panel (A) reproduced with permission from Astolfi M, Péant B, Lateef MA, et al. Micro-dissected tumor tissues on chip: an ex vivo method for drug testing and personalized therapy.* Lab Chip *2016;16(2):312–325. Panel (B) reproduced from Yin L, Du G, Zhang B, et al. Efficient drug screening and nephrotoxicity assessment on co-culture microfluidic kidney chip. Sci. Rep. 2020;10(1):6568, under a Creative Common Attribution 4.0 International License. Panel (C) reproduced with permission from Rismanian M, Saidi MS, Kashaninejad N. A microfluidic concentration gradient generator for simultaneous delivery of two reagents on a millimeter-sized sample. J. Flow Chem. 2020;10:615–625.*

According to the reports analyzing different compounds' clinical developments, oncology drug development's success rate from phase 1, clinical trials to FDA approval is less than 4%.[48] This low success rate in clinical trials demonstrates the deficiencies in existing prescreening methods. In another study, a Tumor Chip system was introduced to deliver cancer–stromal interaction, dynamic perfusion, replicate microvasculature, and high-throughput, as shown in Fig. 5A.[49] The L-TumorChip was a three-layered biochip with spatial control over the reconstruction of microvessels and tumor compartments, with continuous media perfusion. The tumor compartment consisted of human microvascular endothelial cells (HMVEC) and Matrigel-encapsulated breast cancer cells culture layers, separated by a thin membrane to mimic the *in vivo* tumor microenvironment. The tumor biochip was capable of real-time drug response monitoring for high-throughput drug kinetics studies. The drug responses to doxorubicin treatment showed that the presence of cancer-associated fibroblasts affects drug pharmacokinetics, while apoptotic reactions specified by caspase-3 activities are higher in the coculture of cancer cells with normal fibroblasts.[49]

In another research, a chip-based breast cancer equivalent with a tumor-replicating microvasculature was developed to recapitulate the breast tumors' physiology and pathophysiology and assess anticancer drug effectiveness.[50] For resembling high and low flow perfusion regions in the tumor microenvironment, two specific microvascular designs from mouse vascular networks were applied for tumor-mimetic chip production. The two regions with different degrees of fluidic exchange between the microvascular channels and the central tumor chamber represent well-perfused and poorly-perfused breast tumors. Breast cancer cells were cocultured for a prolonged time in 3D hydrogel-based structure with stromal fibroblasts, and the interstitial space between the chambers and endothelium with porous vasculature mimicked the physiological cancer-endothelial cell interactions. Before activating their cytotoxic activity, Provided anticancer drugs must overcome particular rate-controlling steps like the ability to perfuse through the microvasculature and transport to the site of interest. In this work, doxorubicin and paclitaxel were introduced to the breast tumor-on-chip, and the cell viability was assessed. A remarkable reduction in viable tumor areas was observed across all tested conditions, specifically in high perfusion chips, because of higher drug penetration levels in the central tumor chambers.[50]

Glioblastoma multiforme (GBM), an extremely aggressive and high-grade type of brain cancer, has a poor prognosis, with an average survival time of 12–18 months—only 25% of glioblastoma patients survive more

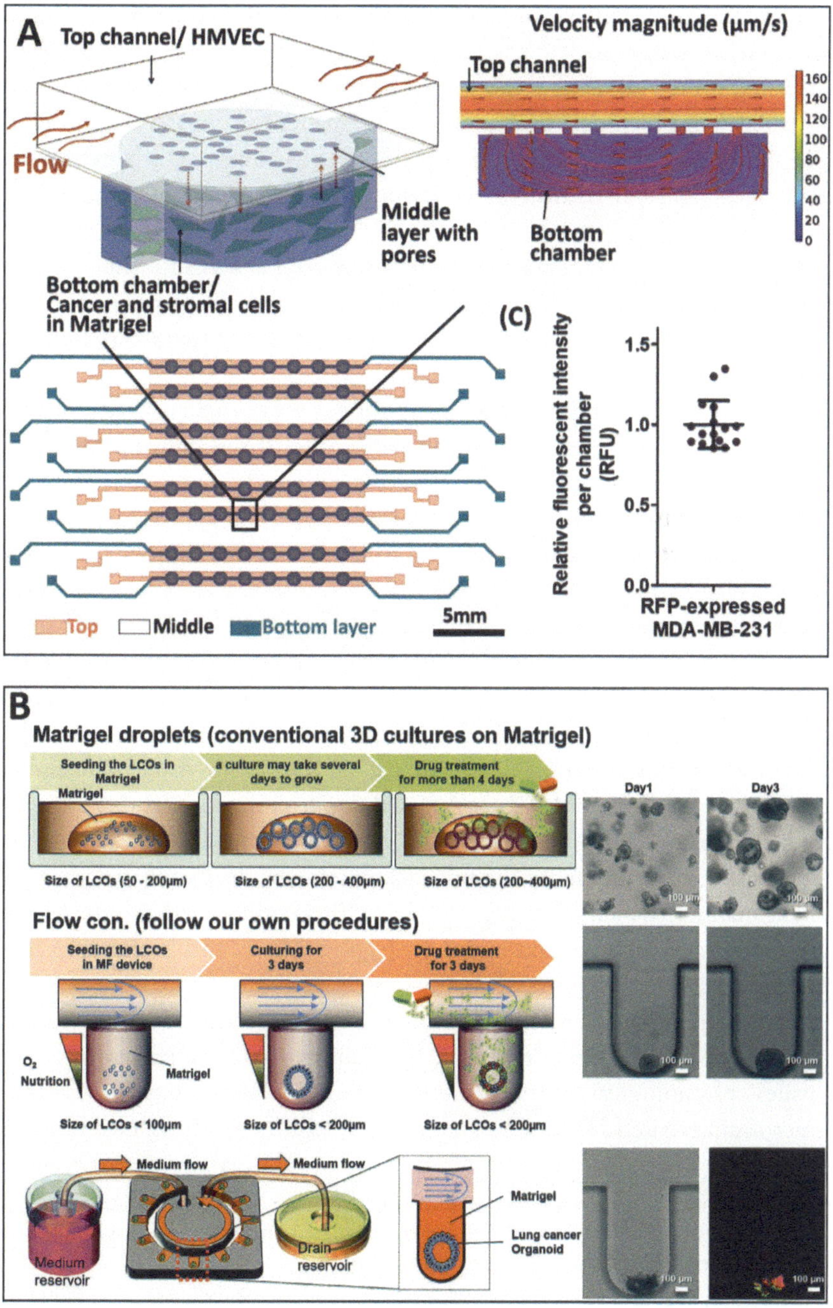

Fig. 5 Chip-based platforms for investigating tumor microenvironment and drug pharmacokinetics. (A) A high-throughput tumor-on-a-chip platform for emulating certain tumor-stroma and tumor–endothelium interactions and the stromal effects on drug *(Continued)*

than 1 year, and only 5% of patients survive more than 5 years—with standard GBM therapies.[51,52] To address this concern, a brain cancer chip for assessing tumor cell drug responses was administered to analyze drugs more efficiently in preclinical trials and improve patient survival.[53] In this work, 3D spheroids obtained from three patient GBM tumor cells were cultured on a hydrogel layer with a concentration gradient generating design to produce two drug concentrations (FDA-approved temozolomide and bevacizumab) simultaneously. The result of drug analysis indicated a common tendency toward higher cytotoxicity of two drugs combined with single-drug treatment and a more robust efficacy of temozolomide than bevacizumab.[53]

Colorectal cancer is now the third major cause of cancer-related death in women and the second in men, primarily due to the high rate of cancer metastasis and chemotherapeutic treatment ineffectiveness.[54,55] There is a dire need for more reliable and predictable screening approaches for colorectal cancer drug development. In one research, a biomimetic tumor-on-a-chip platform was developed to emulate *in vivo* relevant colorectal tumor microenvironment and reconstruct functional microvasculature for precision nanomedicine delivery.[56] Colorectal cancer cells were seeded in matrigel-supported 3D structure, while endothelial cells were cultured in a 3D vessel like architecture. To examine the concentration-dependent response of cells inside the biochip to the antitumor drug, gemcitabine (GEM) was loaded in CMCht/PAMAM dendrimer nanoparticles for delivering the drug to HCT-116 cancer cell line. With introducing GEM-loaded dendrimer nanoparticles to the microchannels, cell death occurred in a gradient manner, which demonstrated that GEM could be efficiently released from the nanoparticles and taken up by the CRCs as a useful anticancer therapeutic.[56]

As a preclinical method, tumor organoids are advantageous for patient-specific treatment thanks to their ability to sustain critical biological tissues' physiological and structural features.[57,58] Jung et al. developed a microfluidic-based 3D lung cancer organoid culture platform enabling drug

Fig. 5—Cont'd responses to doxorubicin treatment. (B) Microfluidic biochip for three-dimensional lung cancer organoid culturing and cisplatin and etoposide apoptosis tests. *Panel (A) reproduced with permission from Chi C-W, Lao Y-H, Ahmed AHR, et al. 2020 High-throughput tumor-on-a-chip platform to study tumor–stroma interactions and drug pharmacokinetics.* Adv. Healthc. Mater. *n/a(n/a):2000880. Panel (B) reproduced with permission from Jung DJ, Shin TH, Kim M, Sung CO, Jang SJ, Jeong GS. A one-stop microfluidic-based lung cancer organoid culture platform for testing drug sensitivity.* Lab Chip *2019;19(17):2854–2865.*

sensitivity tests directly on a microphysiological system, as shown in Fig. 5B.[59] This platform had high control over the size of lung cancer organoids (LCO) and validated the production of LCOs derived from primary small-cell lung cancer tumors with rapid proliferation and exhibition of disease-specific characteristics. The first-line chemotherapeutic for Lung cancer (cisplatin and etoposide) was chosen as the test drugs in this work. While the LCO cores and cells located away from blood vessels were more resistant to chemotherapy and remained viable, apoptosis was shown in cells around the organoid outer region. The drug concentration was associated with reduced rates of cell viability.[59]

Hepatocellular carcinoma has become the second major cause of cancer-related death, with an occurrence rate of about 850,000 new patients each year. As a result of 2D cell culture models' incapability in mimicking complex liver pathophysiology and predicting the toxicity of the compounds, few new anticancer drugs are available now.[60,61] In one research, a biomimetic 3D liver cancer-on-chip was developed, and decellularized liver matrix (DLM) was obtained from rat liver tissue to integrate the ECM proteins and GelMa into the liver chip.[61] This system had an improved competence to maintain hepatocytes viability and functions under perfusion conditions and may be accredited to establish biochemical factors, the maintenance of scaffold proteins, and the restoration of biophysical and biomechanical cues enhanced resemblance of the 3D liver tumor microenvironment. For assessing the platform's drug analyzing capability, acetaminophen and sorafenib were tested for dose-dependent drug toxicity, and linear dose-dependent hepatotoxicity was detected for both drugs in terms of HepG2 cells viability, urea secretion, and albumin production.[61]

3. Microfluidic *in situ* drug delivery at organ level

Recent advances in microfabrication technology revolutionized the drug delivery system. Microneedle (MN) is a noninvasive and painless localized drug delivery system (DDS) that prevents gastrointestinal degradation and bypasses hepatic metabolism. MNs have been used for drug delivery to different organs, including skin, oral mucosa, vaginal mucosa, ocular tissue, nail, anal sphincter, dermal papilla, and cardiac muscle.[62–65] Fig. 6A and B demonstrate the MNs DDS used for a drug-eluting balloon of vascular tissue[66] and intracorneal drug injection,[67] respectively. Various drugs such as insulin,[68,69] calcein,[69] DNA vaccine and nucleic acids,[69] nanomedicine,[69,70] ocular and chronic wounds treated drug,[69,71] peptide or subunit vaccination, and cancer therapeutics[72] have been delivered by MN drug delivery system. MNs are

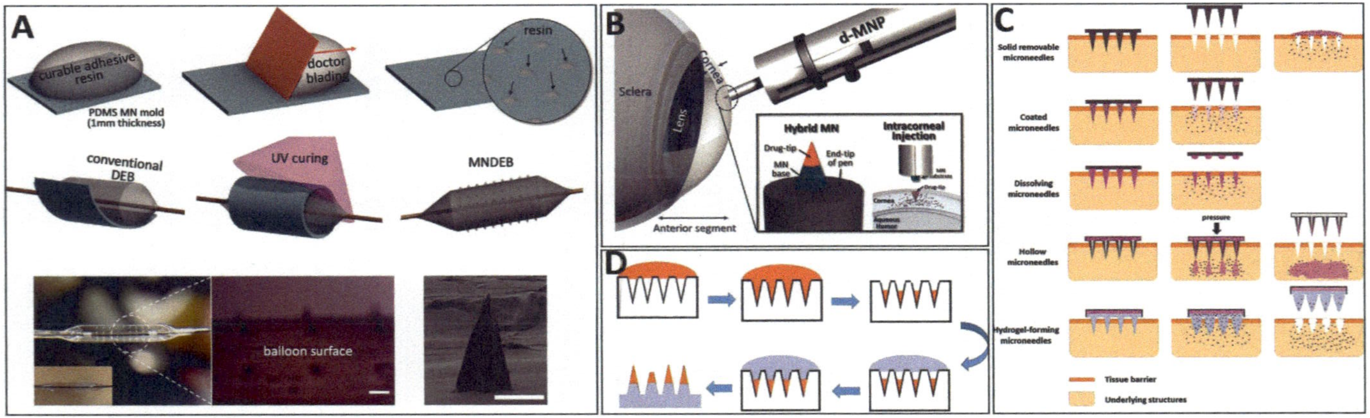

Fig. 6 See figure legend on opposite page.

structures of the arrowhead, bullet, conical, cylindrical, pyramidal, obelisk, octagonal cone shape with the height average between super-short (<100 μm) to super-long (>1000 μm), the width of 50–250 μm, and the tip thickness of 1–25 μm.[73–75] Both sizes and shapes of MNs affect the mechanical strength of the MN and the skin penetration's depth.[76] MNs have been fabricated through various microfabrication techniques such as casting, dry etching, wet etching, micromolding, 3D–printing, electroplating, laser ablation, magnetorheological, and drawing lithography.[75,77] The most-reported microfabrication technique for MNs is the casting method, a low-cost, massproductive, and easy to operation method shown in Fig. 6D. Depending on their applications, MNs are made from different materials, including silicone, metals, polymers,[78] and ceramics.[79] The material that is used for MN fabrication must be safe, nontoxic, and strong enough to penetrate the skin.[80,81] Based on the drug delivery route, MNs can be categorized

Fig. 6 MNs DDS. (A) MNs integrated with eluting balloon for drug delivery to vascular tissue, (a–c) conformal transfer molding steps consisting of premolding with UV curable adhesive resin, (d) alignment the drug eluting balloons (DEB) with flexible planar PDMS mold, (e) UV curing process and (f) completed MNDEB, (g) Stereo- and optical microscopic images of MNDEB and MNs MN onto MNDEB surface (scale bar = 300 μm) (h) the scanning electron microscopic image of the constructed MN in the MNDEB surface, (scale bar = 100 μm). (B) Schematic of sustained drug releasing system using a detachable hybrid microneedle pen system for intracorneal drug injection. (C) Schematic representation of the MNs DDS with five different drug delivery mechanisms, (a) Solid MNs that are used as puncher to create micropores in the skin to increase the permeability, (b) Coated MNs that contain drug on the surface which released after penetration, (c) Dissolvable MNs that released the drug due to dissolution once inserted into the tissue, (d) Hollow MNs that injected the drug into the tissue from the external reservoir by pressure-driven flow or diffusion mechanism. (e) Swollen MNs that absorb the ISF after insertion and administrate the drug. (D) The casting fabrication process of two layers MNs consists of 2 components. (a) pouring component 1 into the mold, (b) vacuumed or centrifuged to release air bubbles, (c) solidification component 1, (d) pouring component 2 as the second layer, (e) vacuumed or centrifuged, (f) Peeled away solidified two-layer MNs. *Panel (A) reproduced with permission from Lee K, Lee J, Lee SG, et al. Microneedle drug eluting balloon for enhanced drug delivery to vascular tissue.* J. Control. Release. *2020;321:174–183. Panel (B) reproduced with permission from Lee K, Song HB, Cho W, Kim JH, Kim JH, Ryu W. Intracorneal injection of a detachable hybrid microneedle for sustained drug delivery.* Acta Biomater. *2018;80:48–57. Panel (C) reproduced with some changes from Rzhevskiy AS, Singh TRR, Donnelly RF, Anissimov YG. Microneedles as the technique of drug delivery enhancement in diverse organs and tissues.* J. Control. Release. *2018;270:184–202. Panel (D) reproduce with permission from Zhuang J, Rao F, Wu D, et al. Study on the fabrication and characterization of tip-loaded dissolving microneedles for transdermal drug delivery. Eur. J. Pharm. Biopharm. 2020;157:66–73.*

into five groups that are solid MN, coated MNs, hollow MN, dissolvable MNs, and swollen MNs that are shown in Fig. 6C. In the following, each group will be discussed in more detail to address the operation mechanism, material, fabrication method, and challenges.

3.1 Solid microneedles

The first generation of MNs is solid MNs. Solid MNs were applied as micro punchers to produce small pores in the skin and increase skin penetration. After solid MN insertion, the drug in the form of liquid, cream, gel, or lotion is applied in created pores, as shown in Fig. 6C(a). Different types of materials, including silicone, metals (stainless steel, titanium), polymers,[78] and ceramics,[79] have been used for solid MNs preparation. Solid MNs have been fabricated through various techniques such as dry etching, wet etching, micro-molding, 3D–printing, electroplating, laser ablation, magnetorheological, and drawing lithography.[82] Various drug formulations such as Nicardipine HCl, Amantadine HCl, Tiagabine HCl, Pramipexole, dihydrochloride, and Carbamazepine have been administered *via* solid MNs.[83] Solid MNs can be made from a wide range of materials that are an excellent factor for accessibility. However, solid MNs fracture under the skin is a challenging task. There also some concerns regarding the two-step drug delivery using solid MNs that increase the odds of contamination.

3.2 Coated microneedles

Coated MNs are generally solid MNs that are coated with matrix drug loaded. After MNs insertion into the skin, the coating layer will be dissolved, and the drug will be released, Fig. 6C(b). This group of MNs was made from metal, silicon, and nondegradable polymers that were coated with biomaterial matrices such as carboxymethyl cellulose sodium salt, hyaluronic acid, methylcellulose, polyvinyl alcohol, polyvinyl pyrrolidone, polylactic acid, polycaprolactone, sucrose, and (2-hydroxypropyl)-β-cyclodextrin.[73] The key parameters of the coated material are hydrophilicity, biocompatibility, and biodegradability. Some stabilizers were usually added to the matrix to decrease the damage to the therapeutics, such as dextran, glucose, inulin, sucrose, and trehalose.[73] There are various coating methods such as dip coating, gas-jet drying, rolling coating, spray coating, electrohydrodynamic atomization coating, and layer-by-layer coating.[73,84] The biggest drawback of the coated MNs is the low drug loading capacity because, in such type of MN, a very small dose of drug material is only coated on the surface of the

MN. For example, for polylactic acid MNs that were coated with a solution of sulforhodamine B, the drug loadings were up 18 ng per needle for 750 μm MNs. It was shown that by increasing the viscosity of the coating solutions, the drug loading could be increased. However, the drug delivery efficiency was reported as 90%.[85]

3.3 Dissolvable microneedles

Dissolvable MNs are made from dissolvable material in aqueous media. After insertion, MNs absorb the interstitial fluid (ISF) and eventually dissolve, leading to the sequential release of the loaded drug, Fig. 6C(c). Dissolvable MNs can be made from a quick or long-time dissolvable material, making them good candidates for controlled-release delivery.[86–88] Biopolymers and sugars material similar to coated MNs were commonly used to prepared dissolvable MNs through micromolding,[75] casting,[89] drawing lithography, droplet-born air blowing methods.[90] As a significant advantage, there is no limitation for the size of therapeutic molecules for applying in dissolvable MNs. Similar to the coated MNs, the efficiency of drug delivery of the dissolved MNs depends on the dissolution of the matrix. However, the agent dosing is much higher for dissolved MNs. This group of MNs is economical and effective, thus attracting many researchers' attention, and promising good tools for MNs DDS.[88,91,92] One of the major problems with the dissolvable MNs were poor mechanical strength and weakness to penetrate the skin, however, recently acceptable strong MNs were reported.[93–96] For example, dissolvable polymeric MNs with a small amount of graphene oxide (GO) was used for the transdermal delivery of the chemotherapeutic, HA15, to melanoma-bearing mouse models. MNs demonstrated good mechanical strength (10–17 times at 500 mg/mL GO), increased moisture resistance, self-sterilization, antibacterial and anti-inflammatory properties, and controlled drug release by near-infrared light-activated.[94] There are two categories of dissolvable MNs, hole body and tip loaded systems. In the first category, the hole MN body was made of dissolvable material, while in the second one, only the tip of the MNs consists of dissolvable drug-loaded material.[67,97,98] An MN with a drug-loaded biodegradable tip and a supporting base was developed to sustained drug release into the cornea of an Acanthamoeba keratitis mouse model. Follow-up administration for 7 days confirmed the therapeutic efficacy of tip-loaded MN.[67] Although the tip-loaded MNs facilitate drug penetration into the deeper region, the dose of the loaded drug is small as the small volume tip.

3.4 Hollow microneedles

Hollow MNs are a miniature version of hypodermic needles. After hollow MNs insertion, the drug is injected under the tissue by pressure-driven flow or diffusion, Fig. 6C(d). Agent dosing is more for hollow MNs in comparison with the other kind of MNs. For example, a single hollow microneedle was integrated with a magnetic polymer composites device and inductive sensing instrument to controlled drug release as a wireless pumping system. The fabricated system was able to dosage sensing.[99] Hollow MNs were made in various configurations with out-off-center pore and side hole. They have been mainly prepared from similar materials for solid MNs, such as silicon metals, glass, polymers,[81] using the multistep process of deep reactive ion etching, wet chemical etching, plasma etching, isotropic etching, laser micromachining, digital light processing-based projection stereolithography, micropipette pulling, and sacrificial micromolding.[79] Due to the complex geometry of the hollow MNs, the manufacturing process of this type of MNs are more complicated and time-consuming than the others.

3.5 Swollen microneedles

A new generation of the MNs is swollen MNs fabricated from swollen materials such as hydrogels. After the insertion, the MNs will swoll due to ISF absorption and subsequently, the loaded agent released, Fig. 6C(e). Swollen MNs were prepared from the combination of the swollen polymers such as methyl vinyl ether and maleic anhydride (PMVE/MAH) and polyethyleneglycol (PEG), PMVE and maleic acid and PEG, 2-hydroxyethyl methacrylate and ethylene glycol dimethacrylate, polyvinyl alcohol, ethyl acrylate, and methacrylate.[73,100] The most common fabrication method of the swollen MNs is the casting method. For example, silk fibroin scaffold was used with poly(ethylene glycol) diacrylate and sucrose as the needle matrix to fabricate swollen MNs with three formulations of drugs, including Rhodamine B, doxorubicin, and indocyanine green.[101] Like hollow MNs, swollen MNs create conduct for drug delivery that can continuously insert the agent from the external reservoir. However, the administration time for the swollen MNs is greater than hollow MNs, which makes the drug delivery time-consuming.

3.6 Challenge of microneedle drug delivery systems

MNs drug delivery has many benefits: MNs realize noninvasive drug delivery, have better patient compliance because of reducing dose frequency, realize the possibility of self-administration, prevent first-pass metabolism and gastrointestinal incompatibility, have minimal or no side effects, can maintain the plasma drug level, have the possibility of large molecules administration, and are suitable for the drug with short biological half-life and narrow therapeutic index.[102] However, there are some limitations and challenges for using MNs for drug delivery, including less accuracy of dose compared to conventional hypodermic needles, variation in delivery due to skin condition, skin-dependent penetration mechanism, the possibility of tissue damage and venues collapse due to repetitive injection, and risk of breaking and remaining under the skin in case of using nonbiodegradable MNs.[103]

4. Conclusion and future perspective

Miniaturization offers significant advantages in drug delivery application that can be realized using microfluidic technology. In this chapter, the focus was given to *in vitro* and *in situ* microfluidic systems for drug delivery. To this aim, first, the latest advances in developing microfluidic concentration gradient generators (MCGGs) that can deliver the drug at cellular level were discussed. In particular, both diffusion-based and flow-based MCGGs as well as gradient generators for high-throughput drug screening, were thoroughly evaluated. These systems have become an essential tool for studying biological phenomena like cell migration, proliferation, chemotaxis, drug screening, and personalized medicine. Next, we evaluated the state-of-the-art lab-on-a-chip systems for *in vitro* drug delivery at tissue level was investigated. Finally, the design and fabrication of various *in situ* microneedle-based drug delivery systems, including solid microneedles, coated microneedles, hollow microneedles, microneedles dissolvable and swollen microneedles, were explained. We concluded each part by briefly identifying the present challenges and presenting the possible solutions.

Although tremendous efforts have been made in the design, fabrication and characterization of microfluidic devices for drug delivery applications, this field, especially *in vivo* microfluidic drug delivery platforms, is still at a preliminary stage. There are still many challenges that need be addressed in order to realize an efficient implantable microfluidic device for controlled drug delivery. Besides biocompatibility and biodegradability, one of the major challenges of the *in vivo* microfluidic systems for drug delivery is the integration of an efficient mechanism for continuous pumping the drug or the therapeutic agent in the body. It should be done autonomously using mainly through capillary action or other self-powered microfluidic pump.[104] Efficient integration of such a self-powered microfluidic pump with microneedles can open up a great avenue in this field, leading to revolutionary drug delivery and vaccine administration platforms. Moreover, droplet-based microfluidics,[105] flow-focusing droplet generators[106] and magnetofluidics[107] are other highly efficient platforms that can potentially be used for both drug synthesis and controlled delivery. There are still plenty of rooms in improving such platforms for *in vivo* drug delivery and personalized medicine.

References

1. Nguyen N-T, Shaegh SAM, Kashaninejad N, Phan D-T. Design, fabrication and characterization of drug delivery systems based on lab-on-a-chip technology. *Adv Drug Deliv Rev*. 2013;65(11):1403–1419.
2. Bhatia SN, Ingber DE. Microfluidic organs-on-chips. *Nat Biotechnol*. 2014;32:760.
3. Esch EW, Bahinski A, Huh D. Organs-on-chips at the frontiers of drug discovery. *Nat Rev Drug Discov*. 2015;14(4):248–260.
4. Zhang B, Korolj A, Lai BFL, Radisic M. Advances in organ-on-a-chip engineering. *Nat Rev Mater*. 2018;3(8):257–278.
5. Shintu L, Baudoin R, Navratil V, et al. Metabolomics-on-a-chip and predictive systems toxicology in microfluidic bioartificial organs. *Anal Chem*. 2012;84 (4):1840–1848.
6. Haddrick M, Simpson PB. Organ-on-a-chip technology: turning its potential for clinical benefit into reality. *Drug Discov Today*. 2019;24(5):1217–1223.
7. Kashaninejad N, Chan WK, Nguyen N-T. Fluid mechanics of flow through rectangular hydrophobic microchannels. In: *Paper Presented at: International Conference on Nanochannels, Microchannels, and Minichannels*; 2011.
8. Wu Q, Liu J, Wang X, et al. Organ-on-a-chip: recent breakthroughs and future prospects. *Biomed Eng Online*. 2020;19(1):9.
9. Lin JM. *Cell Analysis on Microfluidics*. Singapore: Springer; 2017.
10. Rismanian M, Saidi MS, Kashaninejad N. A new non-dimensional parameter to obtain the minimum mixing length in tree-like concentration gradient generators. *Chem Eng Sci*. 2019;195:120–126.
11. Moradi E, Jalili-Firoozinezhad S, Solati-Hashjin M. Microfluidic organ-on-a-chip models of human liver tissue. *Acta Biomater*. 2020;116:67–83.
12. Dravid A, Raos B, Aqrawe Z, Parittotokkaporn S, O'Carroll SJ, Svirskis D. A macroscopic diffusion-based gradient generator to establish concentration gradients of soluble molecules within hydrogel scaffolds for cell culture. *Front Chem*. 2019;7:638.
13. Shi H, Nie K, Dong B, Long M, Xu H, Liu Z. Recent progress of microfluidic reactors for biomedical applications. *Chem Eng J*. 2019;361:635–650.
14. Cabaleiro JM. Flowrate independent 3D printed microfluidic concentration gradient generator. *Chem Eng J*. 2020;382:122742.
15. Ahrens L, Tanaka S, Vonwil D, Christensen J, Iber D, Shastri VP. Generation of 3D soluble signal gradients in cell-laden hydrogels using passive diffusion. *Adv Biosyst*. 2019;3(1):1800237.
16. Barata D, Spennati G, Correia C, et al. Development of a shear stress-free microfluidic gradient generator capable of quantitatively analyzing single-cell morphology. *Biomed Microdevices*. 2017;19(4):81.

17. Jin T, Xu X, Hereld D. Chemotaxis, chemokine receptors and human disease. *Cytokine.* 2008;44(1):1–8.
18. Uzel SG, Amadi OC, Pearl TM, Lee RT, So PT, Kamm RD. Simultaneous or sequential orthogonal gradient formation in a 3D cell culture microfluidic platform. *Small (Weinheim an der Bergstrasse, Germany).* 2016;12(5):612–622.
19. Kilinc D, Schwab J, Rampini S, et al. A microfluidic dual gradient generator for conducting cell-based drug combination assays. *Integr Biol.* 2015;8(1):39–49.
20. Rismanian M, Saidi MS, Kashaninejad N. A microfluidic concentration gradient generator for simultaneous delivery of two reagents on a millimeter-sized sample. *J Flow Chem.* 2020;10:615–625.
21. Parittotokkaporn S, Dravid A, Bansal M, et al. Make it simple: long-term stable gradient generation in a microfluidic microdevice. *Biomed Microdevices.* 2019;21(3):77.
22. Mo SJ, Lee JH, Kye HG, et al. A microfluidic gradient device for drug screening with human iPSC-derived motoneurons. *Analyst.* 2020;145(8):3081–3089.
23. Wolfram CJ, Rubloff GW, Luo X. Perspectives in flow-based microfluidic gradient generators for characterizing bacterial chemotaxis. *Biomicrofluidics.* 2016;10(6):061301.
24. Ebadi M, Moshksayan K, Kashaninejad N, Saidi MS, Nguyen N-T. A tool for designing tree-like concentration gradient generators for lab-on-a-chip applications. *Chem Eng Sci.* 2020;212:115339.
25. Tang Q, Li X, Lai C, et al. Fabrication of a hydroxyapatite-PDMS microfluidic chip for bone-related cell culture and drug screening. *Bioact Mater.* 2021;6(1):169–178.
26. Shimizu A, Goh WH, Itai S, Karyappa R, Hashimoto M, Onoe H. ECM-based microfluidic gradient generator for tunable surface environment by interstitial flow. *Biomicrofluidics.* 2020;14(4):044106.
27. Kang YB, Eo J, Mert S, Yarmush ML, Usta OB. Metabolic patterning on a chip: towards in vitro liver zonation of primary rat and human hepatocytes. *Sci Rep.* 2018;8(1):8951.
28. Hong B, Xue P, Wu Y, Bao J, Chuah YJ, Kang Y. A concentration gradient generator on a paper-based microfluidic chip coupled with cell culture microarray for high-throughput drug screening. *Biomed Microdevices.* 2016;18(1):21.
29. Wei Y, Zhu Y, Fang Q. Nanoliter quantitative high-throughput screening with large-scale tunable gradients based on a microfluidic droplet robot under unilateral dispersion mode. *Anal Chem.* 2019;91(8):4995–5003.
30. Xu Z, Fang P, Xu B, et al. High-throughput three-dimensional chemotactic assays reveal steepness-dependent complexity in neuronal sensation to molecular gradients. *Nat Commun.* 2018;9(1):4745.
31. Zhang Y, Xiao RR, Yin T, et al. Generation of gradients on a microfluidic device: toward a high-throughput investigation of spermatozoa chemotaxis. *PLoS One.* 2015;10(11):e0142555.
32. Ezra Tsur E, Zimerman M, Maor I, Elrich A, Nahmias Y. Microfluidic concentric gradient generator design for high-throughput cell-based studies. *Front Bioeng Biotechnol.* 2017;5:21.
33. Mulholland T, McAllister M, Patek S, et al. Drug screening of biopsy-derived spheroids using a self-generated microfluidic concentration gradient. *Sci Rep.* 2018;8(1):14672.
34. Shen S, Zhang X, Zhang F, Wang D, Long D, Niu Y. Three-gradient constructions in a flow-rate insensitive microfluidic system for drug screening towards personalized treatment. *Talanta.* 2020;208:120477.
35. Lim W, Park S. A microfluidic spheroid culture device with a concentration gradient generator for high-throughput screening of drug efficacy. *Molecules.* 2018;23(12):3355.
36. Nakajima A, Ishida M, Fujimori T, Wakamoto Y, Sawai S. The microfluidic lighthouse: an omnidirectional gradient generator. *Lab Chip.* 2016;16(22):4382–4394.

37. Esch MB, Prot J-M, Wang YI, et al. Multi-cellular 3D human primary liver cell cultures elevate metabolic activity under fluidic flow. *Lab Chip*. 2015;15(10):2269–2277.
38. Bhushan A, Senutovitch N, Bale SS, et al. Towards a three-dimensional microfluidic liver platform for predicting drug efficacy and toxicity in humans. *Stem Cell Res Ther.* 2013;4(Suppl 1):S16.
39. Lin R-Z, Chang H-Y. Recent advances in three-dimensional multicellular spheroid culture for biomedical research. *Biotechnol J.* 2008;3(9–10):1172–1184.
40. Ziółkowska K, Kwapiszewski R, Brzózka Z. Microfluidic devices as tools for mimicking the in vivo environment. *New J Chem*. 2011;35(5):979–990.
41. Astolfi M, Péant B, Lateef MA, et al. Micro-dissected tumor tissues on chip: an ex vivo method for drug testing and personalized therapy. *Lab Chip*. 2016;16(2):312–325.
42. Hay M, Thomas DW, Craighead JL, Economides C, Rosenthal J. Clinical development success rates for investigational drugs. *Nat Biotechnol*. 2014;32(1):40–51.
43. Yin L, Du G, Zhang B, et al. Efficient drug screening and nephrotoxicity assessment on co-culture microfluidic kidney chip. *Sci Rep*. 2020;10(1):6568.
44. Mitxelena-Iribarren O, Zabalo J, Arana S, Mujika M. Improved microfluidic platform for simultaneous multiple drug screening towards personalized treatment. *Biosens Bioelectron*. 2019;123:237–243.
45. Quail DF, Joyce JA. Microenvironmental regulation of tumor progression and metastasis. *Nat Med*. 2013;19(11):1423–1437.
46. Floor SL, Dumont JE, Maenhaut C, Raspe E. Hallmarks of cancer: of all cancer cells, all the time? *Trends Mol Med*. 2012;18(9):509–515.
47. Virumbrales-Muñoz M, Ayuso JM, Olave M, et al. Multiwell capillarity-based microfluidic device for the study of 3D tumour tissue-2D endothelium interactions and drug screening in co-culture models. *Sci Rep*. 2017;7(1), 11998.
48. Wong CH, Siah KW, Lo AW. Estimation of clinical trial success rates and related parameters. *Biostatistics (Oxford, England)*. 2019;20(2):273–286.
49. Chi C-W, Lao Y-H, Ahmed AHR, et al. High-throughput tumor-on-a-chip platform to study tumor–stroma interactions and drug pharmacokinetics. *Adv Healthc Mater.* 2020. n/a(n/a):2000880.
50. Pradhan S, Smith AM, Garson CJ, et al. A microvascularized tumor-mimetic platform for assessing anti-cancer drug efficacy. *Sci Rep*. 2018;8(1):3171.
51. Siegel RL, Miller KD, Jemal A. Cancer statistics, 2017. *CA Cancer J Clin*. 2017;67(1):7–30.
52. Zhu P, Du XL, Lu G, Zhu J-J. Survival benefit of glioblastoma patients after FDA approval of temozolomide concomitant with radiation and bevacizumab: a population-based study. *Oncotarget*. 2017;8(27):44015–44031.
53. Akay M, Hite J, Avci NG, et al. Drug screening of human GBM spheroids in brain cancer chip. *Sci Rep*. 2018;8(1):15423.
54. Vatandoust S, Price TJ, Karapetis CS. Colorectal cancer: metastases to a single organ. *World J Gastroenterol*. 2015;21(41):11767–11776.
55. Carvalho MR, Lima D, Reis RL, Oliveira JM, Correlo VM. Anti-cancer drug validation: the contribution of tissue engineered models. *Stem Cell Rev Rep*. 2017;13(3):347–363.
56. Carvalho MR, Barata D, Teixeira LM, et al. Colorectal tumor-on-a-chip system: a 3D tool for precision onco-nanomedicine. *Sci Adv*. 2019;5(5):eaaw1317.
57. Weeber F, Ooft SN, Dijkstra KK, Voest EE. Tumor organoids as a pre-clinical cancer model for drug discovery. *Cell Chem Biol*. 2017;24(9):1092–1100.
58. Sabhachandani P, Motwani V, Cohen N, Sarkar S, Torchilin V, Konry T. Generation and functional assessment of 3D multicellular spheroids in droplet based microfluidics platform. *Lab Chip*. 2016;16(3):497–505.

59. Jung DJ, Shin TH, Kim M, Sung CO, Jang SJ, Jeong GS. A one-stop microfluidic-based lung cancer organoid culture platform for testing drug sensitivity. *Lab Chip.* 2019;19(17):2854–2865.
60. Llovet JM, Zucman-Rossi J, Pikarsky E, et al. Hepatocellular carcinoma. *Nat Rev Dis Primers.* 2016;2(1):16018.
61. Lu S, Cuzzucoli F, Jiang J, et al. Development of a biomimetic liver tumor-on-a-chip model based on decellularized liver matrix for toxicity testing. *Lab Chip.* 2018;18 (22):3379–3392.
62. Bilal M, Mehmood S, Raza A, Hayat U, Rasheed T, Iqbal H. Microneedles in smart drug delivery. *Adv Wound Care (New Rochelle).* 2020;10:204–219 [ja].
63. Yang J, Liu X, Fu Y, Song Y. Recent advances of microneedles for biomedical applications: drug delivery and beyond. *Acta Pharm Sin B.* 2019;9(3):469–483.
64. Lee K, Goudie MJ, Tebon P, et al. Non-transdermal microneedles for advanced drug delivery. *Adv Drug Deliv Rev.* 2019;165–166:41–59.
65. Rzhevskiy AS, Singh TRR, Donnelly RF, Anissimov YG. Microneedles as the technique of drug delivery enhancement in diverse organs and tissues. *J Control Release.* 2018;270:184–202.
66. Lee K, Lee J, Lee SG, et al. Microneedle drug eluting balloon for enhanced drug delivery to vascular tissue. *J Control Release.* 2020;321:174–183.
67. Lee K, Song HB, Cho W, Kim JH, Kim JH, Ryu W. Intracorneal injection of a detachable hybrid microneedle for sustained drug delivery. *Acta Biomater.* 2018;80:48–57.
68. Ng LC, Gupta M. Transdermal drug delivery systems in diabetes management: a review. *Asian J Pharm Sci.* 2020;15(1):13–25.
69. Azmana M, Mahmood S, Hilles AR, Mandal UK, Al-Japairai KAS, Raman S. Transdermal drug delivery system through polymeric microneedle: a recent update. *J Drug Deliv Sci Technol.* 2020;60, 101877.
70. Chen M, Quan G, Sun Y, Yang D, Pan X, Wu C. Nanoparticles-encapsulated polymeric microneedles for transdermal drug delivery. *J Control Release.* 2020;325:163–175.
71. Barnum L, Samandari M, Schmidt TA, Tamayol A. Microneedle arrays for the treatment of chronic wounds. *Expert Opin Drug Deliv.* 2020;17:1–14.
72. Moreira AF, Rodrigues CF, Jacinto TA, Miguel SP, Costa EC, Correia IJ. Microneedle-based delivery devices for cancer therapy: a review. *Pharmacol Res.* 2019;148:104438.
73. Kang N-W, Kim S, Lee J-Y, et al. Microneedles for drug delivery: recent advances in materials and geometry for preclinical and clinical studies. *Expert Opin Drug Deliv.* 2020;18:929–947.
74. Waghule T, Singhvi G, Dubey SK, et al. Microneedles: a smart approach and increasing potential for transdermal drug delivery system. *Biomed Pharmacother.* 2019;109: 1249–1258.
75. Larraneta E, Lutton RE, Woolfson AD, Donnelly RF. Microneedle arrays as transdermal and intradermal drug delivery systems: materials science, manufacture and commercial development. *Mater Sci Eng: R: Rep.* 2016;104:1–32.
76. Sabri AH, Kim Y, Marlow M, et al. Intradermal and transdermal drug delivery using microneedles–fabrication, performance evaluation and application to lymphatic delivery. *Adv Drug Deliv Rev.* 2020;153:195–215.
77. Kashaninejad N, Munaz A, Moghadas H, Yadav S, Umer M, Nguyen N-T. Microneedle arrays for sampling and sensing skin interstitial fluid. *Chemosensors.* 2021;9(4):83.
78. Al-Japairai KAS, Mahmood S, Almurisi SH, et al. Current trends in polymer microneedle for transdermal drug delivery. *Int J Pharm.* 2020;587:119673.
79. Ita K. Ceramic microneedles and hollow microneedles for transdermal drug delivery: two decades of research. *J Drug Deliv Sci Technol.* 2018;44:314–322.

80. Nagarkar R, Singh M, Nguyen HX, Jonnalagadda S. A review of recent advances in microneedle technology for transdermal drug delivery. *J Drug Deliv Sci Technol.* 2020;59:101923.
81. Ali R, Mehta P, Arshad M, Kucuk I, Chang M, Ahmad Z. Transdermal microneedles—a materials perspective. *AAPS PharmSciTech.* 2020;21(1):12.
82. Yang H, Wu X, Zhou Z, Chen X, Kong M. Enhanced transdermal lymphatic delivery of doxorubicin via hyaluronic acid based transfersomes/microneedle complex for tumor metastasis therapy. *Int J Biol Macromol.* 2019;125:9–16.
83. Parhi R, Divya Supriya N. Review of microneedle based transdermal drug delivery systems. *Int J Pharm Sci Nanotech.* 2019;12(3):4511–4523. https://doi.org/10.37285/ijpsn.2019.12.3.1.
84. Ingrole RS, Gill HS. Microneedle coating methods: a review with a perspective. *J Pharmacol Exp Ther.* 2019;370(3):555–569.
85. Li L, Eyckmans J, Chen CS. Designer biomaterials for mechanobiology. *Nat Mater.* 2017;16(12):1164–1168.
86. Chandrasekharan A, Hwang YJ, Seong K-Y, Park S, Kim S, Yang SY. Acid-treated water-soluble chitosan suitable for microneedle-assisted intracutaneous drug delivery. *Pharmaceutics.* 2019;11(5):209.
87. Lopez-Ramirez MA, Soto F, Wang C, et al. Built-in active microneedle patch with enhanced autonomous drug delivery. *Adv Mater.* 2020;32(1), 1905740.
88. Thakur RRS, Tekko IA, Al-Shammari F, Ali AA, McCarthy H, Donnelly RF. Rapidly dissolving polymeric microneedles for minimally invasive intraocular drug delivery. *Drug Deliv Transl Res.* 2016;6(6):800–815.
89. Chen C-H, Shyu VB-H, Chen C-T. Dissolving microneedle patches for transdermal insulin delivery in diabetic mice: potential for clinical applications. *Materials (Basel).* 2018;11(9):1625.
90. Leone M, Mönkäre J, Bouwstra J, Kersten G. Dissolving microneedle patches for dermal vaccination. *Pharm Res.* 2017;34(11):2223–2240.
91. GhavamiNejad A, Li J, Lu B, et al. Glucose-responsive composite microneedle patch for hypoglycemia-triggered delivery of native glucagon. *Adv Mater.* 2019;31(30):1901051.
92. Ita K. Dissolving microneedles for transdermal drug delivery: advances and challenges. *Biomed Pharmacother.* 2017;93:1116–1127.
93. Nguyen HX, Bozorg BD, Kim Y, et al. Poly (vinyl alcohol) microneedles: fabrication, characterization, and application for transdermal drug delivery of doxorubicin. *Eur J Pharm Biopharm.* 2018;129:88–103.
94. Chen Y, Yang Y, Xian Y, et al. Multifunctional graphene-oxide-reinforced dissolvable polymeric microneedles for transdermal drug delivery. *ACS Appl Mater Interfaces.* 2019;12(1):352–360.
95. Luo Z, Sun W, Fang J, et al. Biodegradable gelatin methacryloyl microneedles for transdermal drug delivery. *Adv Healthc Mater.* 2019;8(3):1801054.
96. Luzuriaga MA, Berry DR, Reagan JC, Smaldone RA, Gassensmith JJ. Biodegradable 3D printed polymer microneedles for transdermal drug delivery. *Lab Chip.* 2018;18(8):1223–1230.
97. Zhu DD, Wang QL, Liu XB, Guo XD. Rapidly separating microneedles for transdermal drug delivery. *Acta Biomater.* 2016;41:312–319.
98. Zhuang J, Rao F, Wu D, et al. Study on the fabrication and characterization of tip-loaded dissolving microneedles for transdermal drug delivery. *Eur J Pharm Biopharm.* 2020;157:66–73.
99. Jayaneththi V, Aw K, Sharma M, Wen J, Svirskis D, McDaid A. Controlled transdermal drug delivery using a wireless magnetic microneedle patch: preclinical device development. *Sensors Actuators B Chem.* 2019;297:126708.

100. Queiroz MLB, Shanmugam S, Santos LNS, et al. Microneedles as an alternative technology for transdermal drug delivery systems: a patent review. *Expert Opin Ther Pat.* 2020;30(6):433–452.
101. Gao Y, Hou M, Yang R, et al. Highly porous silk fibroin scaffold packed in PEGDA/sucrose microneedles for controllable transdermal drug delivery. *Biomacromolecules.* 2019;20(3):1334–1345.
102. Prausnitz MR. Engineering microneedle patches for vaccination and drug delivery to skin. *Ann Rev Chem Biomol Eng.* 2017;8:177–200.
103. Sharma S, Hatware K, Bhadane P, Sindhikar S, Mishra DK. Recent advances in microneedle composites for biomedical applications: advanced drug delivery technologies. *Mater Sci Eng C.* 2019;103:109717.
104. Dal Dosso F, Kokalj T, Belotserkovsky J, Spasic D, Lammertyn J. Self-powered infusion microfluidic pump for ex vivo drug delivery. *Biomed Microdevices.* 2018;20(2):44.
105. Maleki MA, Soltani M, Kashaninejad N, Nguyen N-T. Effects of magnetic nanoparticles on mixing in droplet-based microfluidics. *Phys Fluids.* 2019;31(3):032001.
106. Yaghoobi M, Saidi MS, Ghadami S, Kashaninejad N. An interface–particle interaction approach for evaluation of the co-encapsulation efficiency of cells in a flow-focusing droplet generator. *Sensors.* 2020;20(13):3774.
107. Maleki MA, Zhang J, Kashaninejad N, Soltani M, Nguyen N-T. Magnetofluidic spreading in circular chambers under a uniform magnetic field. *Microfluid Nanofluid.* 2020;24(10):80.

CHAPTER THREE

From organ-on-chip to body-on-chip: The next generation of microfluidics platforms for *in vitro* drug efficacy and toxicity testing

Jerome Lacombe[a,b,*], Maria Soldevila[a], and Frederic Zenhausern[a,b,*]
[a]Center for Applied Nanobioscience and Medicine, Phoenix, AZ, United States
[b]Department of Basic Medical Sciences, College of Medicine-Phoenix, The University of Arizona, Phoenix, AZ, United States
*Corresponding authors: e-mail address: jlacombe@email.arizona.edu; fzenhaus@email.arizona.edu

Contents

Abstract

The high failure rate in drug development is often attributed to the lack of accurate pre-clinical models that may lead to false discoveries and inconclusive data when the compounds are eventually tested in clinical phase. With the evolution of cell culture technologies, drug testing systems have widely improved, and today, with the emergence of microfluidics devices, drug screening seems to be at the dawn of an important revolution. An organ-on-chip allows the culture of living cells in continuously perfused microchambers to reproduce physiological functions of a particular tissue or organ.

Progress in Molecular Biology and Translational Science, Volume 187
ISSN 1877-1173
https://doi.org/10.1016/bs.pmbts.2021.07.019
Copyright © 2022 Elsevier Inc.
All rights reserved.

The advantages of such systems are not only their ability to recapitulate the complex biochemical interactions between different human cell types but also to incorporate physical forces, including shear stress and mechanical stretching or compression. To improve this model, and to reproduce the absorption, distribution, metabolism, and elimination process of an exogenous compound, organ-on-chips can even be linked fluidically to mimic physiological interactions between different organs, leading to the development of body-on-chips. Although these technologies are still at a young age and need to address a certain number of limitations, they already demonstrated their relevance to study the effect of drugs or toxins on organs, displaying a similar response to what is observed *in vivo*. The purpose of this review is to present the evolution from organ-on-chip to body-on-chip, examine their current use for drug testing and discuss their advantages and future challenges they will face in order to become an essential pillar of pharmaceutical research.

1. Introduction

The development of new drugs is a long and expensive process with a high uncertainty of success. Despite a recent increase in success rates in late-stage development, the success in phase II stays relatively modest and the probability of launch from entry to phase I is less than 10%.[1] One percent of these clinical failures are due to operational or technical flaws, 7% for commercial reasons, 13% are the result of strategic realignment, but the largest part (79%) is attributable to safety or efficacy. The reduction of this risk passes through the earliest stages of drug discovery, and thus should focus on a better assessment of efficacy or the elimination of compounds that have mechanism-based toxicity in preclinical phase.[2]

In this phase, efficacy and toxicity of drugs are currently mostly determined and evaluated by the highly used *in vitro* systems or animal models. For example, the efficacy of cancer drugs are usually tested using several immortal cell lines and the toxicity is assessed using cell lines derived from regular tissues.[3] In these models, cells, which grow in monolayer on stiff substrate (plastic flasks/plates, glass), do not reproduce the realistic environment present in the body and their behavior or response to drugs can widely deviate to what is observed *in vivo*. They offer unnatural growth kinetics and cell attachments, lack of cell polarity, misrepresent cell-cell and cell-extracellular matrix (ECM) interactions, oxygen or nutrient gradients, that all ultimately affect their phenotype and function.[4]

To overcome this limitation, different technologies emerged allowing cell culture techniques to evolve to three-dimensional (3D) models.[5,6] Cells cultured in 3D have been shown to display different responses to external stimuli, including drugs, than cells cultured as a monolayer, mainly because of (1) the ability to maintain a normal morphology, (2) the difference in the organization of surface receptors on the cell, (3) the presence of cells at different stages and (4) the cellular shape-induced pH change.[7]

However, 3D models still have limitations. Spheroids/organoids can be variable in size and shape and are limited in their ability to quantify transcellular transport due to common necrotic cores of entrapped cells. These systems may misrepresent multiscale architecture and co-culture tissue interfaces which make up the essential components of organs such as vascular or connective tissue. These models also lack mechanical cues, such as fluid stress, cyclic stain, tension or compression forces, which are imperative for the development and good functioning of the organ. Therefore, animal models are also systematically used to assess drug efficiency and safety in a real *in vivo* environment. However, in addition to all ethical problems risen by animal models, important physiological differences exist between animals and humans and their relevance as model is widely debated.[8–10]

Therefore, to model *in vivo* (patho)physiological functions of human tissues more accurately, organ-on-chip (OOC) rapidly became a technology of interest. These systems combine the presence of physical forces with the biochemical and cellular interactions, permitting analysis of organ-specific response to drugs, toxins or other environmental perturbations.[11] However, when a drug is administrated in the human body, it goes under a complex process that involves its absorption, distribution, metabolism, and elimination (ADME) and its effect cannot be restrained to a specific organ only. Indeed, a drug is first administered to one site (*i.e.*, oral cavity, skin, *etc.*) and then passed into the bloodstream, where it will be carried through the body and distributed to various organs until it is metabolized (mainly in the liver) and its metabolites finally excreted, mainly by urine (kidney) or feces (colon) (Fig. 1). Therefore, several organs are "in contact" with the drug and, naturally, OOCs have been connected to evolve into body-on-chip (BOC) systems that can simulate these multi-organ interactions under near-physiological fluid flow conditions and with realistic tissue-to-tissue size ratios.[3] Used in conjunction with pharmacokinetic (PK) and pharmacodynamic (PD) models, these devices allow an estimation of effective concentrations (*i.e.*, optimizing efficiency to the target while minimizing side effects to the surrounding organs) that can predict the human response.

In this chapter, we will review how the evolution from a simple, dual compartmental OOC to a complex model of BOC led to a better reproduction of the ADME system and improve our understanding of drug toxicity mechanisms. We will also discuss the current limitations of such systems and highlight their future for drug development.

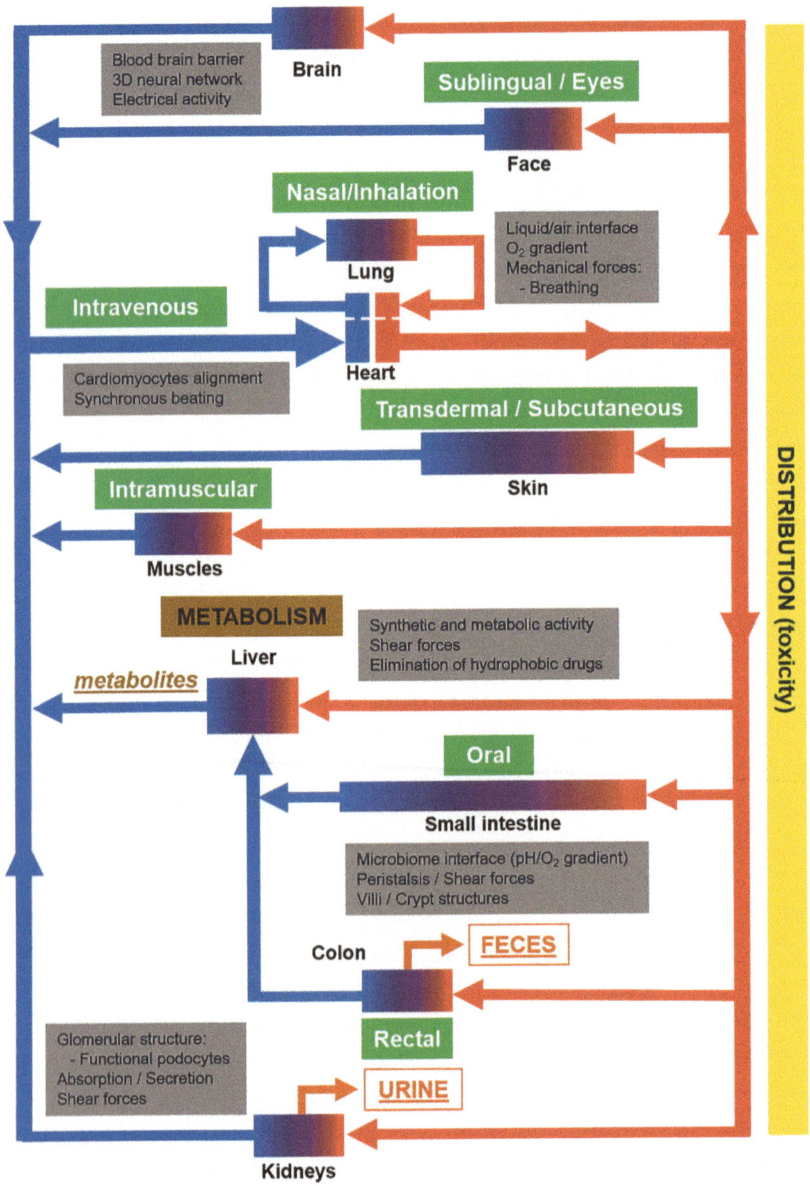

Fig. 1 Schematic representation of human blood circulation and interconnection of major organs involved in absorption, distribution, metabolism, and elimination (ADME) process. Gray boxes inform about the main requirements and current challenges of individual OOC systems to reproduce the main physiological functions involved in ADME process. Blue and red lines represent circulation of deoxygenated and oxygenated blood respectively. Green boxes represent the different administration routes and orange arrows act for elimination routes.

2. Organ-on-chip development: The game changer

2.1 From the first cellular models to the body-on-chips: A brief history

OOC technologies have resulted from a large number of contributed cell culture techniques that have originated over a century ago (Fig. 2). For instance, the foundations of traditional two-dimensional cell culture were developed in 1985 when Wilhelm Roux successfully maintained embryonic chicken cells in warm saline on a flat glass plate for several days.[12] This experimental success led to an overwhelming number of historical developments in the field of medicine based on the principle that cells or tissues could be maintained outside the origin of the host body. This principle was later illustrated by Ross G. Harrison in 1907 during his research into the origin of nerve fibers with a 3D application. Harrison carried out one of the first demonstrations of animal tissue culture using the "hanging-drop" technique to assess the growth and differentiation of neuroblasts outside of the body.[13] 3D models shifted to using host-organism material such as in Ehrmann's case wherein 1956 he utilized a transparent gel of reconstituted rat-tail collagen to achieve a better representation of cells in their native environment.[14] However, the first 3D culture using non-host organism material while paving the way for modern drug therapy testing was made in a soft agar solution in the 1970s.[15] This originally newfound method permitted the characterization of many newly obtainable and characterizable biophysical properties of tumor cells including sedimentation velocity and surface antigenic features. 3D models take many forms and one of which can be distributed within ECM or self-assembled in cellular microstructures (*i.e.*, organoids). These techniques have shown great promise for studying the effects of drugs or radiation on human cells.

However, the test of therapeutic targets and drug candidates in complex biological environment requires advanced technology such as microstructured components. These technological achievements have been made possible from the combination of microfabrication and biology.[16] The microelectronics/semiconductor industry has developed photolithography fabrication techniques in the 1960s to pattern wafers used for the making of processors and integrated circuits. These silicon micromachining techniques have been then adapted to the fabrication of microdevices for biological research. However, these techniques are limited in their applicability due to their expensive nature, restricted control over surface

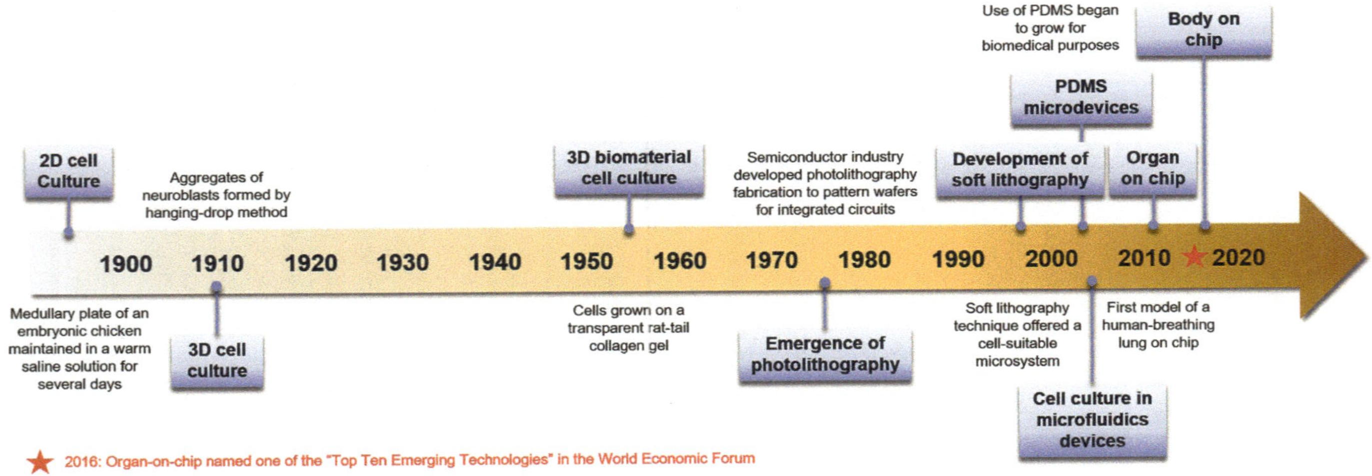

Fig. 2 Timeline of cell culture systems evolution.

properties, and indirect applicability to proteins and cells.[16] For example, dry etching techniques from the semiconductor industry have been repurposed to etch away designs on silicon wafers. The designs left behind from etches mark the foundations for microstructure patterns. Silicon wafers are generally expensive and require precise manufacturing equipment. However, silicon has indirect applicability to cells and the biological system. Only the wafer can be repurposed as a molding technique that gave way for soft lithography. Soft lithography was then developed during the 1990s and offered the ability to fabricate microsystems suitable for cells by molding biochemicals such as organic materials and polymers, typically poly(dimethylsiloxane) (PDMS).[16] The application of soft lithography to biology led to the creation of biocompatible microsystems mimicking physical and mechanical forces to the tissue environment and, ultimately, to the field of microfluidics. Microfluidic technology was developed with the purpose of analyzing small quantities of samples and analytes with high resolution and sensitivity; which has great potential advantage in the advancement of biopharmaceuticals.[17] One of the first documented studies of adherent cell culture in a microfluidic channel system was performed by Tilles et al. where primary rate hepatocytes were seeded in coculture with 3T3-J2 fibroblasts within a polycarbonate and glass microchannel cell culture system.[18] The use of microfluidic channel systems gave rise to the development of multi-compartment and physiological systems often referred now as OOCs.

OOC can be adapted to a wide range of robust cellular systems that offer predictive models of human physiology and disease and can be used as tools for drug discovery and development. One of the first OOC to accurately recapitulate organ-level functionality by creating separate parenchymal and vascular compartments and contain tissue-tissue interfaces with vascular perfusion was a lung alveolus on chip.[19] Recently, OOC systems have evolved in sophistication and are thus allowing a better understanding of the complex interactions that occur as a result of the organs' response to drugs and chemicals. The goal of the OOC is to not replicate an entire organ but the minimal functional qualities of a particular organ enough to mimic physiological interactions that would normally occur *in vivo*. However, OOCs are also limited in that they are not able to reproduce the complex interactions that occur between different organ systems. These interactions are critical as they play essential roles in the maintenance of homeostasis and are subject to several cases of pathologies. This limitation is profound when investigating complex pathologies such as obesity, metabolic diseases, and immunological disorders. Subsequently, BOC systems have been designed to emulate entire human physiological responses to drugs and encapture potential drug toxicity in specific organs. BOC systems are accomplished by integrating several different OOC systems to recapitulate what would be the minimal functional units of the human body and replicate physiological responses to a drug.

2.2 The different types of organ-on-chips and their use in drug testing

In the past few years, a lot of organs have been implemented into microfluidics devices to generate OOCs, including gut, heart, liver, lung, skin, kidney, brain, muscles, bones, female and male reproductive organs, that can recapitulate the mains features of the tissue. The objective of this section is not to detail every technologies and approaches that have been developed for each OOCs (since it has already been extensively reviewed by others (Table 1)) but to provide a rapid state-of-art about the main organs involved in the ADME process and their associated OOCs already used for drug testing.

2.2.1 Liver-on-chip

The liver is the major site of drug metabolism in the body and therefore, is a key organ in the ADME process. Therefore, the conception of a functional liver-on-chip is critical for the design of relevant BOCs. The liver anatomical unit is the lobule, a hexagonal structure with a portal triad (hepatic artery, a portal vein and bile duct) at each corner connecting the central vein through the hepatic sinusoid. The hepatic sinusoid is a specialized capillary of fenestrated liver sinusoidal endothelial cells, wrapped by stellate cells, that also contains macrophages (Kupffer cells) and that is separated from hepatocytes by the space of Disse.[23] The hepatocytes surrounding the sinusoid present numerous microvilli and so, are in direct contact with the blood allowing the exchanges between liver cells and vascular system. In addition to this specific spatial configuration, the gradient of oxygen, growth factors, and hormones that forms along the sinusoid are responsible for the dynamic structural and functional liver heterogeneity, known as liver zonation.[43] Zonation is the precondition for performing all the metabolic functions and maintain homeostasis.[44] Due to this heterogeneity and complex organization, the reproduction of liver physiology in microfluidic devices is therefore a big challenge. However, studies managed to recreate liver lobules[45,46] or the sinusoidal cell composition and architecture[47–49] that

Table 1 List of the OOCs and their features reproducing the main organs involved in the ADME process.

Organ	Features	References[a]
Liver	Hepatic lobule structure; assess liver cell drug toxicity; liver sinusoid composition; oxygen tension	Beckwitt et al.[20] Du et al.[21] Deng et al.[22] Ehrlich et al.[23]
Kidney	Osmotic pressure drug excretion; Filtration/reabsorption	Zanetti[24] Kim and Takayama[25] Paoli and Samitier[26]
Gut	Crypt/villi structure; aerobic/anaerobic microbiome interface; peristalsis	Bein et al.[27]
Lung	Gas exchange; alveoli-capillary barrier; airway mechanical pressure; expansion/contraction	Shrestha et al.[28] Konar et al.[29]
Skin	Inflammation; skin barrier/permeability; epidermal morphogenesis; vascular network; immune system	Zhang et al.[30] van den Broek et al.[31] Sutterby et al.[32] Bal-Öztürk et al.[33]
Heart	Contractility; field potential; spontaneous & synchronous beating	Ribas et al.[34] Kitsara et al.[35] Zhao et al.[36] Zhang et al.[37]
Brain	Electrical/chemical signaling; myelination; blood-brain barrier	Zhao et al.[38] Frimat and Luttge[39] Bang et al.[40] Haring et al.[41] Raimondi et al.[42]

[a]Refer to extensive reviews about each OOCs.

can maintain certain specific features, such as oxygen gradient, cytochrome P450 induction or synthetic and secretory functions, for extended periods of time. Such advances showed the feasibility for the investigation of hepatotoxicity of diverse drug, including acetaminophen (alone or in combination), chlorpromazine, tacrine, caffeine, nimesulide or troglitazone.[50–54] Results showed that the chips were able to detect diverse liver toxicities (*e.g.*, hepatocellular injury, steatosis, cholestasis and fibrosis) or specific mitochondrial damages, and uncover mechanisms unidentifiable by using standard approaches.

2.2.2 Kidney-on-chip

The kidneys serve as a natural blood filter, removing water-soluble waste from the cardiovascular system while reabsorbing useful substances (water, glucose, amino acids, *etc.*).[26] Because of this secretory function, kidney is a central organ in the ADME process and nephrotoxicity has become a common side effect in clinical practice and a major cause of drug development failure. The kidney's function in the maintenance of fluid and osmotic pressure drug excretion is accomplished by its fundamental unit: the nephron. The nephron consists of the renal corpuscle, which includes capillaries (glomerulus) encapsulated in the Bowman's space, and that opens into the renal tubules, formed by three sections: the proximal tubule, the intermediate tubule, and the distal tubule. The nephron contains several cellular types organized into a complex 3D network, including the glomerulus epithelial cells, the podocytes, the glomerular endothelium, mesangial cells, the juxtaglomerular cells, and the tubular epithelial cells.[55] Each nephron's part is highly specialized, the glomerulus is responsible for renal ultrafiltration while the renal tubule reabsorbs water and pre-filtered useful substances toward the circulatory system.[56] Therefore, a kidney-on-chip model would be an useful tool to perform drug toxicity testing, determine the pharmacokinetics (when integrated in BOCs), and increase the precision of drug dose evaluation.[24] So far, models of the glomerulus[57–59] that mimic the glomerular filtration barrier, tissue-tissue interface and the clearance of multiple compounds, and models of proximal[60,61] and distal[62,63] tubules that induce a functional cell polarization and cytoskeletal reorganization, exhibiting biochemical activities and secretory and reabsorptive processes, have been developed, but the creation of a full nephron-on-chip platform remains challenging. However, these devices already showed their interest in the assessment of nephrotoxicity mimicking adriamycin-induced albuminuria and podocyte injury[57] as well as cisplatin toxicity[64] with similar cellular and physiological changes as observed in patients and *in vivo* data.

2.2.3 Gut-on-chip

All oral drug efficacy is substantially determined by its absorption and metabolism in the gut prior to entering the bloodstream, making the intestine a determinant factor in the ADME process. Most of the nutrient transport occurs in the small intestine, whereas the large intestine (colon) is primarily responsible for water and electrolyte transport as well as secretion through the feces.[65] Oral drug absorption is highly influenced by the intestine phenotype, including the structure of mucus network, the presence and regulation of intercellular tight junctions, and the expression of transporters and efflux pumps in epithelial cells.[66] Therefore, there is an important need for the gut-on-chip to recapitulate the intestinal histology as close as possible to the *in vivo* microenvironment. Intestinal villi are a key factor in the absorption of nutrients and their morphology is an essential functional unit to recapitulating drug intake. Several groups managed to recapitulate villi structure by using different approaches (*i.e.*, dynamic flow, cyclic strain, 3D scaffold) on microfluidic chips.[67–71] Results showed that villi/crypt structures are connected through tight junctions, covered by brush borders, and able to secrete mucin. A recent study using a duodenum-intestine-on-chip, made from duodenal organoids that displayed multi-lineage differentiation of native human intestine, showed that rifampicin and 1,25-dihidroxyvitamin D3 were able to induce CYP3A4, an enzyme that impacts the pharmacokinetics and bioavailability of various orally administered drugs and mediate drug-drug interactions.[72] Another important feature of the gut is its close interaction with a large diversity of microorganisms.[73,74] This gut microbiome and the resulting interaction with the host has been shown to be crucial in the maintenance of homeostasis and disease development.[75–78] Importantly, recent studies also showed that two-third of 271 tested oral drugs were metabolized by at least one bacterial strain[79] and that many drugs have unanticipated effects on the composition and metabolic function of the gut microbiota,[80,81] highlighting the non-negligible role of gut microbiome in drug response and the need to be taken into consideration during drug discovery. The controlled culture of microbiome in a chip is challenging but several studies already demonstrated the feasibility to integrate either aerobic or anaerobic microorganisms and to perform a co-culture with intestinal epithelial cells for days to weeks.[82–87]

2.2.4 Lung-on-chip

The lungs are the primary organs of the respiratory system in humans and a major administration route in the ADME process for inhaled drugs or toxins. The functional unit of the lungs is the alveoli, an interface between alveolar epithelial cells and pulmonary microvascular endothelial cells that allows gas exchange.[88] This alveolar-capillary interface has been successfully reproduced on chip,[89] including cyclic stretch capacity to simulate the mechanical stress of breathing.[19,90] The results displayed a functional air-liquid alveolar barrier, with tight and adherens junctions formed by the epithelial and endothelial cells, surfactant production by the epithelium and a low level of protein permeability as observed *in vivo* showing barrier integrity. To note, the presence of cyclic mechanical strain increased the permeability of small hydrophilic molecules without compromising the integrity and influenced the metabolic activity and the cytokine secretion of epithelial cells.[90] Interestingly, this strain also accentuated toxic and inflammatory responses of the lung to silica nanoparticles.[19] The same type of device was used to investigate toxicity of interleukin-2 (IL-2) treatment.[91] The results showed that the addition of IL-2 in the vascular compartment leaked into the air compartment and induced edema, the breathing movement even worsening the leakage. Interestingly, this device also allowed the identification of new drugs against IL-2-induced lung toxicity, highlighting the great potential of lung-on-chip for drug screening and toxicity assessment. A similar lung alveolus-on-a-chip, but with perfused whole blood instead of culture media in the vascular compartment, reproduced *in vivo* thrombus formation, including platelet-endothelial dynamics, pro-inflammatory signals secretion, and revealed that lipopolysaccharide (LPS) endotoxin indirectly stimulates intravascular thrombosis by activating the alveolar epithelium, rather than acting directly on endothelium.[92] Microfluidics devices have also been used to reproduce the human airway (bronchial epithelium-endothelium interface). Sellgren et al. developed a primary cells tri-culture on a chip, comprising lung fibroblasts, microvascular endothelial and tracheo-bronchial epithelial cells.[93] The results demonstrated that the chip could maintain viable cells up to 35 days, displaying mucociliary differentiation and barrier function. A simplified model also showed that the exposure of the airway epithelium to IL-13 reconstituted the goblet cell hyperplasia, cytokine hypersecretion and decreased ciliary function of asthmatics.[94] Altogether, these studies demonstrate that *in vitro* analysis using OOCs offer a great potential for enhancing pulmonary research and drug development.

2.2.5 Skin-on-chip

As the largest organ in our body, skin offers a promising pathway for drug delivery. The long-time use of ointments, gels and potions and the development of transdermal drug delivery, for either local or systemic effect, as an alternative to injections and oral delivery have seen significant success.[95,96] However, the skin is a complex organ, made by three majors layers: (1) the epidermis, primarily comprising highly stratified keratinocytes, (2) the highly vascularized dermis, connected to the epidermis through the basement membrane and comprising heterogeneous fibroblasts, and

(3) the hypodermis that includes reticular fibroblasts and adipocytes, key parameters for intercellular communication.[32] The most common skin-on-chip models integrate either one layer[97,98] or epidermis and dermis together[99–103] while more advanced models also incorporate blood vessels[104–106] and even the hypodermis layer.[107] These models allow epidermal morphogenesis and differentiation, an enhanced barrier function, and autocrine and paracrine communication between the different cell types. In addition, the dynamic perfusion of media has been showed to improve tight junction formation and promote a faster maturation. These chips already showed their great potential for drug testing. In a vascularized epidermis-on-chip, the chemical irritants steartrimonium chloride (SC) and lauryl sulfate (SLS) have been tested and results showed that SLS disturbed the keratinocyte cell proliferation and significantly affect angiogenic performance.[106] Interestingly, SC also promoted vessel sprouting and lumen formation, despite its classification as nonirritating and its use in supposedly safe formulations. Ramadan and Ting exposed a skin-on-chip allowing co-culture of keratinocyte and monocytes to LPS and demonstrated a significant decrease of IL-6 and IL-β secretion compared to the culture of monocytes alone, suggesting that keratinocyte layer formed a robust barrier against LPS invasion.[97] A epidermis-dermis-endothelium model also proposed to test the effect of tumor necrosis factor alpha (TNFα) and its countermeasure dexamethasone (Dex). Results showed that TNFα exposure induced cytokine production, disrupted tight junctions (*i.e.*, zonula occludens 1) and increased endothelial permeability, thus mimicking fluid accumulation in the interstitium and creating a new model of edema.[105] Interestingly, the device allowed to show that Dex counteracted the effect of TNFα by decreasing inflammation and edema. On a human skin equivalents-on-chip, comprising both the epidermal and dermal compartments, Abaci et al. showed that doxorubicin treatment caused a spatial detachment of the basal layer along the epidermal–dermal interface, stopped the proliferation of the basal keratinocytes and induced the expression of Loricrin, a stratum corneum marker, in the basal and suprabasal layers.[101] Overall, this data demonstrated that the skin-on-chip was able to recapitulate the toxic effect of a drug on keratinocyte proliferation and differentiation. By using a three chambers microfluidic device (*i.e.*, drug delivery compartment, skin preparations from rodents, and receptor compartment), another study assessed the characteristics, degree, and speed of absorption of caffeine through the dermal barrier.[108] Despite some slight discrepancies probably due to different backgrounds, the authors showed that the results

obtained with the device are similar to the conventional Franz diffusion cell systems. Similarly, another study demonstrated the utility of microfluidics devices to assess the antiaging effect of *Curcuma longa* leaf extract, a natural product cosmetic ingredient, on the skin barrier function of the skin epidermis layer.[109] However, although skin-on-chip technology demonstrated great promise for drug testing, improvements are still necessary for repeated dose toxicity testing.[31]

2.2.6 Heart-on-chip

Many drugs and drug regimens can lead to heart failure by direct myocardial toxicity and/or drug-drug interactions.[110,111] Therefore, heart-on-chips, recapitulating either contractility or electrophysiology (or both) have been developed over the last years in order to provide a new tool for cardiotoxicity-inducing drug screening. These chips already demonstrated tissues growth, transmission of electrical potential facilitated by the presence of well-formed gap junctions and synchronous contraction. One of the first study assessed the effect of haloperidol by using on-chip cardiomyocyte cell network.[112] Authors showed that beating interval of a four-cell network is affected by drug exposure but returned to normal 10 min after the haloperidol was washed out of the chip. Interestingly, they also showed that using a nine-cell network, the beating interval was able to return to normal faster after exposure than the four-cell network, highlighting the importance of the community size in cell networks used for drug screening. A similar study also showed that epinephrine exposure increased the average frequency of contraction of multiple tissues, recapitulating previously reported epinephrine dose response curves from isolated atria strips of adult rats.[113] For a high throughput perspective, Agarwal et al. developed an heart-on-chip with 35 tissues, measured their diastolic and systolic stresses and thus, could show the isoproterenol-induced changes on cardiac contractility.[114] In order to improve the key physical and biological cues of the native myocardium, Marsano and al. developed a microfluidic device able to induce homogeneous uniaxial cyclic strains to 3D cardiac microtissues.[115] Authors showed an increase in junction complexes in stimulated microtissues compared to control, leading to a superior cardiac differentiation, as well as electrical and mechanical coupling. In addition, these cardiac microtissues were exposed to isoprenaline, a β1- and β2-adrenoreceptor agonist commonly used to treat bradycardia or heart block, in the presence or the absence of an external electrical pacing signal. Results showed that spontaneously contracting human cardiac microtissues responded to low concentrations of isoprenaline with a positive chronotropic effect, as previously described in large- and micro-scale human embryonic cells. Similarly, another study showed the interest of heart-on-chip to test the tachycardia effect of β-adrenergic agonist (isoproterenol).[116] The same study also assessed the effect of terfenadine, a cardiotoxic pro-drug that may cause life-threatening cardiotoxicity, and its active but innocuous antihistamine metabolite, fexofenadine. Results showed that fexofenadine and low dose of terfenadine do not induce significant prolongation in rate-corrected field potential duration. However, high dose of terfenadine prolonged rate-corrected field potential duration and even cessation of cardiac field potentials activity, demonstrating that the device is sensitive enough to differentiate electrophysiological properties of a pro-drug and its metabolite. With a novel cardiac platform that can provide real-time, non-invasive data acquisition of both cardiac electrophysiology and contractility under physiological conditions of human induced pluripotent stem cell-derived cardiomyocytes, Qian et al. also showed the effect of norepinephrine (*i.e.*, enhanced action potential propagation speed, increased beat rate and altered activation).[117] Overall, heart-on-chips, incorporating both electrical and mechanical stimulation, start to show their efficacy for testing new compounds. However, additional developments are still required to efficiently translate this technology into new drug screening platforms.

2.2.7 Brain-on-chip

The brain is a complex organ, responsible for many tasks, in response to environmental demands, that include the processing and interpretation of sensory information to make coordinated decisions, learning and memory, motor control, the secretion of chemicals, the consciousness, the sexual functions, *etc.*[118] The brain is structurally mainly composed of neuron and glial cells. The neuron is responsible for receiving and sending, sometime over long distances, nerve impulses or signals while glial cells (*e.g.*, astrocytes, microglia, and oligodendrocytes) provide support and nutrition, maintain homeostasis, form myelin and facilitate signal transmission in the nervous system. Given this complex structure, the current brain-on-a-chips do not recapitulate the full functions but rather focus on specific areas of the brain, including: unidirectional neural network, myelination process, blood-brain barrier and structure of the spinal cord.[40] The development of devices with stacked microchannels of different heights and widths to compartmentalize different tissue cultures allowed the separation of the axon from the soma of a neuron, thus guiding the growth of the axon and controlling its development or connection with others. The growth and culture of neurons on chip (innervation-on-chip) is detailed in Section 2.3.3. For instance, such devices allowed the study of axon regeneration after lesions[119] or the effect of mepivacaine, which blocks Na+ channels, on the velocity of the propagating signals.[120] Following an identical design but mimicking the neuromuscular junction, microfluidics devices also managed to reproduce the connection between muscle cells and motoneurons[121,122] and showed the presence of potential action and a significant increase in contraction frequency after glutamate stimulation.[122] Interestingly, the addition of D-tubocurarine, a neuromuscular blocker that binds to and inactivates acetylcholine receptors, was shown to reduce glutamate-stimulated contraction frequencies to baseline levels, suggesting that such systems respond to drug stimulation and could be used for drug screening. To note, because of the isolation of axon, these systems can also allow the study of myelination,[123,124] the formation of a myelin membrane around the axon that increases the efficiency of electrical transmission. The selective transport ability of the blood-brain barrier, that allows the passage of nutrients and wastes but prevents the influx of pathogens and large hydrophilic molecules, plays a significant role in the failure of many drugs tested for brain diseases. Hence, there is an important need to reproduce this selective semipermeable border *in vitro* in order to improve drug discovery and neurotoxicity screening. Many microfluidics devices with different types of model design have been already developed and showed that, although permeability levels as low as *in vivo* has yet to be achieved, they can reproduce appropriate shear stress, transcytosis, tight junction upregulation and functional efflux pumps.[125,126] Thus, such models showed their efficiency to mimic the effect of drugs on permeability, like for example, the effect of intravascular administration of the psychoactive drug methamphetamine, that induced disruption of cell–cell junctions and lowered expression of VE-cadherin, leading to barrier breakdown (reversible after drug withdrawal), as observed *in vivo*.[127] Finally, and as mentioned above, microfluidics devices were also able to reproduce a spinal cord-on-a-chip where induced pluripotent stem cell (iPSCs)-derived ventral spinal neurons were cocultured with brain endothelial cells to form a vascular-neural interaction.[128] Authors discovered that by using this system, the gene expression was like that of a fetus in comparison to monoculture or 2D systems. Although a complete brain-on-chip, or at least a technology gathering several of the main brain structures and functions, is far from being achieved, these recent developments have led to a great potential for advancing drug therapy treatments for neurological diseases.

2.3 The next generation of organ-on-chip

2.3.1 3D cell culture integration

The majority of the OOCs described above rest upon the same configuration and mainly consist on perfused microfluidic chamber that contains a single monolayer of cells, thus exhibiting limited functions of one single tissue type, and that can be sometime separated by a porous membrane from another compartment to recreate artificial interfaces between different tissues.[11] Therefore, in terms of cellular heterogeneity, phenotype fidelity, and physiologically relevant complexity, these OOCs still suffer important limitations.[129] Organoids (and to a lesser degree, spheroids) have the main advantage of a structural sophistication made of multiple lineages with the ability of self-organization,[130,131] thus recapitulating the microanatomy and key physiological aspects of the targeted tissue. The combination of organoids technology with OOCs is therefore a promising approach to improve our current *in vitro* models and has seen an increased interest these last years.[132] To date, several microfluidics devices have been developed either to promote spheroids formation or maintain spheroids in culture[133] and helped to better understand the behavior of 3D cellular structures in such environment. This already led to the development of multi-organoids-on-chip,[134] in which several tissue are cultured as organoids under perfusion and their interactions are investigated to study a variety of different mechanisms including drug toxicity or metastasis dissemination.[135,136] In an evolved model, Choi et al. developed a microsystem that enabled co-culture of breast tumor spheroids on a monolayer of human mammary ductal epithelial cells, separated by a porous membrane from mammary fibroblasts to replicate the 3D microenvironment of breast ductal carcinoma *in situ*.[137] By using this model, they showed the anticancer effect of paclitaxel on breast cancer spheroids and its low toxicity on normal cells. Another example of a full integration of organoids in a vasculature-like perfused chip is the recent conception of a new microphysiological model of the human retina integrating more than seven different essential retinal cell types derived from human induced pluripotent stem cells and recapitulating for the first time the interaction of mature photoreceptor segments with retinal pigment epithelium *in vitro*.[138] Interestingly, authors also demonstrated the applicability of the chip for drug testing, by reproducing the retinopathic side-effects of the anti-malaria drug chloroquine and the antibiotic gentamicin. Despite the traditional challenges faced by such approaches, including organoids/spheroids retrieval, imaging capability, high-throughput screening, microfabrication processes or biosensors integration, the current devices paved the way to a promising technology development, combining organoids and OOCs, that should mirroring the complex inner workings of human organs in easily accessible and controllable model systems.[139]

2.3.2 OOC vascularization

In order to better reproduce organ physiology, the vascularization and innervation of the tissue are vital and need to be integrated in the OOC. Vascularization is responsible of tissue oxygenation and nutriment supply, maintains tissue homeostasis, regulates the interaction with immune cells and its dysfunction can lead to critical diseases.[140] Two main types of vascularized OOCs can be differentiated: OOCs with endothelialized channels and angiogenesis/vasculogenesis-on-a-chip.[141] OOCs with an endothelium barrier are the most common and usually consist on monolayers of endothelial cells and parenchymal cells seeded on the opposite sides of a porous membrane.[142] This approach is probably so far the simplest to provide a naïve vascularization system to the chip. The attachment of endothelial cells on membrane and the culture in an independent compartment perfused by the appropriate cell medium is well controlled. However, although easy to implement and able to recapitulate the main endothelium-mediated mechanisms, this structure does not really reproduce the organ-specific microvasculature. Therefore, news chips have been developed to induce angiogenesis/vasculogenesis and promote the formation of 3D vascular networks. These 3D vessel structures add morphological complexity and thus can better reproduce the vascular delivery function. For example, Zhang et al. constructed a myocardium-on-chip that they vascularized by bioprinting endothelial cells within microfibrous hydrogel scaffolds that gradually migrated toward the peripheries of the microfibers to form a layer of confluent endothelium.[143] The cardiomyocytes seeded on the hydrogel showed contraction ability and when combined with a microfluidic perfusion bioreactor, the model could investigate the response of both cardiomyocytes (decrease of beating rate) and endothelial cells (decrease of vWF level) to doxorubicin. The main limitation of such model is that the microfibrous structures are not perfusable and consequently, despite the 3D structure, blood circulation cannot be reproduced. Further studies would need to address this point to merge the advantages of endothelialized channels and 3D networks.

To note, lymphangiogenesis should also be considered as the lymphatic system also plays an important role in the regulation of tissue homeostasis. Without a functional lymphatic microvasculature, a tissue or organ cannot efficiently drain the fluid that filters out of the capillaries along with vital proteins and the recruitment of immune cells would not be optimal.[144] Although microdevices have been developed to culture lymphatic vessels and promote lymphangiogenesis,[145,146] showing in passing the synergy with blood vessels and vascular angiogenesis, the integration of a lymphatic circulation within a more evolved OOC is still to be demonstrated. Recently, a study assessed the effect of doxorubicin on breast cancer cells by using a tumor-on-chip pairing 3D-printed blood and lymphatic vessels.[147] The results revealed that the drug diffusion and breast cancer cells viability varied following the different combinations of these bioprinted blood/lymphatic vessels (*i.e.*, presence of blood circuit only, or both blood and lymphatic circuits), highlighting the important role of the lymphatic system in PK/PD response and the necessity to take it also in consideration for drug testing-on-chip.

2.3.3 OOC innervation

The central nervous system (*i.e.*, brain and spinal cord) commands the rest of nerves of the body (peripheral nervous system) that connect to the targeted tissue or organs, through axonal connections from neurons, and ultimately regulate their development, maturation, function, regeneration or pathology.[148] The peripheral nervous system can be divided into two systems: the somatic nervous system that voluntary controls the motor functions *via* skeletal muscles, and the autonomic nervous system that regulates organs functions, such as the heart rate and contraction force of cardiac muscle, bladder wall contraction and relaxation, regulation of intestinal transit, absorption, secretion and immune response, pupillary function or respiratory rate.[149] To date, only a few types of innervated tissues have been explored on a chip and this needs to be developed further to reproduce better the *in vivo* functions. One of the first innervated organ-on-chip was the heart.[150–152] In a microfluidics device, neurons were seeded in one compartment while ventricular myocytes or pluripotent cell-derived cardiomyocytes were cultured in another chamber. Results showed that neurons were able to connect to cardiac cells in the other compartment *via* the formation of neurites and that their electrical stimulation induced changes in the beat rate, correlated to the pulse frequency. Interestingly, the addition of propranolol, a β-adrenoreceptor antagonist, in culture medium prevented the increase of beat rate after stimulation. Together, these data demonstrated the possibility of reproducing an organ-autonomic nervous system connection. Following an identical approach by promoting the sprouting of neurons and their connection to target cells in a microfluidic device, recent studies also demonstrated proof-of-concept for modeling the blood–retinal barrier,[153] the neuronal-induced osteogenic differentiation capacity of mesenchymal stem cells by the mediation of several molecular events (including the increase of alkaline phosphatase activity, the upregulation of osteoblast-specific genes, cytoplasmic accumulation of β-catenin translocation into the nucleus, *etc.*),[154] the neuromuscular junctions[155] and the repulsive or attractive effects of teeth on trigeminal innervation mimicking the *in vivo* development of orofacial tissues and organs.[156]

In order to understand better the innervation-on-chip and the mechanisms that govern such systems, several studies also focus on the growth and isolation of axons and the implementation of synaptic connection.[157,158] Thus, by using the properties of PDMS and collagen, a recent study developed a microfluidic platform that allowed the assembly and alignment of fibrous structures in a multi-modular 3D conglomerate to reconstruct a hippocampal neural network.[159] Data showed that CA3 neurons extended parallel axons to and synapsed with CA1 neurons, establishing a functional connectivity. A similar synapse-on-chip reconstituted an *in vivo* corticostriatal network to reproduce healthy and Huntington's disease (HD) conditions.[160] The device highlighted the major defects in the different compartments of the HD corticostriatal circuit, in particular the role of presynaptic neurons. Another model investigated synaptic competition using a 3-compartment microfluidic device in which neurons seeded into two chambers innervated target neurons placed in a central compartment through microgrooves.[161] Interestingly, the inhibition of one side chamber, using the $GABA_A$ receptor agonist muscimol, led to an increased synapse number and axon elongation in the opposing untreated chamber and microgrooves to connect to the central compartment. Altogether, these studies demonstrated the importance of innervated OOCs for investigating pathogenic mechanisms and for identifying drugs.

2.3.4 Mechanobiology-on-chip

During the last two decades, increasing evidences showed that cell function is regulated not only by biochemical signals (*i.e.*, growth factors, cytokines, and chemotactic molecules, *etc.*) but also by physical and mechanical forces applied to the tissue environment (shear stress,[162,163] matrix stiffness,[164,165] tension and compression,[166,167] *etc.*).[168] These mechanical cues appeared to control multiples biological processes such as gene expression,[169] cell fate switching, pattern formation, and tissue development in the embryo,[170] as well as the onset and progression of multiple diseases including atherosclerosis, fibrosis, cardiac hypertrophy, muscular dystrophy and cancer.[171–173] The conversion of these mechanical signals into a cellular response is called mechanotransduction. From this concept has emerged the new fields of mechanopharmacology[174,175] and mechanomedicine that focus on "therapeutically targeting tissue mechanics, either by directly altering the mechanical cues presented to cells or by disrupting the cellular response to mechanics."[176]

Therefore, the implementation of mechanical cues in addition to the standard biochemical regulation between cells is of importance to reproduce more accurately the *in vivo* microenvironment and needs to be taken into consideration for OOCs design. Since the perfusion of liquid is integral part of the OOC functioning, most of them can easily recapitulate fluid stress. However, few OOC studies used liquid flow to study a direct effect of shear stress on tissue/organ or drug delivery efficacy but instead, mainly to move liquid from one compartment to another, thus neglecting the effect of liquid movement on seeded cells. The effect of shear stress on the vascular compartment and endothelial cells has probably been the most studied phenomenon to date, however, there is a need to investigate also the effect of fluid flow on others cells to understand better others mechanisms such as those regulated by interstitial fluid pressure and movement.[177]

Using deformability property of PDMS, Ingber group developed a multi-compartmentalized vacuum-connected device that, in addition to fluid flow (*i.e.*, air flow at the air–epithelium interface and blood flow at the endothelium–blood interface), reproduced the mechanical stretching of breathing movements on a lung-on-chip.[19] Data showed that the inclusion of this mechanical stretching had a significant effect on ROS generation and nanoparticles absorption compared to a static device, highlighting the importance of mechanical constraints to reproduce *in vivo* physiology. Following an identical design, the same group managed to develop a glomerulus-on-chip that also models the dynamic mechanical strain observed in living glomeruli due to the cyclic pulsations of renal blood flow.[57] The application of cyclic mechanical strain showed to enhance podocytes spreading and to increase their levels of nephrin expression, suggesting that mechanical forces can influence podocyte differentiation by enhancing expression of lineage-specific markers. The current OOCs implementing mechanical stimuli have been extensively reviewed by Ergir et al.[178] and Kaarj and Yoon.[179]

3. Interconnected organ-on-chips: The emergence of body-on-chips

As described in Section 2, OOCs rapidly evolved and can now mimic healthy organ physiology or pathophysiological conditions, and, therefore, have been used to test different drugs. However, to reproduce ADME system and PK/PD response of different drugs (Fig. 1) for generating reliable results that could be extrapolated and compared to human *in vivo* data, the concept of fluidically linking OOCs has emerged,[180] giving birth to multi-organ or whole-body systems, referred in this paper as "body-on-chips" (BOCs).

3.1 The first macroscale body-on-chip systems

As for OOCs, first BOC concepts were relatively simple and consisted of a limiting number of organs, represented as one kind of cultured cell and linked through a unique fluidic circuit. Among the pioneer works, the *macroscale* cell culture analog platforms[181–183] can be considered as one of the first attempt for establishing *in vitro* PK/PD based testing methods. These platforms consisted of two cell culture glass bottles reproducing liver (H4IIE rat hepatoma cell line) and lung (L2 rat lung cell line) compartments, linked together through a third cell-free "other tissues" compartment. These experiments demonstrated that reactive metabolites of naphthalene (naphthoquinone and naphthalene epoxide) are produced by H4IIE cells in the liver compartment and circulate to the lung compartment where they can induce L2 cell death, suggesting the promising interest of such multi-organ physiological systems for studying the action of compounds with reactive and toxic metabolites. However, one of the main limitation of these macroscale systems is that they did not establish physiologically realistic ratios of liquid to cell volumes. Indeed, most of the signals involved in cell and organ communication are dependent on the effector concentration. A large volume of liquid compared to the amount of cells present in the device dilutes the drug metabolites and soluble ligands, leading to incorrect PK/PD data and potential failure to detect certain drug effects.[184] Thus, in addition to their advantage of using minimal amount of cell or tissue material and potentially expensive reagents, *micro*systems have been developed to better mimic the appropriate physiological scale.

3.2 The emergence of the microscale recirculating platforms with a single fluidic circuit

Microscale microfluidics devices and their recent advances in fluidic control, platform materials, platform throughput, and compatibility of circulating cells have been extensively reviewed in exhaustive papers from Shuler group.[184,185] In these papers, the authors classified BOC systems into four categories: (1) static microscale platforms, (2) single-pass microfluidic platforms, (3) pump-driven and (4) pumpless recirculating microfluidic platforms. As specified by their name, static microscale platforms do not involve any active fluid movement, and interactions between tissue are achieved through direct physical contact and/or by passive diffusion of soluble signal *via* a common medium connecting the different compartments. Single-pass microfluidic platforms integrate different tissue compartments connected through a shared medium perfused in unidirectional fluid routes. Recirculating multiorgan platforms connect individual organ compartments in a closed-loop system that allows reciprocal organ interactions though the continuous circulation of medium driven by either pumps or gravity. To note, single-pass or recirculating platforms can integrate one single or multiple fluid routes depending on whether these systems reproduce non-barriers or barriers organs. Although they involve complex fluidic components, these latest should be prioritized for drug toxicity screening as they better recapitulate the blood distribution and the exchanges between the parenchyma and vascular compartments that participate to drug dilution and PD/PK.

Among the first microscale BOCs, Viravaidya et al. conceived a device based on the PD/PK model of a rat,[186] that flowed medium through a first chamber mimicking lung tissue and that is then distributed to a fat (6%), liver (25%) and other tissue (66%) chambers before being combined and pumped back to the lung compartment.[187,188] Using this system, authors studied bioaccumulation, distribution, and toxicity of different compounds for a maximum period of 6 h. Results revealed that hydrophobic compounds preferentially accumulated in the fat chamber. Similar to the macroscale systems previously described, naphthalene was converted into reactive metabolites (*i.e.*, 1,2-naphthalenediol and 1,2-naphthoquinone) in the "liver" compartment, which then circulated to the other compartments and favorized glutathione (GSH) depletion in lung cells. Interestingly, the presence of differentiated adipocytes reduced this depletion and thus the lung toxicity. The same group later developed a platform that allowed the culture of four interconnected chambers (liver (HepG2/C3A), bone marrow (MEG-01), uterine cancer (MESSA), and a multidrug resistant variant of uterine cancer (MESSA/DX-5)) for 3 days, which have been scaled to match human tissue residence times as well as cell-to-organ volume ratio. This system has been used to evaluate exposure to doxorubicin (DOX) in combination with drug modulators cyclosporine (CSP) and nicardipine (NCP).[189] Results showed that acute exposure of DOX induces a significant decrease of MESSA and HepG2/C3A viability but has a limited effect on MESSA/DX5 cells. The presence of both drug modulators had no-to-few effect compared to DOX alone but when CSP and NCP are combined with DOX, toxicity is enhanced toxicity in the MESSA/DX-5 but not in MESSA and MEG-01 cells. This synergic effect can be partially explained by the enhancement of DOX uptake in presence of drug modulators. Interestingly, when these results were compared with static culture in 96-well plate, discrepancy was shown as modulators significantly enhanced DOX toxicity on all cell lines, demonstrating here an additive effect rather than synergistic. Although the particular mechanism of this effect is unknown, authors suggested that the "exchange of metabolites and response to flow are two key differences" between the two models, highlighting the significant advantage of BOC to better reproduce *in vivo* organ-to-organ exchanges and PD/PK than standard approaches. The same team confirmed later these observations by using a three-compartments BOC, reproducing liver, colon and bone marrow tissue, to assess the cytotoxic effect of the oral prodrug Tegafur and the metabolism of its product 5-fluorouracil (5-FU).[190,191] The addition of Tegafur within the BOC showed a distinct cytotoxic effect on colon cancer cells in a time-dependent manner while no toxicity had been detected in static 96-well plate culture as well as in BOC in absence of liver cells, suggesting that the BOC allowed the conversion of Tegafur to 5-FU in the liver compartment and its circulation to tumor cells. In addition, the integration of PK/PD model showed that in the dynamic BOC model cell death kinetics is faster than conventional cell culture suggesting that cells were more sensitive in a lower concentration range in BOC. Also, the rate of natural cell death (without drug) is different for all cell types in BOC while relatively similar for the static culture, implying that cells have differential adaptability to the microfluidic environment. Authors observed that "this dependency of cell responses on cell types and culture environment demonstrates the fact that conventional *in vitro* cell based assays may not necessarily reflect the drug effects that would be observed in the same environment under dynamic conditions."

An evolution of this microscale system has been made with the integration of Transwell inserts for cell culture in the platform. Although the upper compartment of the insert is not perfused (and so it must still be considered as a recirculating platform with one single fluidic circuit), this allows a physical separation between the basal and apical compartments of the same tissue. Using this approach, Wagner et al. managed to maintain stable co-culture of skin tissue at the air-liquid interface with 3D liver constructs for 28 days.[192] In addition, liver microtissue respond in a dose-dependent manner to a long exposure to troglitazone as shown by an increased expression of P450 3A4, indicating a typical reproduction of the toxic mechanism *in vivo*. This upregulated biotransformation activity following troglitazone treatment has also been reproduced in an endothelialized chip although the response was lower than expected.[193] The authors did not discuss if such difference could be due to a barrier-effect from the endothelial cells that may prevent free diffusion of the drug or an additional cell-cell communication that may modify liver cells behavior, but this shows the importance of vasculature in ADME process and highlights the need to include endothelial tissue in BOC whenever possible. Using the same technology, the group performed the co-culture of liver and neurospheres and demonstrated the dose-dependent effect of the repeated exposure to the neurotoxic 2,5-hexanedione (over 2 weeks) as marked by the significant increase in LDH activity starting at day 7 (for low dose) and day 9 (high dose).[194] Interestingly, drug exposure of single tissue culture (either neurospheres or liver cells alone in the chip) revealed that co-cultivated tissues were more sensitive to 2,5-hexanedione. Although additional experiments would be required to elucidate this phenomenon, authors advanced the hypothesis of a transmission of necrotic signals from one tissue to the another, highlighting again the importance of tissue-tissue communication in drug response and the need of recapitulating such interactions *in vitro*.

3.3 Microscale systems complexification with the integration of multiple fluidic circuits

Among the first models with two independent fluidic circuits, Esch et al. constructed a microfluidic "GI tract–liver–other tissue" BOC device in which they cultured enterocytes (Caco-2) and mucin-producing cells (HT29-MTX) as well as hepatocytes (HepG2/C3A) connected through a first closed circuit that served the apical side of the Caco-2/HT29-MTX cultures and a second that served the basolateral side of the Caco-2/

HT29-MTX culture, the HepG2/C3A culture and the "other tissue" compartment.[195] To test the permeability of the intestinal layer to compounds and a possible resulting downstream toxicity, authors used carboxylated polystyrene nanoparticles (NPs) injected in the first fluidic circuit. Data showed that intestinal cells act as a barrier preventing most of NPs to cross and only single NPs and small aggregates were able to pass to the basolateral compartment. However, these small NPs were shown to induce aspartate aminotransferase (AST) release, an intracellular liver enzyme indicating cell injury. Interestingly, AST release was only observed for high NPs concentration when liver cells were cultured alone whereas low concentration was enough to increase AST concentration when co-cultured with intestinal cells. This paradoxical intestinal effect (*i.e.*, intestinal cells presents a significant barrier but their presence increases liver sensitivity to NPs) brings another proof of the importance of the tissue-tissue communication that BOC can recapitulate. The presence of a second fluidic circuit mimicking the interaction of two independent compartments adds a significant feature that better represents the absorption and distribution of the compounds.

An et al. designed a complex multi-layer BOC, vertically assembled, that includes a gut compartment (Caco-2 cells) separated from endothelial cells (HUVECs) by a porous membrane, located above a liver compartment, which is again separated from breast cancer cells (MCF7) by another HUVECs cells layer.[196] Below the MCF7 cells are disposed lung, heart and adipose tissues. In a first attempt to introduce the kidney compartment and the filtration process, a dialysis membrane is located below these tissues. This device is run by three fluidic circuits recapitulating the "digestive juice" (through the gut compartment), the blood (through the MCF7/HUVECs chamber) and the urine (through the empty chamber below the dialysis membrane). These circuits are independent, and recirculate only through their own chamber. Drug is diffused passively through the different membranes, from the gut compartment to the last chamber after the dialysis membrane. For these reasons, this system has limitation to reproduce the actual *in vivo* fluid circulation, but may provide valuable information on the ADME process. Thus, authors showed that concentration of propranolol, thiopentone and pentobarbital increased during absorption to reach different maximums due to different solubility properties. The concentration then decreased during metabolism and distribution, and the drugs were completely eliminated at the end of the processes. The device can recapitulate the interactions of these compounds with different organs as showed by propranolol highly distributed in lung and heart compared to adipose tissue, while pentobarbital was detected only in the fat tissue. The test of anti-tumor compound also showed that liver compartment could metabolized cyclophosphamide into aldehyde phosphoboramide leading to a toxic activity on MCF7 cells. In addition, antitumor activity of taxol, injected in the gut compartment, was lower than that in a 96-well plate due to the poor intestinal permeability, reproducing *in vivo* data showing that taxol must only be administered by injection (blood) and not orally.

The integration of the kidney tissue, a key organ in the ADME process, responsible of the elimination, is an important challenge. An et al. used a simplified model with a dialysis membrane but an implementation with kidney cells would be more relevant. Vernetti et al. proposed one of the first model with renal drug clearance.[197] The system comprised four modules reproducing the liver, the gut, the blood-brain barrier and the kidney compartment. The kidney tissue is made with renal proximal tubule epithelial cells and HUVECs to form a two lumen, dual perfusion system. However, the system is not fully integrated and each module are run by different laboratories and after flowing in the first module, the shared medium is collected and shipped to the other labs following the physiological order. Consequently, although the fluid circulation is preserved, caution has to be taken because of the lack of real-time and the risk of degradation of secreted metabolites and compounds in the medium. Nevertheless, the study showed that compounds such as terfenadine or trimethylamine could be metabolized to fexofenadine or trimethylamine-N-oxide respectively by the intestinal epithelium and liver, cleared through the liver, and secreted into the kidney lumen. Fexofenadine did not pass the blood-brain barrier but the study showed for the first time that trimethylamine-N-oxide could penetrate it. Although we cannot consider this system as a complete BOC, this study showed the feasibility of kidney integration in multi-organ systems. Since, several studies have used kidney modules in BOC to provide new information about excretion process during drug testing.[198,199]

3.4 The new generation of body-on-chips

As showed by these previous studies, the integration of multiples fluidic circuits to reproduce fluid circulation between key organs such as gut, liver or kidney is not an easy task. To overcome these problems, the Ingber group built a robotic system able to connect and maintain viable for 3 weeks eight vascularized OOCs (intestine, liver, kidney, heart, lung, skin, blood–brain barrier and brain) by intermittently flowing a common blood substitute through their reservoirs of medium and endothelium-lined vascular channels.[200] First, each coupled OOCs were made with a dual-channel architecture, including both parenchyma and endothelium, thus eliminating the lack of a separate vascular compartment with associated flux between the parenchymal and vascular spaces that could limit comparison with human data. Therefore, the system comprised multiple fluidic circuits for each chamber, whose linking control between each module is ensured by the robotic hardware and software following computational PD-PK modeling. Second, each OOCs were individually more developed than a simple culture of monolayer cells or organoids in a unique medium but instead, in addition to the endothelium compartment, also integrated specific physiological functions for each organ, such as gut peristalsis, liver microstructure or lung air-liquid interface. Authors demonstrated the utility of this system to predict pharmacokinetic responses of orally administered nicotine (using linked gut, liver and kidney chips) and for intravenously injected cisplatin (using linked bone marrow, liver and kidney chips).[201] Results showed that the nicotine could be effectively metabolized as detected by the presence of cotinine in the effluent from the apical hepatocyte channel of the liver chip. In addition, the maximum level of nicotine, the time needed for nicotine to reach the different tissue compartments, and the clearance rates in the liver chip matched closely what had been measured previously in patients. For cisplatin administration, the results also mirrored patient data recapitulating its bioavailability as well as a blood and kidney toxicity while the liver did not show any signs of injury.

In addition to all previous described systems, other BOCs have been used to test the efficacy and/or toxicity of several drugs, including doxorubicin, atorvastatin, valproic acid, N-Acetyl-p-aminophenol, N-Acetyl-m-aminophenol, epinephrine, bleomycin, propranolol, cyclosporine A, epirubicin, 17-β estradiol, soy isoflavone, aflatoxin B1 and diclofenac,[198,199,202–205] and to uncover new mechanisms that could not have been identified with standard models. All these studies are recapitulated in Table 2.

Table 2 Non-exhaustive list of the main microscale body-on-chips used for assessing drug efficacy and toxicity.

References	Organs	Cell lines	Category	Drug tested	Time	Results
Viravaidya et al.[187]	Liver	HepG2/C3A	Recirculating platform with one single fluidic circuit	Naphtalene	6 h	Naphthalene was converted into reactive metabolites in the liver compartment, which then circulated to the lung and induced toxicity that is reduced by the presence of the differentiated 3T3-L1 cells
	Lung	L2 rat lung Type II epithelial cells				
	Fat	3T3-L1 adipocytes				
Tatosian and Shuler[189]	Liver	HepG2/C3A	Recirculating platform with one single fluidic circuit	Doxurubicin (DOX) with drug modulators cyclosporine and/or nicardipine	24–72 h	DOX with modulators had better efficiency than DOX or modulators alone in the device compared to standard cell culture, suggesting cell-specific synergistic response of DX-5 cells in the device
	Bone marrow	MEG-01 (megakaryoblast)				
	Uterus	MES-SA (cancer) & MES-SA/DX-5 (doxorubicin resistant)				
Sung and Shuler[191]	Liver	HepG2/C3A	Recirculating platform with one single fluidic circuit	Tegafur	24–72 h	Tegafur was metabolized by HepG2/C3A cells into 5-fluorouracil which induced HCT-116 and Kasumi-1 cell death. Results not visible with standard cell culture model
	Colon	HCT-116				
	Marrow	Kasumi-1 (myeloblast)				
Imura et al.[204]	Liver	HepG2	Single-pass platform	Cyclophosphamide (CTX), epirubicin (EPI), 17-β estradiol (E2), and soy isoflavone (IF)	72 h (48 h with drugs)	CTX crossed the intestine layer but not EPI, E2 and IF. HepG2 cells metabolized CTX, E2 and IF but not EPI. MCF-7 cell proliferation was stimulated by E2 and IF (stimulation higher in absence of HepG2)
	Gut	Caco-2				
	Breast	MCF7				
Sung et al.[190]	Liver	HepG2/C3A	Recirculating platform with one single fluidic circuit	5-fluorouracil (5-FU)	24–72 h	Each cell type responded differentially to 5-FU in the device and compared to standard cell culture model
	Colon	HCT-116				
	Marrow	Kasumi-1 (myeloblast)				

Continued

Table 2 Non-exhaustive list of the main microscale body-on-chips used for assessing drug efficacy and toxicity.—cont'd

References	Organs	Cell lines	Category	Drug tested	Time	Results
Wagner et al.[192]	Liver	HepaRG + HSCs	Recirculating platform with one single fluidic circuit	Troglitazone	Up to 28 days (7 days with drug)	Repeated exposure to troglitazone induced hepatotoxicity with an increase of LDH activity and CYP3A4 in a dose dependent manner
	Skin	Biopsy				
Mashmeyer et al.[193]	Liver	HepaRG + HSCs	Recirculating platform with one single fluidic circuit	Troglitazone	2 weeks (3 days exposure)	Troglitazone was able to cross the intestinal barrier, and then was pumped to the liver compartment where it was transformed into quinone and epoxide
	Skin	Prepuce biopsy				
	Gut	Primary small intestine epithelial cells				
	Endothelium	HDMECs from biopsy				
Materne et al.[194]	Liver	HepaRG + HSCs	Recirculating platform with one single fluidic circuit	2,5-hexanedione	2 weeks	Drug induced apoptosis in liver and neurospheres. Toxicity profiles of two different doses could be discriminated. Co-cultures were more sensitive to the drug compared to respective single-tissue cultures in the BOC
	Brain	Neurospheres differentiated from NT2 cells				
Esch et al.[195]	Liver	HepG2/C3A	Recirculating platform with two fluidic circuits	50 nm carboxylated polystyrene nanoparticles (NPs)	24 h	Caco-2/HT29-MTX barrier presents an effective barrier for the NPs. At high concentration, single NPs and small aggregates crossed Caco2/HT29-MTX barrier and induced liver cell toxicity
	Gut	Caco-2 + TH29-MTX				

An et al.[196]	Gut	Caco-2	Recirculating platform with three fluidic circuits ("digestive juice, blood and urine")	propranolol (PRO), thiopentone and pentobarbital; N-Acetyl-p-aminophenol (APAP); CTX, taxol and 5-FU	72 h	The concentration of pentobarbital, thiopentone and PRO first increased during the absorption process and then declined during the metabolism, distribution and elimination processes. PRO was preferentially deposited in the lung and heart tissues and pentobarbital in fat, matching animal testing. The primary hepatocytes metabolized CTX that ultimately showed antitumor activity on MCF7 cells
	Breast	MCF7				
	Liver	Primary hepatocytes (rat)				
	Lung	Primary tissue (rat)				
	Heart	Primary tissue (rat)				
	Adipose tissue	Primary tissue (rat)				
	Endothelium	HUVECs				
Oleaga et al.[202]	Liver	HepG2/C3A	Recirculating platform with one single fluidic circuit	DOX, Atorvastatin (ATR), Valproic acid (VPA), APAP, N-Acetyl-m-aminophenol	5 days +2 days drug exposure	DOX induced liver and cardiac toxicity, and ATR muscle, neural and liver toxicity. Both showed reduction in muscle functionality. VPA affected the liver compartment while showing neuroprotective effects. APAP overexposure decreased liver viability while preserving cardiac functionality. All these results are consistent with published toxicity results from human and animal data
	Heart	iPSC CMs				
	Brain	Motoneurons from hSCSC				
	Skeletal muscle	Myofibers from hSKMSCs				
Vernetti et al.[197]	Gut	Jejunum enteroids, Goblet, Paneth and enteroendocrine cells	Single-pass platform with multiple compartments	Terfenadine, trimethylamine (TMA), vitamin D3	6–120 h for each organ	Results are consistent with reported *in vivo* data. Terfenadine was metabolized to fexofenadine by the gut and liver and secreted into the kidney lumen. Fexofenadine did not penetrate the BBB. TMA was transported across the jejunum, metabolized by the liver to TMAO and then secreted into the kidney lumen. Vitamin D3 was transported without metabolism across the jejunal monolayer, metabolized to 25-(OH) vitamin D3 by the liver and penetrated the BBB
	Liver	Primary hepatocytes, Kupffer, Stellate and endothelial cells				
	BBB/ neurovascular unit	iPSC-derived human neurons, pericytes, astrocytes and endothelial cells				
	Kidney	RPTEC and endothelial cells				

Continued

Table 2 Non-exhaustive list of the main microscale body-on-chips used for assessing drug efficacy and toxicity.—cont'd

References	Organs	Cell lines	Category	Drug tested	Time	Results
Skardal et al.[203]	Liver	Primary hepatocytes, hHSCs and Kupffer cells	Recirculating platform with one single fluidic circuit	APAP (liver); PRO and epinephrine for liver-heart; bleomycin for liver-heart-lung	Up to 14 days (9 days for Liver-Lung-Heart)	APAP induced liver toxicity, countermeasured by the addition of *N*-acetyl-L-cysteine. PRO is metabolized by liver increasing heart beating rate upon administration of epinephrine. Bleomycin induced lung cytokine releases leading to cardiac toxicity
	Heart	iPSC CMs and primary cardiac fibroblasts				
	Lung	Lung endothelial, airway stromal mesenchymal and bronchial epithelial cells				
Edington et al.[198]	Liver	Primary hepatocytes and Kupffer cells	Recirculating platform with one single fluidic circuit	Diclofenac (DCF)	4 weeks (2 days with drugs)	The DCF concentrations in each compartment are similar to the PBPK model predictions. DCF on the apical side of the gut increased in the basal compartment and propagated to the liver. At 48 h, concentration in the liver declined while concentration of its metabolite (4-OH-DCF) increased. 4-OH-DCF was metabolized in liver compartment, circulated throughout the platform and was detected in all the other compartments
	Gut	C2BBe1 and HT29-MTX-E21				
	Lung	NHBE				
	Uterus	Ishikawa human endometrial epithelial adenocarcinoma cells and tHESCs				
	Brain	iPSC-derived astrocytes and neurons				
	Heart	iCell CMs 2				

	Pancreas	Pancreatic islets from biopsy (rat)				
	Muscle	Primary skeletal muscle myoblasts				
	Skin	Normal keratinocytes				
	Kidney	RPTEC				
Lin et al.[199]	Liver	HepaRG and HSCs	Recirculating platform with one single fluidic circuit	cyclosporine A (CsA) alone or in combination with rifampicin (RFP)	14 days	High CsA concentrations increased the epithelial barrier permeability and nephrotoxicity, reversed by RFP
	Kidney	RPTEC				
Schimek et al.[205]	Liver	HepaRG and HSCs	Recirculating platform with one single fluidic circuit	Aflatoxin B1	14 days (24 h with drug)	Aflatoxin B1 impaired functionality in lung tissue when culture alone but showed a protective effect when the tissues were co-cultured with liver spheroids
	Lung	Bronchial MucilAir culture				
Herland et al.[201]	Liver	Primary hepatocytes and sinusoidal endothelial cells	Recirculating platform with multiple fluidic circuits	Cisplatin (bone marrow) and nicotine (gut)	Up to 6 days	The maximum nicotine concentrations, the time needed to reach the different compartments, and the clearance rates in the liver are consistent with patient data. Nicotine altered gut chip barrier function and kidney transporter. Cisplatin pharmacodynamics matches clinical data, including the reduction in blood cell count, the high expression of biomarkers for kidney injury and the lack of hepatotoxicity
	Kidney	RPTECs and GMVECs				
	Gut	Caco-2 Bbe and HUVECs				
	Bone marrow	CD34+ progenitor cells and HUVECs				

CMs, cardiomyocytes; GMVECs, glomerular microvascular endothelial cells; HDMECs, human dermal microvascular endothelial cells; HSCs, hepatic stellate cells; hSCSC, human spinal cord stem cell; HSKMSCs, human skeletal muscle satellite cells; HUVECs, human umbilical vein endothelial cells; iPSCs, induced pluripotent stem cells; NHBE, normal human bronchial epithelial cells; RPTECs, primary renal proximal tubule epithelial cells; tHESCs, hTERT-immortalized human endometrial stromal cells

4. Advantages and limitations of OOCs and BOCs

BOCs and OOCs are a relatively recent technology, that emerged for their potential to combine the strengths of animal models and cell culture (Table 3). Standard cell culture models are easy to handle, cost-effective,

Table 3 Comparison of standard cell culture, organ- and body-on-chips and animal models features.

	Standard cell culture	OOCs/BOCs	Animal model
Time	Cells need to be passed regularly. Limited by *in vitro* senescence/ derivation	Setup (~1–2 week) + Experiment (up to 1 month)[a]	Animal lifespan
Cost	Low	Require large equipment but low price per chip	Low for wild-type (small animal) but high for transgenic and/or humanized animals (and large animals)[b]
High-throughput	Yes	No	No
Knowledge/ technical skills	No	Yes, for chip conception but limited for routine use	Require training in animal care
Co-culture	Limited to 2 or 3 cell types	Up to 10 organs (BOC)[c]	Unlimited
3D structure	Limited to Spheroids/ Organoids	Limited to Spheroids/ Organoids	Yes
Biophysical environment	No. Static and stiff surface	Yes. Fluid, compressive and/or stretching stress	Yes
Scaling	Very high	Moderate	Ideal

[a]Setup usually includes fabrication, assembly, sterilization, cell seeding and adaptation. Microfluidic devices allow relatively long-term maintenance cell culture (usually 2–4 weeks, but an automated microfluidic cell culture allowed culture of differentiated neurons up to 100 days[206]).
[b]In addition to numerous ethical problems.
[c]A recent study demonstrated the possibility of linking 10 organs,[198] but may be higher in the future.

allow the usage of human material and, especially, can be used for high throughput screening. However, they lack physiological relevance: they do not reproduce the 3D structure of the tissue, the appropriate organ scaling, the interfaces between different tissues and they do not incorporate the physical constraint present in the *in vivo* environment. Animal models can overcome these problems but at a higher cost and despite limited translational results to Humans due to species difference. OOC and BOC systems appear to be able to merge these two approaches, although they still face several challenges.

First, and as discussed in Section 2.1, most of the chips and/or platforms are made with soft lithography using material such as PDMS. PDMS has multiples advantages: it is biocompatible, gas permeable, flexible (providing tunable mechanic features) and transparent, thus allowing performant cell culture with a possible direct visualization.[207,208] However, the main limitation of PDMS is its hydrophobic surface and tendency to absorb many hydrophobic molecules,[209] that can highly impact drug screening and ADME studies.[210] In addition, this hydrophobicity also requires pretreating (with extracellular matrix proteins or paraffin, for example) the surface to facilitate cellular attachment. Furthermore, the unselective absorption and lack of precise models for drug absorption rates can complicate the localization of target compounds in the devices resulting in analytical misinterpretation. Other issues such as high evaporation, leaching of uncured oligomers and unstable surface treatment contribute to limited scalability for production. An alternative could be to use materials such as stainless-steel, glass, Teflon, polystyrene, poly(methyl methacrylate) (PMMA), and polycarbonate that are considered to be inert to most chemicals, display slow degradation and allow mass production despite more complicated fabrication techniques.[211]

OOCs and BOCs allow human *in vitro* experiments in a highly controlled environment, reproducing as close as possible the *in vivo* environment. However, this control comes with a cost of large equipment (pumps, incubators, sensors, imaging system, *etc.*) and challenges in the connectivity and setting up. Indeed, OOC and BOC platforms require a fluid delivery system (pressure driven pumps, peristaltic pumps, syringe pumps, *etc.*), which connects the chips to the medium reservoirs, leading to a complex interface with multiple tubing and connectors. The temperature and gas levels in the chips also need to be controlled and so the entire platform is often placed in an incubator. To make the OOCs and BOCs accessible to a maximum of biologists, technology development is required to improve design, minimize platform size and volume, simplify steps in the setting up to facilitate the user experience. In addition, this simplification should naturally lead to an increase of the throughput which is still currently limited.

Indeed, most studies using OOC and BOC systems employ individual chips containing one or only a few tissues per chip.[212] Recent microfluidic OOC systems have been developed for large-scale drug screening,[213,214] however, their configurations are still naive and do not integrate the complete features allowing a similar performance to complex OOC or BOC platforms that can integrate mechanical constraints, multi-tissue compartments and multiple fluidic circuits.

In addition to the engineering challenges, there also are biological needs to be addressed by BOC platforms. First, the optimization of cell culture with multiple cell types sharing the same growing medium is essential for progressing toward more developed BOC systems.[185] This medium, often referred as blood surrogate, flows through all the compartments and is the main component for the organ-organ interaction, carrying nutriments, drugs and secreted factors. The adaptation of cells to a new and unique medium often leads to phenotypic changes that can falsify the results and has to be performed with caution after a good characterization to ensure that the cells still behave as expected.[215] The conception of BOC linked through the vascular compartments of several OOCs (with independent fluidic circuit to feed parenchyma compartments) seems to be a reasonable approach to overcome this challenge, the unique medium shared by all the chips being the "vascular" medium and feeding only the endothelial cells. However, the introduction of immune cells, another critical parameter to improve current BOC systems as immune responses can affect drug dose responses,[216] will require an optimization of this medium to adapt both endothelial and immune cells. To note, this medium can also be modified to match the physical parameters of whole blood (*e.g.*, viscosity), by adding for example, polyacrylamide, glycerol or xanthan gum,[217] in order to better reproduce tension and shear stress in vascular compartment.

The source of biological tissue is also one of the most important parameters in OOC and BOC systems,[218] a different genetic background of the tissues during the cocultivation being a major drawback when investigating organ-organ interactions. Although cell lines are the most convenient source to work with, they do not represent normal physiological conditions (often derived from cancer cells), come from multiple sources, and may have a limited impact. Primary cells constitute a reasonable option, as they derive directly from normal tissue, but they are difficult to access (by biopsies, that can even come from different donors as well), have a very low proliferative *in vitro* activity that limits cell culture time and promotes instability, and are, overall, an expensive resource. A good alternative is based on the use of stem cells and especially induced pluripotent stem cells (iPSCs) that spread enthusiasm for the development of new models of human disease and enhanced platforms for drug discovery these last years.[219] iPSCs are more accessible than primary cells, are derived from one single donor and can be differentiated into any tissue. Although this differentiation is not straightforward and there is a substantial need to improve the protocols for reprogramming the cells in the tissue of choice, the use of iPSCs will likely become the gold standard for OOC/BOC systems.

Finally, as described above, an additional challenge includes the important scaling issue ensuring a correct liquid/cell number ratio to mimic accurately organ size and medium volume. These microscale conditions are required to have a physiologically relevant model, in particular in a context of drug testing/screening, and extrapolate reliable PK/PD data. Moreover, miniaturization would allow the performance of multiple tests by minimizing expensive reagents and cells/tissue material, and microfabrication technology would provide identical units at low cost, facilitating rapid parallel testing.[220]

Because of these multiple parameters, reproducibility is still difficult to achieve for OOCs and BOCs. However, we believe that addressing the engineering (through automation, miniaturization, material choice, *etc.*) and biological (through the use of iPSCs, standardized protocols, *etc.*) challenges should eventually decrease variability and improve accuracy of such systems.

5. Conclusion

The perspective of human OOCs and BOCs being able to reproduce the complex biochemical and physical interactions present at tissue and organ levels open great promises for drug development, but also for other applications such as disease modeling, precision medicine and even clinical-trial-on-a-chip[221] (Fig. 3). As a result, OOC and BOC start-ups have already started to emerge from the academic research to propose commercial solutions for the drug discovery industry,[222] bringing these systems in a new era but critical phase.

However, the development of such technologies is not complete yet, and many improvements are still necessary to provide a reliable alternative

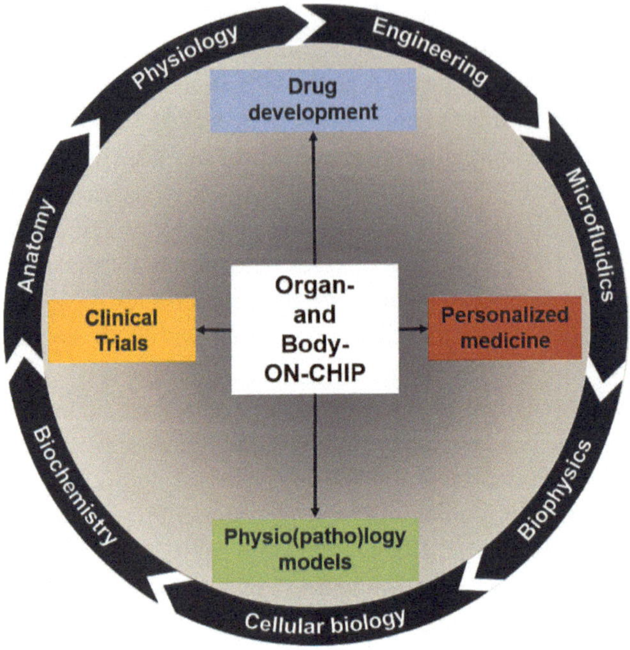

Fig. 3 OOCs and BOCs are at the interface of multiple scientific disciplines with great promises for several applications.

to animal models. The conception of OOC and BOC systems is not an easy task and requires a total collaboration and synergy between many scientific fields, including physiology, anatomy, cellular biology but also engineering and biophysics (Fig. 3). OOCs and BOCs have already benefit for miniaturization, automatization and parallelization; and improved cell culture techniques will have a vital role in all future cell-based assays,[223] especially those using stem cells[224] and/or gene editing.[225]

To demonstrate the great interest of such platforms, the goal of OOCs and BOCs development should not be too ambitious and claim immediate animal replacement or the feasibility of linking tens (or more) of different organs, but focus on the need of pharmaceutical industry and address first the main objectives defined for efficient human drug efficacy and toxicity testing. In view of the speed of technology development and biological knowledge improvement, this goal could happen faster than expected.

References

1. Dowden H, Munro J. Trends in clinical success rates and therapeutic focus. *Nat Rev Drug Discov*. 2019;18(7):495–496. https://doi.org/10.1038/d41573-019-00074-z.
2. Kola I, Landis J. Can the pharmaceutical industry reduce attrition rates? *Nat Rev Drug Discov*. 2004;3(8):711–716. https://doi.org/10.1038/nrd1470.
3. Esch MB, King TL, Shuler ML. The role of body-on-a-chip devices in drug and toxicity studies. *Annu Rev Biomed Eng*. 2011;13(1):55–72. https://doi.org/10.1146/annurev-bioeng-071910-124629.
4. Kapałczyńska M, Kolenda T, Przybyła W, et al. 2D and 3D cell cultures—a comparison of different types of cancer cell cultures. *Arch Med Sci*. 2018;14(4):910–919. https://doi.org/10.5114/aoms.2016.63743.
5. Langhans SA. Three-dimensional in vitro cell culture models in drug discovery and drug repositioning. *Front Pharmacol*. 2018;9:6. https://doi.org/10.3389/fphar.2018.00006.
6. Fang Y, Eglen RM. Three-dimensional cell cultures in drug discovery and development. *SLAS Discov*. 2017;22(5):456–472. https://doi.org/10.1177/1087057117696795.
7. Jensen C, Teng Y. Is it time to start transitioning from 2D to 3D cell culture? *Front Mol Biosci*. 2020;7:33. https://doi.org/10.3389/fmolb.2020.00033.
8. Godlee F. We need better animal research, better reported. *BMJ*. 2018;360:k124. https://doi.org/10.1136/bmj.k124.
9. Hackam DG, Redelmeier DA. Translation of research evidence from animals to humans. *JAMA*. 2006;296(14):1727–1732. https://doi.org/10.1001/jama.296.14.1731.
10. van der Worp HB, Howells DW, Sena ES, et al. Can animal models of disease reliably inform human studies? *PLoS Med*. 2010;7(3):e1000245. https://doi.org/10.1371/journal.pmed.1000245.
11. Bhatia SN, Ingber DE. Microfluidic organs-on-chips. *Nat Biotechnol*. 2014;32(8):760–772. https://doi.org/10.1038/nbt.2989.
12. Sander K. *Landmarks in Developmental Biology 1883–1924: Historical Essays from Roux's Archives*. Springer-Verlag; 1997. https://doi.org/10.1007/978-3-642-60492-8.
13. Harrison RG, Greenman MJ, Mall FP, Jackson CM. Observations of the living developing nerve fiber. *Anat Rec*. 1907;1(5):116–128. https://doi.org/10.1002/ar.1090010503.
14. Ehrmann RL, Gey GO. The growth of cells on a transparent gel of reconstituted rat-tail collagen. *J Natl Cancer Inst*. 1956;16(6):1375–1403.
15. Hamburger AW, Salmon SE. Primary bioassay of human tumor stem cells. *Science*. 1977;197(4302):461–463. https://doi.org/10.1126/science.560061.
16. Whitesides GM, Ostuni E, Takayama S, Jiang X, Ingber DE. Soft lithography in biology and biochemistry. *Annu Rev Biomed Eng*. 2001;3:335–373. https://doi.org/10.1146/annurev.bioeng.3.1.335.
17. Whitesides GM. The origins and the future of microfluidics. *Nature*. 2006;442(7101):368–373. https://doi.org/10.1038/nature05058.
18. Tilles AW, Baskaran H, Roy P, Yarmush ML, Toner M. Effects of oxygenation and flow on the viability and function of rat hepatocytes cocultured in a microchannel flat-plate bioreactor. *Biotechnol Bioeng*. 2001;73(5):379–389. https://doi.org/10.1002/bit.1071.
19. Huh D, Matthews BD, Mammoto A, Montoya-Zavala M, Hsin HY, Ingber DE. Reconstituting organ-level lung functions on a chip. *Science*. 2010;328(5986):1662–1668. https://doi.org/10.1126/science.1188302.
20. Beckwitt CH, Clark AM, Wheeler S, et al. Liver "organ on a chip". *Exp Cell Res*. 2018;363(1):15–25. https://doi.org/10.1016/j.yexcr.2017.12.023.

21. Du Y, Li N, Long M. Liver sinusoid on a chip. *Methods Cell Biol.* 2018; 146:105–134. https://doi.org/10.1016/bs.mcb.2018.06.002.
22. Deng J, Wei W, Chen Z, et al. Engineered liver-on-a-chip platform to mimic liver functions and its biomedical applications: a review. *Micromachines.* 2019;10(10):676. https://doi.org/10.3390/mi10100676.
23. Ehrlich A, Duche D, Ouedraogo G, Nahmias Y. Challenges and opportunities in the design of liver-on-chip microdevices. *Annu Rev Biomed Eng.* 2019;21:219–239. https://doi.org/10.1146/annurev-bioeng-060418-052305.
24. Zanetti F. Chapter 7—kidney-on-a-chip. In: Hoeng J, Bovard D, Peitsch MC, eds. *Organ-on-a-Chip.* Academic Press; 2020:233–253. https://doi.org/10.1016/B978-0-12-817202-5.00007-3.
25. Kim S, Takayama S. Organ-on-a-chip and the kidney. *Kidney Res Clin Pract.* 2015;34 (3):165–169. https://doi.org/10.1016/j.krcp.2015.08.001.
26. Paoli R, Samitier J. Mimicking the kidney: a key role in organ-on-chip development. *Micromachines.* 2016;7(7):126. https://doi.org/10.3390/mi7070126.
27. Bein A, Shin W, Jalili-Firoozinezhad S, et al. Microfluidic organ-on-a-chip models of human intestine. *Cell Mol Gastroenterol Hepatol.* 2018;5(4):659–668. https://doi.org/10.1016/j.jcmgh.2017.12.010.
28. Shrestha J, Bazaz SR, Es HA, et al. Lung-on-a-chip: the future of respiratory disease models and pharmacological studies. *Crit Rev Biotechnol.* 2020;40(2):213–230. https://doi.org/10.1080/07388551.2019.1710458.
29. Konar D, Devarasetty M, Yildiz DV, Atala A, Murphy SV. Lung-on-a-chip technologies for disease modeling and drug development. *Biomed Eng Comput Biol.* 2016;7 (Suppl. 1):17–27. https://doi.org/10.4137/BECB.S34252.
30. Zhang Q, Sito L, Mao M, He J, Zhang YS, Zhao X. Current advances in skin-on-a-chip models for drug testing. *Microphysiol Syst.* 2018;2:4. Accessed September 21, 2020. http://mps.amegroups.com/article/view/4737.
31. van den Broek LJ, Bergers LIJC, Reijnders CMA, Gibbs S. Progress and future prospectives in skin-on-chip development with emphasis on the use of different cell types and technical challenges. *Stem Cell Rev.* 2017;13(3):418–429. https://doi.org/10.1007/s12015-017-9737-1.
32. Sutterby E, Thurgood P, Baratchi S, Khoshmanesh K, Pirogova E. Microfluidic skin-on-a-chip models: toward biomimetic artificial skin. *Small.* 2020;16(39):2002515. https://doi.org/10.1002/smll.202002515.
33. Bal-Öztürk A, Miccoli B, Avci-Adali M, et al. Current strategies and future perspectives of skin-on-a-chip platforms: innovations, technical challenges and commercial outlook. *Curr Pharm Des.* 2018;24(45):5437–5457. https://doi.org/10.2174/1381612825666190206195304.
34. Ribas J, Sadeghi H, Manbachi A, et al. Cardiovascular organ-on-a-chip platforms for drug discovery and development. *Appl In Vitro Toxicol.* 2016;2(2):82–96. https://doi.org/10.1089/aivt.2016.0002.
35. Kitsara M, Kontziampasis D, Agbulut O, Chen Y. Heart on a chip: micro-nanofabrication and microfluidics steering the future of cardiac tissue engineering. *Microelectron Eng.* 2019;203–204:44–62. https://doi.org/10.1016/j.mee.2018.11.001.
36. Zhao Y, Rafatian N, Wang EY, et al. Towards chamber specific heart-on-a-chip for drug testing applications. *Adv Drug Deliv Rev.* 2020;165–166:60–76. https://doi.org/10.1016/j.addr.2019.12.002.
37. Zhang YS, Aleman J, Arneri A, et al. From cardiac tissue engineering to heart-on-a-chip: beating challenges. *Biomed Mater.* 2015;10(3):034006. https://doi.org/10.1088/1748-6041/10/3/034006.
38. Zhao Y, Demirci U, Chen Y, Chen P. Multiscale brain research on a microfluidic chip. *Lab Chip.* 2020;20(9):1531–1543. https://doi.org/10.1039/C9LC01010F.

39. Frimat J-P, Luttge R. The need for physiological micro-nanofluidic systems of the brain. *Front Bioeng Biotechnol.* 2019;7:100. https://doi.org/10.3389/fbioe.2019.00100.
40. Bang S, Jeong S, Choi N, Kim HN. Brain-on-a-chip: a history of development and future perspective. *Biomicrofluidics.* 2019;13(5):051301. https://doi.org/10.1063/1.5120555.
41. Haring AP, Sontheimer H, Johnson BN. Microphysiological human brain and neural systems on a chip: potential alternatives to small animal models and emerging platforms for drug discovery and personalized medicine. *Stem Cell Rev.* 2017;13(3):381–406. https://doi.org/10.1007/s12015-017-9738-0.
42. Raimondi I, Izzo L, Tunesi M, Comar M, Albani D, Giordano C. Organ-on-a-chip in vitro models of the brain and the blood-brain barrier and their value to study the microbiota-gut-brain axis in neurodegeneration. *Front Bioeng Biotechnol.* 2020;7:435. https://doi.org/10.3389/fbioe.2019.00435.
43. Gebhardt R, Matz-Soja M. Liver zonation: novel aspects of its regulation and its impact on homeostasis. *World J Gastroenterol: WJG.* 2014;20(26):8491–8504. https://doi.org/10.3748/wjg.v20.i26.8491.
44. Ben-Moshe S, Itzkovitz S. Spatial heterogeneity in the mammalian liver. *Nat Rev Gastroenterol Hepatol.* 2019;16(7):395–410. https://doi.org/10.1038/s41575-019-0134-x.
45. Ma X, Qu X, Zhu W, et al. Deterministically patterned biomimetic human iPSC-derived hepatic model via rapid 3D bioprinting. *Proc Natl Acad Sci U S A.* 2016;113(8):2206–2211. https://doi.org/10.1073/pnas.1524510113.
46. Ma C, Zhao L, Zhou E-M, Xu J, Shen S, Wang J. On-chip construction of liver lobule-like microtissue and its application for adverse drug reaction assay. *Anal Chem.* 2016;88(3):1719–1727. https://doi.org/10.1021/acs.analchem.5b03869.
47. Kang YB(A), Sodunke TR, Lamontagne J, et al. Liver sinusoid on a chip: Long-term layered co-culture of primary rat hepatocytes and endothelial cells in microfluidic platforms. *Biotechnol Bioeng.* 2015;112(12):2571–2582. https://doi.org/10.1002/bit.25659.
48. Du Y, Li N, Yang H, et al. Mimicking liver sinusoidal structures and functions using a 3D-configured microfluidic chip. *Lab Chip.* 2017;17(5):782–794. https://doi.org/10.1039/C6LC01374K.
49. Li X, George SM, Vernetti L, Gough AH, Taylor DL. A glass-based, continuously zonated and vascularized human liver acinus microphysiological system (vLAMPS) designed for experimental modeling of diseases and ADME/TOX. *Lab Chip.* 2018;18(17):2614–2631. https://doi.org/10.1039/c8lc00418h.
50. Bavli D, Prill S, Ezra E, et al. Real-time monitoring of metabolic function in liver-on--chip microdevices tracks the dynamics of mitochondrial dysfunction. *Proc Natl Acad Sci U S A.* 2016;113(16):E2231–E2240. https://doi.org/10.1073/pnas.1522556113.
51. Delalat B, Cozzi C, Ghaemi SR, et al. Microengineered bioartificial liver chip for drug toxicity screening. *Adv Funct Mater.* 2018;28(28):1801825. https://doi.org/10.1002/adfm.201801825.
52. Deng J, Zhang X, Chen Z, et al. A cell lines derived microfluidic liver model for investigation of hepatotoxicity induced by drug-drug interaction. *Biomicrofluidics.* 2019;13(2):024101. https://doi.org/10.1063/1.5070088.
53. Jang K-J, Otieno MA, Ronxhi J, et al. Reproducing human and cross-species drug toxicities using a Liver-Chip. *Sci Transl Med.* 2019;11(517). https://doi.org/10.1126/scitranslmed.aax5516.
54. Vernetti LA, Senutovitch N, Boltz R, et al. A human liver microphysiology platform for investigating physiology, drug safety, and disease models. *Exp Biol Med (Maywood).* 2016;241(1):101–114. https://doi.org/10.1177/1535370215592121.
55. Oxburgh L. Kidney nephron determination. *Annu Rev Cell Dev Biol.* 2018;34(1):427–450. https://doi.org/10.1146/annurev-cellbio-100616-060647.

56. Pollak MR, Quaggin SE, Hoenig MP, Dworkin LD. The glomerulus: the sphere of influence. *Clin J Am Soc Nephrol.* 2014;9(8):1461–1469. https://doi.org/10.2215/CJN.09400913.
57. Musah S, Mammoto A, Ferrante TC, et al. Mature induced-pluripotent-stem-cell-derived human podocytes reconstitute kidney glomerular-capillary-wall function on a chip. *Nat Biomed Eng.* 2017;1(5):1–12. https://doi.org/10.1038/s41551-017-0069.
58. Wang L, Tao T, Su W, Yu H, Yu Y, Qin J. A disease model of diabetic nephropathy in a glomerulus-on-a-chip microdevice. *Lab Chip.* 2017;17(10):1749–1760. https://doi.org/10.1039/c7lc00134g.
59. Zhou M, Zhang X, Wen X, et al. Development of a functional glomerulus at the organ level on a chip to mimic hypertensive nephropathy. *Sci Rep.* 2016;6(1):31771. https://doi.org/10.1038/srep31771.
60. Sciancalepore AG, Sallustio F, Girardo S, et al. A bioartificial renal tubule device embedding human renal stem/progenitor cells. *PLoS One.* 2014;9(1):e87496. https://doi.org/10.1371/journal.pone.0087496.
61. Weber EJ, Chapron A, Chapron BD, et al. Development of a microphysiological model of human kidney proximal tubule function. *Kidney Int.* 2016;90(3):627–637. https://doi.org/10.1016/j.kint.2016.06.011.
62. Baudoin R, Griscom L, Monge M, Legallais C, Leclerc E. Development of a renal microchip for in vitro distal tubule models. *Biotechnol Prog.* 2007;23(5):1245–1253. https://doi.org/10.1021/bp0603513.
63. Jang K-J, Suh K-Y. A multi-layer microfluidic device for efficient culture and analysis of renal tubular cells. *Lab Chip.* 2010;10(1):36–42. https://doi.org/10.1039/B907515A.
64. Jang K-J, Mehr AP, Hamilton GA, et al. Human kidney proximal tubule-on-a-chip for drug transport and nephrotoxicity assessment. *Integr Biol (Camb).* 2013;5(9):1119–1129. https://doi.org/10.1039/c3ib40049b.
65. Kiela PR, Ghishan FK. Physiology of intestinal absorption and secretion. *Best Pract Res Clin Gastroenterol.* 2016;30(2):145–159. https://doi.org/10.1016/j.bpg.2016.02.007.
66. Vertzoni M, Augustijns P, Grimm M, et al. Impact of regional differences along the gastrointestinal tract of healthy adults on oral drug absorption: an UNGAP review. *Eur J Pharm Sci.* 2019;134:153–175. https://doi.org/10.1016/j.ejps.2019.04.013.
67. Shim K-Y, Lee D, Han J, Nguyen N-T, Park S, Sung JH. Microfluidic gut-on-a-chip with three-dimensional villi structure. *Biomed Microdevices.* 2017;19(2):37. https://doi.org/10.1007/s10544-017-0179-y.
68. Maurer M, Gresnigt MS, Last A, et al. A three-dimensional immunocompetent intestine-on-chip model as in vitro platform for functional and microbial interaction studies. *Biomaterials.* 2019;220:119396. https://doi.org/10.1016/j.biomaterials.2019.119396.
69. Kim HJ, Ingber DE. Gut-on-a-Chip microenvironment induces human intestinal cells to undergo villus differentiation. *Integr Biol (Camb).* 2013;5(9):1130–1140. https://doi.org/10.1039/c3ib40126j.
70. Chi M, Yi B, Oh S, Park D-J, Sung JH, Park S. A microfluidic cell culture device (μFCCD) to culture epithelial cells with physiological and morphological properties that mimic those of the human intestine. *Biomed Microdevices.* 2015;17(3):9966. https://doi.org/10.1007/s10544-015-9966-5.
71. Kasendra M, Tovaglieri A, Sontheimer-Phelps A, et al. Development of a primary human Small Intestine-on-a-Chip using biopsy-derived organoids. *Sci Rep.* 2018;8(1):2871. https://doi.org/10.1038/s41598-018-21201-7.
72. Kasendra M, Luc R, Yin J, et al. Duodenum Intestine-Chip for preclinical drug assessment in a human relevant model. *Elife.* 2020;9:e50135. https://doi.org/10.7554/eLife.50135.

73. Schloissnig S, Arumugam M, Sunagawa S, et al. Genomic variation landscape of the human gut microbiome. *Nature*. 2013;493(7430):45–50. https://doi.org/10.1038/nature11711.
74. Eckburg PB, Bik EM, Bernstein CN, et al. Diversity of the human intestinal microbial flora. *Science*. 2005;308(5728):1635–1638. https://doi.org/10.1126/science.1110591.
75. Bäckhed F, Ley RE, Sonnenburg JL, Peterson DA, Gordon JI. Host-bacterial mutualism in the human intestine. *Science*. 2005;307(5717):1915–1920. https://doi.org/10.1126/science.1104816.
76. Hooper LV, Midtvedt T, Gordon JI. How host-microbial interactions shape the nutrient environment of the mammalian intestine. *Annu Rev Nutr*. 2002;22(1):283–307. https://doi.org/10.1146/annurev.nutr.22.011602.092259.
77. Kho ZY, Lal SK. The human gut microbiome—potential controller of wellness and disease. *Front Microbiol*. 2018;9:1835. https://doi.org/10.3389/fmicb.2018.01835.
78. Shreiner AB, Kao JY, Young VB. The gut microbiome in health and in disease. *Curr Opin Gastroenterol*. 2015;31(1):69–75. https://doi.org/10.1097/MOG.0000000000000139.
79. Zimmermann M, Zimmermann-Kogadeeva M, Wegmann R, Goodman AL. Mapping human microbiome drug metabolism by gut bacteria and their genes. *Nature*. 2019;570(7762):462–467. https://doi.org/10.1038/s41586-019-1291-3.
80. Maier L, Pruteanu M, Kuhn M, et al. Extensive impact of non-antibiotic drugs on human gut bacteria. *Nature*. 2018;555(7698):623–628. https://doi.org/10.1038/nature25979.
81. Vich Vila A, Collij V, Sanna S, et al. Impact of commonly used drugs on the composition and metabolic function of the gut microbiota. *Nat Commun*. 2020;11(1):362. https://doi.org/10.1038/s41467-019-14177-z.
82. Kim HJ, Huh D, Hamilton G, Ingber DE. Human gut-on-a-chip inhabited by microbial flora that experiences intestinal peristalsis-like motions and flow. *Lab Chip*. 2012;12(12):2165–2174. https://doi.org/10.1039/c2lc40074j.
83. Jalili-Firoozinezhad S, Gazzaniga FS, Calamari EL, et al. A complex human gut microbiome cultured in an anaerobic intestine-on-a-chip. *Nat Biomed Eng*. 2019;3(7):520–531. https://doi.org/10.1038/s41551-019-0397-0.
84. Shah P, Fritz JV, Glaab E, et al. A microfluidics-based in vitro model of the gastrointestinal human–microbe interface. *Nat Commun*. 2016;7(1):11535. https://doi.org/10.1038/ncomms11535.
85. Kim HJ, Li H, Collins JJ, Ingber DE. Contributions of microbiome and mechanical deformation to intestinal bacterial overgrowth and inflammation in a human gut-on-a-chip. *Proc Natl Acad Sci*. 2016;113(1):E7–E15. https://doi.org/10.1073/pnas.1522193112.
86. Shin W, Kim HJ. Pathomimetic modeling of human intestinal diseases and underlying host-gut microbiome interactions in a gut-on-a-chip. *Methods Cell Biol*. 2018;146:135–148. https://doi.org/10.1016/bs.mcb.2018.05.006.
87. Eain MMG, Baginska J, Greenhalgh K, Fritz JV, Zenhausern F, Wilmes P. Engineering solutions for representative models of the gastrointestinal human-microbe interface. *Engineering*. 2017;3(1):60–65. https://doi.org/10.1016/J.ENG.2017.01.011.
88. Knudsen L, Ochs M. The micromechanics of lung alveoli: structure and function of surfactant and tissue components. *Histochem Cell Biol*. 2018;150(6):661–676. https://doi.org/10.1007/s00418-018-1747-9.
89. Zhang M, Xu C, Jiang L, Qin J. A 3D human lung-on-a-chip model for nanotoxicity testing †. *Toxicol Res*. 2018;7(6):1048–1060. https://doi.org/10.1039/c8tx00156a.
90. Stucki AO, Stucki JD, Hall SRR, et al. A lung-on-a-chip array with an integrated bio-inspired respiration mechanism. *Lab Chip*. 2015;15(5):1302–1310. https://doi.org/10.1039/C4LC01252F.

91. Huh D, Leslie DC, Matthews BD, et al. A human disease model of drug toxicity-induced pulmonary edema in a lung-on-a-chip microdevice. *Sci Transl Med*. 2012;4(159):159ra147. https://doi.org/10.1126/scitranslmed.3004249.
92. Jain A, Barrile R, van der Meer AD, et al. Primary human lung alveolus-on-a-chip model of intravascular thrombosis for assessment of therapeutics. *Clin Pharmacol Ther*. 2018;103(2):332–340. https://doi.org/10.1002/cpt.742.
93. Sellgren KL, Butala EJ, Gilmour BP, Randell SH, Grego S. A biomimetic multicellular model of the airways using primary human cells. *Lab Chip*. 2014;14(17):3349–3358. https://doi.org/10.1039/C4LC00552J.
94. Benam KH, Villenave R, Lucchesi C, et al. Small airway-on-a-chip enables analysis of human lung inflammation and drug responses in vitro. *Nat Methods*. 2016;13(2):151–157. https://doi.org/10.1038/nmeth.3697.
95. Benson HAE, Grice JE, Mohammed Y, Namjoshi S, Roberts MS. Topical and transdermal drug delivery: from simple potions to smart technologies. *Curr Drug Deliv*. 2019;16(5):444–460. https://doi.org/10.2174/1567201816666190201143457.
96. Prausnitz MR, Langer R. Transdermal drug delivery. *Nat Biotechnol*. 2008;26(11):1261–1268. https://doi.org/10.1038/nbt.1504.
97. Ramadan Q, Ting FCW. In vitro micro-physiological immune-competent model of the human skin. *Lab Chip*. 2016;16(10):1899–1908. https://doi.org/10.1039/C6LC00229C.
98. O'Neill AT, Monteiro-Riviere NA, Walker GM. Characterization of microfluidic human epidermal keratinocyte culture. *Cytotechnology*. 2008;56(3):197–207. https://doi.org/10.1007/s10616-008-9149-9.
99. Kim JJ, Ellett F, Thomas CN, et al. A microscale, full-thickness, human skin on a chip assay simulating neutrophil responses to skin infection and antibiotic treatments. *Lab Chip*. 2019;19(18):3094–3103. https://doi.org/10.1039/c9lc00399a.
100. Lim HY, Kim J, Song HJ, et al. Development of wrinkled skin-on-a-chip (WSOC) by cyclic uniaxial stretching. *J Ind Eng Chem*. 2018;68:238–245. https://doi.org/10.1016/j.jiec.2018.07.050.
101. Abaci HE, Gledhill K, Guo Z, Christiano AM, Shuler ML. Pumpless microfluidic platform for drug testing on human skin equivalents. *Lab Chip*. 2015;15(3):882–888. https://doi.org/10.1039/C4LC00999A.
102. Sriram G, Alberti M, Dancik Y, et al. Full-thickness human skin-on-chip with enhanced epidermal morphogenesis and barrier function. *Mater Today*. 2018;21(4):326–340. https://doi.org/10.1016/j.mattod.2017.11.002.
103. Lee S, Jin S-P, Kim YK, Sung GY, Chung JH, Sung JH. Construction of 3D multicellular microfluidic chip for an in vitro skin model. *Biomed Microdevices*. 2017;19(2):22. https://doi.org/10.1007/s10544-017-0156-5.
104. Mori N, Morimoto Y, Takeuchi S. Skin integrated with perfusable vascular channels on a chip. *Biomaterials*. 2017;116:48–56. https://doi.org/10.1016/j.biomaterials.2016.11.031.
105. Wufuer M, Lee G, Hur W, et al. Skin-on-a-chip model simulating inflammation, edema and drug-based treatment. *Sci Rep*. 2016;6(1):37471. https://doi.org/10.1038/srep37471.
106. Jusoh N, Ko J, Jeon NL. Microfluidics-based skin irritation test using in vitro 3D angiogenesis platform. *APL Bioeng*. 2019;3(3):036101. https://doi.org/10.1063/1.5093975.
107. Kim BS, Gao G, Kim JY, Cho D-W. 3D cell printing of perfusable vascularized human skin equivalent composed of epidermis, dermis, and hypodermis for better structural recapitulation of native skin. *Adv Healthc Mater*. 2019;8(7):e1801019. https://doi.org/10.1002/adhm.201801019.
108. Lukács B, Bajza Á, Kocsis D, et al. Skin-on-a-chip device for ex vivo monitoring of transdermal delivery of drugs—design, fabrication, and testing. *Pharmaceutics*. 2019;11(9):445. https://doi.org/10.3390/pharmaceutics11090445.

109. Kim K, Jeon HM, Choi KC, Sung GY. Testing the effectiveness of curcuma longa leaf extract on a skin equivalent using a pumpless skin-on-a-chip model. *Int J Mol Sci*. 2020;21(11):3898. https://doi.org/10.3390/ijms21113898.
110. Mladěnka P, Applová L, Patočka J, et al. Comprehensive review of cardiovascular toxicity of drugs and related agents. *Med Res Rev*. 2018;38(4):1332–1403. https://doi.org/10.1002/med.21476.
111. Page Robert L, O'Bryant Cindy L, Davy C, et al. Drugs that may cause or exacerbate heart failure. *Circulation*. 2016;134(6):e32–e69. https://doi.org/10.1161/CIR.0000000000000426.
112. Kaneko T, Kojima K, Yasuda K. An on-chip cardiomyocyte cell network assay for stable drug screening regarding community effect of cell network size. *Analyst*. 2007;132(9):892–898. https://doi.org/10.1039/B704961G.
113. Grosberg A, Alford PW, McCain ML, Parker KK. Ensembles of engineered cardiac tissues for physiological and pharmacological study: heart on a chip. *Lab Chip*. 2011;11(24):4165–4173. https://doi.org/10.1039/C1LC20557A.
114. Agarwal A, Goss JA, Cho A, McCain ML, Parker KK. Microfluidic heart on a chip for higher throughput pharmacological studies. *Lab Chip*. 2013;13(18):3599–3608. https://doi.org/10.1039/C3LC50350J.
115. Marsano A, Conficconi C, Lemme M, et al. Beating heart on a chip: a novel microfluidic platform to generate functional 3D cardiac microtissues. *Lab Chip*. 2016;16(3):599–610. https://doi.org/10.1039/C5LC01356A.
116. Kujala VJ, Pasqualini FS, Goss JA, Nawroth JC, Parker KK. Laminar ventricular myocardium on a microelectrode array-based chip. *J Mater Chem B*. 2016;4(20):3534–3543. https://doi.org/10.1039/C6TB00324A.
117. Qian F, Huang C, Lin Y-D, et al. Simultaneous electrical recording of cardiac electrophysiology and contraction on chip. *Lab Chip*. 2017;17(10):1732–1739. https://doi.org/10.1039/C7LC00210F.
118. Raichle ME. Two views of brain function. *Trends Cogn Sci*. 2010;14(4):180–190. https://doi.org/10.1016/j.tics.2010.01.008.
119. Tong Z, Segura-Feliu M, Seira O, Homs-Corbera A, Río JAD, Samitier J. A microfluidic neuronal platform for neuron axotomy and controlled regenerative studies. *RSC Adv*. 2015;5(90):73457–73466. https://doi.org/10.1039/C5RA11522A.
120. Dworak BJ, Wheeler BC. Novel MEA platform with PDMS microtunnels enables the detection of action potential propagation from isolated axons in culture. *Lab Chip*. 2009;9(3):404–410. https://doi.org/10.1039/B806689B.
121. Southam KA, King AE, Blizzard CA, McCormack GH, Dickson TC. Microfluidic primary culture model of the lower motor neuron–neuromuscular junction circuit. *J Neurosci Methods*. 2013;218(2):164–169. https://doi.org/10.1016/j.jneumeth.2013.06.002.
122. Smith AST, Long CJ, Pirozzi K, Hickman JJ. A functional system for high-content screening of neuromuscular junctions in vitro. *Technology (Singap World Sci)*. 2013;1(1):37–48. https://doi.org/10.1142/S2339547813500015.
123. Kerman BE, Kim HJ, Padmanabhan K, et al. In vitro myelin formation using embryonic stem cells. *Development*. 2015;142(12):2213–2225. https://doi.org/10.1242/dev.116517.
124. Lee HU, Nag S, Blasiak A, Jin Y, Thakor N, Yang IH. Subcellular optogenetic stimulation for activity-dependent myelination of axons in a novel microfluidic compartmentalized platform. *ACS Chem Nerosci*. 2016;7(10):1317–1324. https://doi.org/10.1021/acschemneuro.6b00157.
125. Jiang L, Li S, Zheng J, Li Y, Huang H. Recent progress in microfluidic models of the blood-brain barrier. *Micromachines*. 2019;10(6):375. https://doi.org/10.3390/mi10060375.
126. Oddo A, Peng B, Tong Z, et al. Advances in microfluidic blood–brain barrier (BBB) models. *Trends Biotechnol*. 2019;37(12):1295–1314. https://doi.org/10.1016/j.tibtech.2019.04.006.

127. Maoz BM, Herland A, FitzGerald EA, et al. A linked organ-on-chip model of the human neurovascular unit reveals the metabolic coupling of endothelial and neuronal cells. *Nat Biotechnol.* 2018;36(9):865–874. https://doi.org/10.1038/nbt.4226.
128. Sances S, Ho R, Vatine G, et al. Human iPSC-derived endothelial cells and micro-engineered organ-chip enhance neuronal development. *Stem Cell Rep.* 2018;10 (4):1222–1236. https://doi.org/10.1016/j.stemcr.2018.02.012.
129. Takebe T, Zhang B, Radisic M. Synergistic engineering: organoids meet organs-on-a-chip. *Cell Stem Cell.* 2017;21(3):297–300. https://doi.org/10.1016/j.stem.2017.08.016.
130. Rossi G, Manfrin A, Lutolf MP. Progress and potential in organoid research. *Nat Rev Genet.* 2018;19(11):671–687. https://doi.org/10.1038/s41576-018-0051-9.
131. Schutgens F, Clevers H. Human organoids: tools for understanding biology and treating diseases. *Annu Rev Pathol: Mech Dis.* 2020;15(1):211–234. https://doi.org/10.1146/annurev-pathmechdis-012419-032611.
132. Wang Z, He X, Qiao H, Chen P. Global trends of organoid and organ-on-a-chip in the past decade: a bibliometric and comparative study. *Tissue Eng Part A.* 2019;26 (11–12):656–671. https://doi.org/10.1089/ten.tea.2019.0251.
133. Moshksayan K, Kashaninejad N, Warkiani ME, et al. Spheroids-on-a-chip: recent advances and design considerations in microfluidic platforms for spheroid formation and culture. *Sens Actuators B.* 2018;263:151–176. https://doi.org/10.1016/j.snb.2018.01.223.
134. Skardal A, Shupe T, Atala A. Organoid-on-a-chip and body-on-a-chip systems for drug screening and disease modeling. *Drug Discov Today.* 2016;21(9):1399–1411. https://doi.org/10.1016/j.drudis.2016.07.003.
135. Skardal A, Devarasetty M, Forsythe S, Atala A, Soker S. A reductionist metastasis-on-a-chip platform for in vitro tumor progression modeling and drug screening. *Biotechnol Bioeng.* 2016;113(9):2020–2032. https://doi.org/10.1002/bit.25950.
136. Niu Y, Bai J, Kamm RD, Wang Y, Wang C. Validating antimetastatic effects of natural products in an engineered microfluidic platform mimicking tumor microenvironment. *Mol Pharm.* 2014;11(7):2022–2029. https://doi.org/10.1021/mp500054h.
137. Choi Y, Hyun E, Seo J, et al. A microengineered pathophysiological model of early-stage breast cancer. *Lab Chip.* 2015;15(16):3350–3357. https://doi.org/10.1039/c5lc00514k.
138. Achberger K, Probst C, Haderspeck J, et al. Merging organoid and organ-on-a-chip technology to generate complex multi-layer tissue models in a human retina-on-a-chip platform. *Elife.* 2019;8:e46188. https://doi.org/10.7554/eLife.46188.
139. Park SE, Georgescu A, Huh D. Organoids-on-a-chip. *Science.* 2019;364(6444):960–965. https://doi.org/10.1126/science.aaw7894.
140. Kim S, Kim W, Lim S, Jeon JS. Vasculature-on-a-chip for in vitro disease models. *Bioengineering.* 2017;4(1):8. https://doi.org/10.3390/bioengineering4010008.
141. Park D, Lee J, Chung JJ, Jung Y, Kim SH. Integrating organs-on-chips: multiplexing, scaling, vascularization, and innervation. *Trends Biotechnol.* 2020;38(1):99–112. https://doi.org/10.1016/j.tibtech.2019.06.006.
142. Huh D, Kim HJ, Fraser JP, et al. Microfabrication of human organs-on-chips. *Nat Protoc.* 2013;8(11):2135–2157. https://doi.org/10.1038/nprot.2013.137.
143. Zhang YS, Arneri A, Bersini S, et al. Bioprinting 3D microfibrous scaffolds for engineering endothelialized myocardium and heart-on-a-chip. *Biomaterials.* 2016;110:45–59. https://doi.org/10.1016/j.biomaterials.2016.09.003.
144. Auger FA, Gibot L, Lacroix D. The pivotal role of vascularization in tissue engineering. *Annu Rev Biomed Eng.* 2013;15(1):177–200. https://doi.org/10.1146/annurev-bioeng-071812-152428.

145. Osaki T, Serrano JC, Kamm RD. Cooperative effects of vascular angiogenesis and lymphangiogenesis. *Regen Eng Transl Med.* 2018;4(3):120–132. https://doi.org/10.1007/s40883-018-0054-2.
146. Sato M, Sasaki N, Ato M, Hirakawa S, Sato K, Sato K. Microcirculation-on-a-chip: a microfluidic platform for assaying blood- and lymphatic-vessel permeability. *PLoS One.* 2015;10(9):e0137301. https://doi.org/10.1371/journal.pone.0137301.
147. Cao X, Ashfaq R, Cheng F, et al. A tumor-on-a-chip system with bioprinted blood and lymphatic vessel pair. *Adv Funct Mater.* 2019;29(31):1807173. https://doi.org/10.1002/adfm.201807173.
148. Das S, Gordián-Vélez WJ, Ledebur HC, et al. Innervation: the missing link for biofabricated tissues and organs. *NPJ Regen Med.* 2020;5(1):1–19. https://doi.org/10.1038/s41536-020-0096-1.
149. Keast JR. Innervation of developing and adult organs. *Organogenesis.* 2013;9(3):168. https://doi.org/10.4161/org.25886.
150. Takeuchi A, Nakafutami S, Tani H, et al. Device for co-culture of sympathetic neurons and cardiomyocytes using microfabrication. *Lab Chip.* 2011;11(13):2268–2275. https://doi.org/10.1039/C0LC00327A.
151. Takeuchi A, Shimba K, Mori M, et al. Sympathetic neurons modulate the beat rate of pluripotent cell-derived cardiomyocytes in vitro. *Integr Biol.* 2012;4(12):1532–1539. https://doi.org/10.1039/c2ib20060k.
152. Sakai K, Shimba K, Ishizuka K, et al. Functional innervation of human induced pluripotent stem cell-derived cardiomyocytes by co-culture with sympathetic neurons developed using a microtunnel technique. *Biochem Biophys Res Commun.* 2017;494(1):138–143. https://doi.org/10.1016/j.bbrc.2017.10.065.
153. Yeste J, García-Ramírez M, Illa X, et al. A compartmentalized microfluidic chip with crisscross microgrooves and electrophysiological electrodes for modeling the blood–retinal barrier. *Lab Chip.* 2017;18(1):95–105. https://doi.org/10.1039/C7LC00795G.
154. Silva DI, dos Santos BP, Leng J, Oliveira H, Amédée J. Dorsal root ganglion neurons regulate the transcriptional and translational programs of osteoblast differentiation in a microfluidic platform. *Cell Death Dis.* 2017;8(12):1–14. https://doi.org/10.1038/s41419-017-0034-3.
155. Uzel SGM, Platt RJ, Subramanian V, et al. Microfluidic device for the formation of optically excitable, three-dimensional, compartmentalized motor units. *Sci Adv.* 2016;2(8):e1501429. https://doi.org/10.1126/sciadv.1501429.
156. Pagella P, Neto E, Jiménez-Rojo L, Lamghari M, Mitsiadis TA. Microfluidics co-culture systems for studying tooth innervation. *Front Physiol.* 2014;5:326. https://doi.org/10.3389/fphys.2014.00326.
157. Taylor AM, Dieterich DC, Ito HT, Kim SA, Schuman EM. Microfluidic local perfusion chambers for the visualization and manipulation of synapses. *Neuron.* 2010;66(1):57–68. https://doi.org/10.1016/j.neuron.2010.03.022.
158. Taylor AM, Blurton-Jones M, Rhee SW, Cribbs DH, Cotman CW, Jeon NL. A microfluidic culture platform for CNS axonal injury, regeneration and transport. *Nat Methods.* 2005;2(8):599–605. https://doi.org/10.1038/nmeth777.
159. Kim SH, Im S-K, Oh S-J, et al. Anisotropically organized three-dimensional culture platform for reconstruction of a hippocampal neural network. *Nat Commun.* 2017;8(1):14346. https://doi.org/10.1038/ncomms14346.
160. Virlogeux A, Moutaux E, Christaller W, et al. Reconstituting corticostriatal network on-a-chip reveals the contribution of the presynaptic compartment to Huntington's disease. *Cell Rep.* 2018;22(1):110–122. https://doi.org/10.1016/j.celrep.2017.12.013.

161. Coquinco A, Kojic L, Wen W, et al. A microfluidic based in vitro model of synaptic competition. *Mol Cell Neurosci*. 2014;60:43–52. https://doi.org/10.1016/j.mcn.2014.03.001.
162. Ando J, Yamamoto K. Vascular mechanobiology: endothelial cell responses to fluid shear stress. *Circ J*. 2009;73(11):1983–1992. https://doi.org/10.1253/circj.cj-09-0583.
163. Dahl KN, Kalinowski A, Pekkan K. Mechanobiology and the microcirculation: cellular, nuclear and fluid mechanics. *Microcirculation*. 2010;17(3):179–191. https://doi.org/10.1111/j.1549-8719.2009.00016.x.
164. Humphrey JD, Dufresne ER, Schwartz MA. Mechanotransduction and extracellular matrix homeostasis. *Nat Rev Mol Cell Biol*. 2014;15(12):802–812. https://doi.org/10.1038/nrm3896.
165. Vogel V. Unraveling the mechanobiology of extracellular matrix. *Annu Rev Physiol*. 2018;80(1):353–387. https://doi.org/10.1146/annurev-physiol-021317-121312.
166. Charras G, Yap AS. Tensile forces and mechanotransduction at cell-cell junctions. *Curr Biol*. 2018;28(8):R445–R457. https://doi.org/10.1016/j.cub.2018.02.003.
167. Chen Y, Li Z, Ju LA. Tensile and compressive force regulation on cell mechanosensing. *Biophys Rev*. 2019;11(3):311–318. https://doi.org/10.1007/s12551-019-00536-z.
168. Wang N. Review of cellular mechanotransduction. *J Phys D Appl Phys*. 2017;50(23):233002. Accessed October 12, 2020 https://www.ncbi.nlm.nih.gov/pmc/articles/PMC5662120/.
169. Uhler C, Shivashankar GV. Regulation of genome organization and gene expression by nuclear mechanotransduction. *Nat Rev Mol Cell Biol*. 2017;18(12):717–727. https://doi.org/10.1038/nrm.2017.101.
170. Mammoto T, Mammoto A, Ingber DE. Mechanobiology and developmental control. *Annu Rev Cell Dev Biol*. 2013;29(1):27–61. https://doi.org/10.1146/annurev-cellbio-101512-122340.
171. Panciera T, Azzolin L, Cordenonsi M, Piccolo S. Mechanobiology of YAP and TAZ in physiology and disease. *Nat Rev Mol Cell Biol*. 2017;18(12):758–770. https://doi.org/10.1038/nrm.2017.87.
172. Martino F, Perestrelo AR, Vinarský V, Pagliari S, Forte G. Cellular mechanotransduction: from tension to function. *Front Physiol*. 2018;9. https://doi.org/10.3389/fphys.2018.00824.
173. Broders-Bondon F, Nguyen Ho-Bouldoires TH, Fernandez-Sanchez M-E, Farge E. Mechanotransduction in tumor progression: the dark side of the force. *J Cell Biol*. 2018;217(5):1571–1587. https://doi.org/10.1083/jcb.201701039.
174. Liao F, Li M, Han D, Cao J, Chen K. Biomechanopharmacology: a new borderline discipline. *Trends Pharmacol Sci*. 2006;27(6):287–289. https://doi.org/10.1016/j.tips.2006.04.001.
175. Krishnan R, Park J-A, Seow CY, Lee PV-S, Stewart AG. Cellular biomechanics in drug screening and evaluation: mechanopharmacology. *Trends Pharmacol Sci*. 2016;37(2):87–100. https://doi.org/10.1016/j.tips.2015.10.005.
176. Lampi MC, Reinhart-King CA. Targeting extracellular matrix stiffness to attenuate disease: from molecular mechanisms to clinical trials. *Sci Transl Med*. 2018;10(422):eaao0475. https://doi.org/10.1126/scitranslmed.aao0475.
177. Yao W, Li Y, Ding G. Interstitial fluid flow: the mechanical environment of cells and foundation of meridians. *Evid Based Complement Alternat Med*. 2012;2012:853516. https://doi.org/10.1155/2012/853516.
178. Ergir E, Bachmann B, Redl H, Forte G, Ertl P. Small force, big impact: next generation organ-on-a-chip systems incorporating biomechanical cues. *Front Physiol*. 2018;9:1417. https://doi.org/10.3389/fphys.2018.01417.
179. Kaarj K, Yoon J-Y. Methods of delivering mechanical stimuli to organ-on-a-chip. *Micromachines*. 2019;10(10):700. https://doi.org/10.3390/mi10100700.

180. Fowler S, Chen WLK, Duignan DB, et al. Microphysiological systems for ADME-related applications: current status and recommendations for system development and characterization. *Lab Chip*. 2020;20(3):446–467. https://doi.org/10.1039/C9LC00857H.
181. Sweeney LM, Shuler ML, Babish JG, Ghanem A. A cell culture analogue of rodent physiology: application to naphthalene toxicology. *Toxicol In Vitro*. 1995;9(3):307–316. https://doi.org/10.1016/0887-2333(95)00007-U.
182. Ghanem A, Shuler ML. Combining cell culture analogue reactor designs and PBPK models to probe mechanisms of naphthalene toxicity. *Biotechnol Prog*. 2000;16(3):334–345. https://doi.org/10.1021/bp9901522.
183. Shuler ML, Ghanem A, Quick D, Wong MC, Miller P. A self-regulating cell culture analog device to mimic animal and human toxicological responses. *Biotechnol Bioeng*. 1996;52(1):45–60. https://doi.org/10.1002/(SICI)1097-0290(19961005)52:1<45::AID-BIT5>3.0.CO;2-Z.
184. Wang YI, Carmona C, Hickman JJ, Shuler ML. Multiorgan microphysiological systems for drug development: strategies, advances, and challenges. *Adv Healthc Mater*. 2018;7(2):1701000. https://doi.org/10.1002/adhm.201701000.
185. Sung JH, Wang YI, Sriram NN, et al. Recent advances in body-on-a-chip systems. *Anal Chem*. 2019;91(1):330–351. https://doi.org/10.1021/acs.analchem.8b05293.
186. Sin A, Chin KC, Jamil MF, Kostov Y, Rao G, Shuler ML. The design and fabrication of three-chamber microscale cell culture analog devices with integrated dissolved oxygen sensors. *Biotechnol Prog*. 2004;20(1):338–345. https://doi.org/10.1021/bp034077d.
187. Viravaidya K, Shuler ML. Incorporation of 3T3-L1 cells to mimic bioaccumulation in a microscale cell culture analog device for toxicity studies. *Biotechnol Prog*. 2004;20(2):590–597. https://doi.org/10.1021/bp034238d.
188. Viravaidya K, Sin A, Shuler ML. Development of a microscale cell culture analog to probe naphthalene toxicity. *Biotechnol Prog*. 2004;20(1):316–323. https://doi.org/10.1021/bp0341996.
189. Tatosian DA, Shuler ML. A novel system for evaluation of drug mixtures for potential efficacy in treating multidrug resistant cancers. *Biotechnol Bioeng*. 2009;103(1):187–198. https://doi.org/10.1002/bit.22219.
190. Sung JH, Kam C, Shuler ML. A microfluidic device for a pharmacokinetic-pharmacodynamic (PK-PD) model on a chip. *Lab Chip*. 2010;10(4):446–455. https://doi.org/10.1039/b917763a.
191. Sung JH, Shuler ML. A micro cell culture analog (microCCA) with 3-D hydrogel culture of multiple cell lines to assess metabolism-dependent cytotoxicity of anti-cancer drugs. *Lab Chip*. 2009;9(10):1385–1394. https://doi.org/10.1039/b901377f.
192. Wagner I, Materne E-M, Brincker S, et al. A dynamic multi-organ-chip for long-term cultivation and substance testing proven by 3D human liver and skin tissue co-culture. *Lab Chip*. 2013;13(18):3538–3547. https://doi.org/10.1039/c3lc50234a.
193. Maschmeyer I, Hasenberg T, Jaenicke A, et al. Chip-based human liver-intestine and liver-skin co-cultures—a first step toward systemic repeated dose substance testing in vitro. *Eur J Pharm Biopharm*. 2015;95(Pt A):77–87. https://doi.org/10.1016/j.ejpb.2015.03.002.
194. Materne E-M, Ramme AP, Terrasso AP, et al. A multi-organ chip co-culture of neurospheres and liver equivalents for long-term substance testing. *J Biotechnol*. 2015;205:36–46. https://doi.org/10.1016/j.jbiotec.2015.02.002.
195. Esch MB, Mahler GJ, Stokol T, Shuler ML. Body-on-a-chip simulation with gastrointestinal tract and liver tissues suggests that ingested nanoparticles have the potential to cause liver injury. *Lab Chip*. 2014;14(16):3081–3092. https://doi.org/10.1039/c4lc00371c.

196. An F, Qu Y, Luo Y, et al. A laminated microfluidic device for comprehensive preclinical testing in the drug ADME process. *Sci Rep*. 2016;6(1):25022. https://doi.org/10.1038/srep25022.
197. Vernetti L, Gough A, Baetz N, et al. Functional coupling of human microphysiology systems: intestine, liver, kidney proximal tubule, blood-brain barrier and skeletal muscle. *Sci Rep*. 2017;7(1):42296. https://doi.org/10.1038/srep42296.
198. Edington CD, Chen WLK, Geishecker E, et al. Interconnected microphysiological systems for quantitative biology and pharmacology studies. *Sci Rep*. 2018;8(1):4530. https://doi.org/10.1038/s41598-018-22749-0.
199. Lin N, Zhou X, Geng X, et al. Repeated dose multi-drug testing using a microfluidic chip-based coculture of human liver and kidney proximal tubules equivalents. *Sci Rep*. 2020;10(1):8879. https://doi.org/10.1038/s41598-020-65817-0.
200. Novak R, Ingram M, Marquez S, et al. Robotic fluidic coupling and interrogation of multiple vascularized organ chips. *Nat Biomed Eng*. 2020;4(4):407–420. https://doi.org/10.1038/s41551-019-0497-x.
201. Herland A, Maoz BM, Das D, et al. Quantitative prediction of human pharmacokinetic responses to drugs via fluidically coupled vascularized organ chips. *Nat Biomed Eng*. 2020;4(4):421–436. https://doi.org/10.1038/s41551-019-0498-9.
202. Oleaga C, Bernabini C, Smith AST, et al. Multi-organ toxicity demonstration in a functional human in vitro system composed of four organs. *Sci Rep*. 2016;6(1): 20030. https://doi.org/10.1038/srep20030.
203. Skardal A, Murphy SV, Devarasetty M, et al. Multi-tissue interactions in an integrated three-tissue organ-on-a-chip platform. *Sci Rep*. 2017;7(1):8837. https://doi.org/10.1038/s41598-017-08879-x.
204. Imura Y, Sato K, Yoshimura E. Micro total bioassay system for ingested substances: assessment of intestinal absorption, hepatic metabolism, and bioactivity. *Anal Chem*. 2010;82(24):9983–9988. https://doi.org/10.1021/ac100806x.
205. Schimek K, Frentzel S, Luettich K, et al. Human multi-organ chip co-culture of bronchial lung culture and liver spheroids for substance exposure studies. *Sci Rep*. 2020; 10(1):7865. https://doi.org/10.1038/s41598-020-64219-6.
206. Kane KIW, Moreno EL, Hachi S, et al. Automated microfluidic cell culture of stem cell derived dopaminergic neurons. *Sci Rep*. 2019;9(1):1796. https://doi.org/10.1038/s41598-018-34828-3.
207. Sosa-Hernández JE, Villalba-Rodríguez AM, Romero-Castillo KD, et al. Organs-on-a-chip module: a review from the development and applications perspective. *Micromachines*. 2018;9(10):536. https://doi.org/10.3390/mi9100536.
208. Torino S, Corrado B, Iodice M, Coppola G. PDMS-based microfluidic devices for cell culture. *Inventions*. 2018;3(3):65. https://doi.org/10.3390/inventions3030065.
209. Halldorsson S, Lucumi E, Gómez-Sjöberg R, Fleming RMT. Advantages and challenges of microfluidic cell culture in polydimethylsiloxane devices. *Biosens Bioelectron*. 2015;63:218–231. https://doi.org/10.1016/j.bios.2014.07.029.
210. van Meer BJ, de Vries H, Firth KSA, et al. Small molecule absorption by PDMS in the context of drug response bioassays. *Biochem Biophys Res Commun*. 2017;482(2):323–328. https://doi.org/10.1016/j.bbrc.2016.11.062.
211. Ding C, Chen X, Kang Q, Yan X. Biomedical application of functional materials in organ-on-a-chip. *Front Bioeng Biotechnol*. 2020;8:823. https://doi.org/10.3389/fbioe.2020.00823.
212. Probst C, Schneider S, Loskill P. High-throughput organ-on-a-chip systems: current status and remaining challenges. *Crit Rev Biomed Eng*. 2018;6:33–41. https://doi.org/10.1016/j.cobme.2018.02.004.

213. Phan DTT, Wang X, Craver BM, et al. A vascularized and perfused organ-on-a-chip platform for large-scale drug screening applications. *Lab Chip*. 2017;17(3):511–520. https://doi.org/10.1039/C6LC01422D.
214. Wevers NR, van Vught R, Wilschut KJ, et al. High-throughput compound evaluation on 3D networks of neurons and glia in a microfluidic platform. *Sci Rep*. 2016;6 (1):38856. https://doi.org/10.1038/srep38856.
215. Yao T, Asayama Y. Animal-cell culture media: history, characteristics, and current issues. *Reprod Med Biol*. 2017;16(2):99–117. https://doi.org/10.1002/rmb2.12024.
216. Kidd BA, Wroblewska A, Boland MR, et al. Mapping the effects of drugs on the immune system. *Nat Biotechnol*. 2016;34(1):47–54. https://doi.org/10.1038/nbt.3367.
217. van den Broek CN, Pullens RAA, Frøbert O, Rutten MCM, den Hartog WF, van de Vosse FN. Medium with blood-analog mechanical properties for cardiovascular tissue culturing. *Biorheology*. 2008;45(6):651–661.
218. Wu Q, Liu J, Wang X, et al. Organ-on-a-chip: recent breakthroughs and future prospects. *Biomed Eng Online*. 2020;19(1):9. https://doi.org/10.1186/s12938-020-0752-0.
219. Rowe RG, Daley GQ. Induced pluripotent stem cells in disease modelling and drug discovery. *Nat Rev Genet*. 2019;20(7):377–388. https://doi.org/10.1038/s41576-019-0100-z.
220. Gourley PL. Brief overview of BioMicroNano technologies. *Biotechnol Prog*. 2005; 21(1):2–10. https://doi.org/10.1021/bp0498239.
221. Ronaldson-Bouchard K, Vunjak-Novakovic G. Organs-on-a-chip: a fast track for engineered human tissues in drug development. *Cell Stem Cell*. 2018;22(3):310–324. https://doi.org/10.1016/j.stem.2018.02.011.
222. Zhang B, Radisic M. Organ-on-a-chip devices advance to market. *Lab Chip*. 2017;17 (14):2395–2420. https://doi.org/10.1039/c6lc01554a.
223. Neužil P, Giselbrecht S, Länge K, Huang TJ, Manz A. Revisiting lab-on-a-chip technology for drug discovery. *Nat Rev Drug Discov*. 2012;11(8):620–632. https://doi.org/10.1038/nrd3799.
224. Geraili A, Jafari P, Hassani MS, et al. Controlling differentiation of stem cells for developing personalized organ-on-chip platforms. *Adv Healthc Mater*. 2018;7(2):1700426. https://doi.org/10.1002/adhm.201700426.
225. van den Berg A, Mummery CL, Passier R, van der Meer AD. Personalised organs-on-chips: functional testing for precision medicine. *Lab Chip*. 2019;19(2):198–205. https://doi.org/10.1039/C8LC00827B.

CHAPTER FOUR

Micro and nanofluidics for high throughput drug screening

Bonolo S.P. Mathekga[a], Zandile Nxumalo[b], and Deepak B. Thimiri Govinda Raj[b,*]
[a]Drizzle Health LLC, Baltimore, MD, United States
[b]Synthetic Nanobiotechnology and Biomachines Group, Synthetic Biology and Precision Medicine Centre, CSIR, Pretoria, South Africa
*Corresponding author: e-mail address: dgovindaraj@csir.co.za

Contents

Progress in Molecular Biology and Translational Science, Volume 187
ISSN 1877-1173
https://doi.org/10.1016/bs.pmbts.2021.07.020
Copyright © 2022 Elsevier Inc.
All rights reserved.

Abstract

In this book chapter, we elaborate on the state-of-the-art technology developments in high throughput screening, microfluidics and nanofluidics. This book chapter further elaborated on the application of microfluidics and nanofluidics for high throughput drug screening with respect to communicable diseases and non-communicable diseases such as cancer. As a future perspective, there is tremendous potential for microfluidics and nanofluidics to be applied in high throughput drug screening which could be applied for various biotechnology applications such as in cancer precision medicine, point-of-care diagnostics and imaging. With the integration of Fourth industrial revolution (4IR) technologies with micro and nanofluidics technologies, it envisioned that such integration along with digital health would enable next generation technology development in medical field.

1. Introduction

Nanotechnology is an emerging technology that has enabled research, development and industrial innovation across the various field such as life sciences, medical sciences, chemical sciences and physical sciences.[1] Applications of nanotechnology in physical and chemical sciences are based on two major manufacturing approaches such as top-down and bottom-up approach.[2] Top-down approach with manufacturing involves generation of nano-size materials or particles by breaking-down of bulk material. Precision engineering and lithography has been used and optimized for industrial applications. Particularly, precision engineering supported production process of the microelectronics industry. Some of the examples included use of diamond-based nanostructures and sensors for advanced servo-drive technologies.[3] Bottom-up manufacturing approach is based on bottom-up used of atom-by atom or molecule-by-molecule based building-up of nanostructures using physical and chemical approaches. Such approaches resulted in generation of nanomaterial in with size range of 1–100 nm through self-assembly or positional or controlled assembly of molecules and atoms. With respect to application of nanotechnology in life and medical sciences includes use of nanotechnology is cell biology, organelle isolation, hyperthermia, clinical research, proteomics and drug screening.[4–7] We used the magnetic nanoparticle-based methodology for plasma membrane, endosome and lysosome proteomics.[8–10] Recently, we also established magnetic nanoparticles functionalized with gold nanoparticles that can be potentially be applied in molecular cell biology, medical science and clinical research applications.[11] We also reported applications of nanotechnology in diagnostics, system biology and synthetic biology.[12] In addition, several teams are exploring use of nanotechnology in functional precision medicine.

Currently, several nanotechnology applications in medicine are implemented in clinical setting. List of the nanotechnology applications includes hyperthermia,[12,13] magnetothermia,[14] precision nanomedicine,[15,16] point-of-care diagnostics,[17] drug delivery[18] systems and drug discovery.[19] Hyperthermia applications includes use of gold nanoparticles in cancer treatment wherein gold nanoparticles has the ability to absorb near infrared light that results in ablation and heating of tumors. In addition, the gold nanoparticle has been used for enhanced X-ray dose to tumors. Synergistic combinations of hyperthermia and radiotherapy has shown to produce improved therapeutic results with reduced X-ray dose and side effects thus making radiotherapy treatment more effective. Similarly, magnetic nanoparticle with iron oxide core are used for magnetothermia for cancer treatment. In addition, magneto-therapy techniques have been successfully applied for pain control, for blood pressure normalization, reduction of oedemata and strengthening of blood vessels. Iron oxide superparamagnetic nanoparticle are also applied for magnetothermia, stem cell tracking and Magnetic Resonance imaging (MRI) analysis in clinical setting. For example, Stem cells can be labeled with superparamagnetic nanoparticles (SPMNPs) in vivo. Using externally induced magnetic field, internalized nanoparticles are magnetized which are used for MRI contrast agents for stem cell tracking. Precision nanomedicine is another application research area where nanotechnology has been applied. Some of the applications includes development of Nano formulations where the drug candidates has been incorporated and optimized with respect to route of administration.[15] In point-of-care diagnostics application, nanotechnology used in cost-effective detection of communicable and non-communicable diseases such as Tuberculosis (TB), Malaria and cancer using Surface Enhanced Raman Spectroscopy (SERS) or later-flow. Nanotechnology has been extensively applied in drug delivery and drug discovery systems where nanoparticles enhanced controlled drug release applied along with applications of numerous biopolymeric materials such as chitosan, alginate, cellulose, and liposomes. In this review article, we focused on current state of the art in micro and nanofluidics and use of the technology has been integrated for life science applications, followed by current and recent development in high throughput systems and applications of micro and nanofluidics in high-throughput systems and drug screening. As an example, we focused on the applications of micro and nanofluidics for TB diagnostics and cancer precision medicine. Finally, we also elaborated on future perspective and our inference on the applications of micro and nanofluidics in biotechnology, biomedical and clinical research, particularly in precision medicine research.

2. State of the art in micro and nanofluidics

Drug screening is a long process (reported to take approximately 12–15 years and costing up to USD 700 million) to bring a potential drug candidate to market.[20–22] Fig. 1 is a depiction of the drug development process. Due to the high risk and long duration of the drug screening process, the pharmaceutical industry is interested in the use of micro and nanofluidics technology to bring down costs and increase the throughput of the drug development process.

In recent years, the use of organ on a chip during the pre-clinical phase of drug development has grown in popularity.[23] The latest microfabrication techniques used in the construction of biomimetic tissues on a chip are discussed. Furthermore, the integration of microfluidics paper based analytical devices with smartphone technology is explored.

2.1 3D bio-printing techniques

The emergence of advanced 3D bio printing techniques has made it possible to create assays that closely mimic in-vivo conditions for use in clinical applications as well as studying underlying disease mechanisms and screening drug compounds.[23,24] In addition, these models can be personalized to individual patients and their unique disease states by utilizing the patient's biological materials in the construction of biomimetic tissue models.[25] Currently, there are three primary 3D bio printing methods; namely inkjet, Extrusion based and Light assisted printing.

Inkjet based bio printing is similar to conventional printing, as it constructs 3D organs on chips by dispensing droplets of biomaterial at predetermined locations on the printing stage. This method is used for constructing small scale models with high resolution demand.[25] Inkjet bio

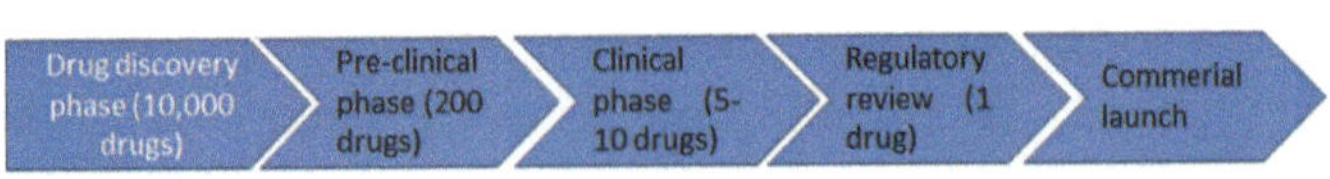

Fig. 1 Drug development phases. The testing and optimization of thousands of drug candidates results in ultimately one drug compound that reaches market. *Adapted from Brancato V, Oliveira JM, Correlo VM, Reis RL, Kundu SC. Could 3D models of cancer enhance drug screening?* Biomaterials. *2020;232:119744.*

printers have a resolution of approximately 20 μm and achieve printing speeds in the order of hundreds of millimeters per second.[26] However, the inkjet printer is limited by its nozzle head, which determines the resolution and does not allow for a wide variety of biomaterials to be printed (due to clogging issues).

Extrusion-based bio printers work by continuously dispensing biomaterials onto the printing stage. By doing so, these printers are compatible with a greater selection of biomaterials, enabling them to be used in the construction of intricate heterogeneous organ models.[25] The downfall of extrusion based bio printing is that it has a slow printing speed of approximately 10–50 μm/s, while achieving resolutions of over 100 μm.[26]

Lastly, light assisted bio printing works by curing photosensitive bio-ink. These printers are commonly used due to their favorable resolution, high printing speeds and minimal shear-stress induced damage to cells.[24] The nozzle free printer is used in the construction of large-scale tissues.[25] Popular light assisted bio printers include the Digital Light Processing (DLP) based and the Two-Photon Polymerization (TPP) bioprinters.[24]

When it comes to the selection of biomaterials, a trade-off is made between mechanical stability and biological compatibility as offered by synthetic and natural biomaterials, respectively. Natural biomaterials are beneficial since their structure recapitulates the extracellular matrix leading to 3D constructs which can support cell growth.[24] While synthetic materials have the advantage of offering mechanical stability and strength to scaffolds. To overcome the shortcomings of both synthetic and natural biomaterials, organs on chips are currently being constructed by combining both biomaterials in multilateral 3D printing. Synthetic materials such as Poly-dimethylsiloxane (PDMS) are used to provide stability, and natural biomaterials such as fibrin and collagen are printed in between the PDMS filaments.[24]

2.2 Microfluidic paper-based analytical devices and smartphone technology

The use of microfluidic paper-based analytical devices (μPADs) in medical diagnostics has grown since the invention of immune-chromatographic assays in the 1980s.[27] Paper is low-cost, disposable and allows for mass production, making it attractive for use in chemical assays. The urine dipstick and pregnancy test kit are examples of commercially available μPADs.[28]

Lateral flow Assays (LFAs) are commonly used μPADs in clinical applications. These tests typically have fast turnaround times of approximately

10 min, are disposable, and do not require sophisticated lab-based equipment or machinery.[28] Hence, LFAs are widely used in point of care settings.[28]

LFAs give yes or no answers based on the detection of target analytes. The biological sample to be tested moves along the length of the immune-chromatographic strip, and as it does so, it is exposed to test lines which comprise of specific biomarkers and a control line. The biological sample reacts with the test lines and change its color based on the absence or presence of the target analyte. Inconsistencies in reading the colorimetric signal (due to varying shades of positive/negative test lines) remains a technical challenge. Mobile technology is being leveraged to circumvent this by giving out digital readings of the colorimetric signals.[29,30] Smartphone digital readers increase the diagnostic utility of μPADS at point of care settings by providing a quantitative reading of the colorimetric signal, while reducing the possibility of human error in result interpretation. Smartphones also have the added benefit of data storage and digital recording of results.[31,32]

The integration of mobile technology with microfluidics has substantially increased due to the penetration of smartphones across global markets. One of the earliest examples of this is the Cellscope, a microscope attachment for mobile devices that is capable of bright field and fluorescence microscopy.[33,34] In 2019, Wang et al.[35] developed a smartphone based biosensor system to detect Salmonella. Their device was based on immuno-magnetic separation, fluorescence labeling and smartphone video processing, and was reported to achieve the limit of detection (LoD) of 58 CFU/mL. In other work, the colorimetric analysis of blood hematocrit was conducted using disposable microfluidic devices.[32] A smartphone was utilized for data collection, image processing and subsequent classification of blood hematocrit levels (ranging between 10% and 65%).

3. Current and recent development in high throughput systems

In recent years, the ability to micro-engineer complex chemical and biological science systems has transformed the manner at which high throughput research is conducted. High throughput screening (HTS) is regarded as a widely used method of choice in the research and development of drug discovery and other relevant fields of biology and chemist.[36] The incorporation of robotics and high-speed computer technology makes HTS more feasible. More recently microfluidics systems have emerged as potential technologies for the application into high throughput systems.[37]

Nevertheless, running up an HTS facility is expensive and requires highly specialized screening tools, and therefore research institutions wishing to establish an HTS facility are advised to initially use existing HTS facilities, rather than setting up one. As this innovative field of science still requires highly specialized and expensive screening laboratory technologies to run an HTS operation successfully. In the following section, we will illustrate a brief high throughput screening process and briefly discuss some of the recent and current impacts of micro-engineered systems as a high throughput screening tools in drug development.

3.1 An overview of a high throughput screening (HTS) process

High throughput systems and usage are not relatively, over the last decade the usage of high throughput systems is regarded as a widely used method of choice in the research and development of drug discovery and other relevant fields of biology and chemistry. To illustrate, compound synthesis and drug discovery processes often rely on the purification and the sensitivity of highly integrated micro-engineered high throughput systems.[36] This HTS process often involves a screening test; a screening test is an experiment that is designed to investigate the presence or absence of specific bio-molecular markers or drug compounds.[36] Once the high throughput system has been validated and the conditions optimized, data is generated and analyzed for the development of targeted biological markers and or drug development. Moreover, the data generated is also stored as libraries or datasets. In Fig. 2 is an overview illustration of the approaches and designs that are employed in high throughput screening research and development processes.

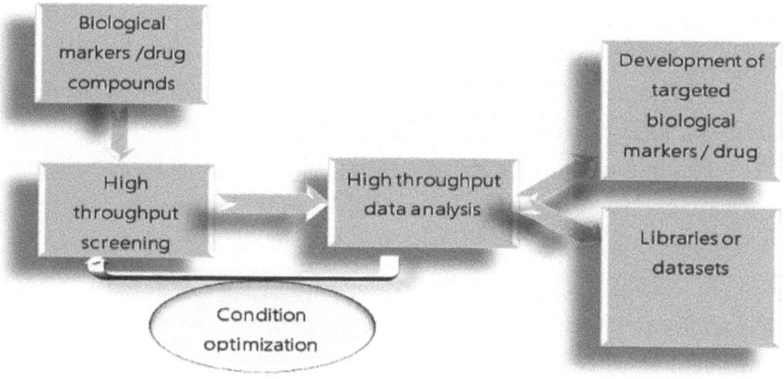

Fig. 2 An overview of the high-throughput approach design for the discovery and development of biological markers or drug compounds.

3.2 Recent and current high throughput systems in drug development

Most commonly known technologies that use micro-engineered techniques is the development of microfluidics devices and micro-reactor devices.[23,38] Traditionally microfluidics devices and micro-reactors are used to manipulate fluid in channels with dimensions of tens of micrometers.[23,37,38] However, more recently this technology has emerged in several distinct new areas of research, all credits can be given to its complex micro-engineering, fabrication techniques and more importantly the diverse number of application fields such as biology, chemistry, medicine, and physical sciences.[38] For these systems to function efficiently, they need to be integrated via the micro-engineered processes with the solvents of interest and must be compatible with the detection methods.[38] Some of the processes employed in the micro-engineering of these systems include but is not limited to device fabrication, injection molding techniques and the incorporation of highly sensitive biochemical assays via a purification step. The fabrication techniques mainly consist of thermoforming or simply the application of heating onto polydimethylsiloxane (PDMS) or glass and a Teflon material to make it soft.[38] Whereas, injection molding or hot embossing is made by the introduction of thermoplastic pellets into a heated mold to create the microfluidic chip or channels.[38]

4. Micro and nanofluidics for high throughput in diagnostics application

The present SARS-CoV-2 global pandemic has made the apparent need for high throughput screening with short turnaround times urgent. Laboratories across the globe are currently swamped with the increased demand for molecular tests and immunoassays for detecting SARS-CoV-2.[39,40] Consequently, national health departments have had to shift their focus from other major diseases (e.g., diabetes, hypertension, maternal and neonatal health, tuberculosis, HIV/AIDs, etc.) to address the on-going COVID-19 pandemic. This re-purposing of laboratory facilities has resulted in disruptions to programs for other health priorities. For example, in South Africa Tuberculosis (TB) testing is extensively done by means of the Gene Xpert MTB/RIF. Using COVID-19 specific cartridges, some of the Gene Xpert machines have been re-purposed to accelerate SARS-CoV-2 testing.[41,42] This is one of the primary reasons why it is estimated that there has been at least a 25% decline in global TB detection during the first 3 months of the pandemic.[40]

Micro and nanofluidics devices offer countless advantages (including low-cost, high speed, portability, and automation) over traditional bulk assays.[43] Thus, lending the technology useful in increasing the throughput of diagnostic applications. Droplet microfluidics has become an important field in biological research because it has enabled disease interrogation to be carried out at a cellular level. Since the analysis is conducted at a single-cell level, information that is typically overlooked in bulk assays is uncovered, owing to the increased signal to noise to ratio of droplet microfluidics.[44] Moreover, the detection of rare biomarkers (e.g., cancer mutations which are known to occur at a single-cell level) by the high throughput analysis of millions of micro droplets is highly sensitive compared to conventional methods.[45]

In this section, we explore different droplet generation techniques and quantitatively evaluate their throughput based on droplet size and the frequency of droplet generation. In addition, we discuss single cell analysis of nucleic acid and protein-based biomarkers for the detection for various diseases. Lastly, the diagnostic applications of microfluidic flow cytometry are explored.

4.1 High throughput droplet microfluidics

4.1.1 Generation of microfluidic droplets

The discretization of single cells from biological fluids is necessary for cells to be interrogated individually. This is achieved using either active or passive droplet generation techniques.[46] A trade-off between control and throughput must be made when selecting the between the droplet formation methods.[45] Active droplet generation techniques provide greater control of encapsulation size, motion, and content at the cost of greater throughput. While passive techniques can generate droplets at high frequencies, however, lack the precision offered by their counterpart.

In its typical form active droplet generation utilizes an external force (e.g., electrical, magnetic, optical, centrifugal, etc.) to assemble the material into droplets.[47] The intensity and duration of the exerted force is used to control the morphology of the resulting droplet. While in passive techniques, the droplet is generated by means of introducing a dispersed immiscible fluid into another continuously flowing fluid.[46] Table 1 shows a high-level throughput analysis of the various droplet generation methods. Once the droplets are encapsulated, they are incubated to allow the generation of a detectable signal, before analyzing the molecular target as depicted in Fig. 3. Commercial sequencing devices (e.g., Drop-Seq) typically require high frequency droplet generation and deterministic encapsulation of

Table 1 Throughput analysis of droplet generation methods.

Classification	Mechanism	Device configuration	Droplet radius	Throughput	References
Passive	–	Flow focusing	~nm	Hz–10 kHz	45,48–51
	–	Cross flow	Limited by channel dimension	Hz	52,53
	–	Co-flow	~μm	Hz–kHz	54,55
	–	Step emulsification	~μm	Hz	56–58
Active	Hydraulic/ pneumatic	Flow focusing	~μm	Hz	59,60
	Optical	Flow focusing	~nm	kHz	61
	Mechanical vibration	Flow focusing	~nm	Hz	62
	Electrical	Cross flow	~nm	Continuous	63

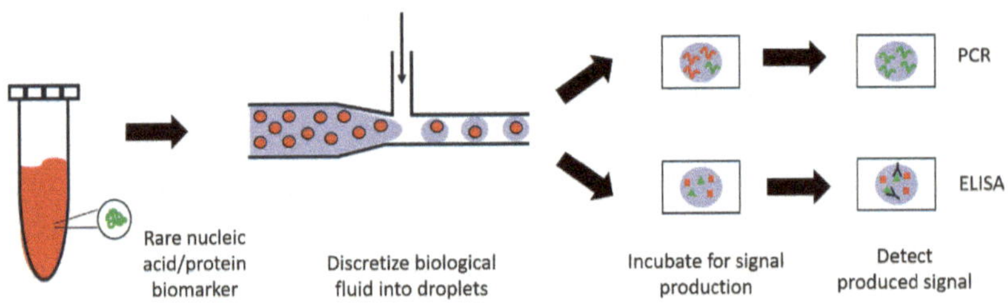

Fig. 3 Overview of droplet microfluidics.

biomarkers.[48] Hence, passive droplet formation in the flow focusing geometry is preferred for diagnostic applications.

4.1.2 Detection of nucleic acid biomarker

For nucleic acid biomarkers the isolated cell is lysed prior to amplification and detection. PCR analysis of droplets has grown in popularity for the detection of cancer related biomarkers. The technology is used to monitor tumor growth and burden by analyzing Cell free circulating tumor DNA (ctDNA), which is challenging for traditional bulk assays as ctDNA is typically low in concentration.[64]

The RainDrop System (RainDance Technologies, Billerica, US) was used by Zonta et al.[65] to detect lung cancer gene mutations (EGFR, KRAS and TP53), and by Heritier et al.[66] to investigate BRAFV600E mutations in clinical samples of patients with childhood Langerhans cell histiocytosis. Another commonly used digital droplet PCR (ddPCR) device is the QX200 ddPCR system (Bio-Rad Laboratories, Hercules, CA, USA). This has been used in cancer research to detect gene mutations. ctDNA samples were emulsified into 20,000 nanoliter droplets prior to mixing with primers/probes. Using this configuration, van Ginkel et al.[64] were able to successfully detect tumor specific TP53 mutations in low level ctDNA from six patients with head and neck squamous cell carcinoma. Sacher et al.[67] used the same QX200 ddPCR system to detect common EGFR and KRAS mutations and the acquired EGFR T790M mutation from blood plasma of 180 patients with non-small cell lung cancer. They found that plasma ddPCR had the shortest turnaround time (TAT) of 3 days. While tissue genotyping had a median TAT of 12 and 27 days for newly diagnosed and acquired resistance patients, respectively.

In other applications, the QX200 ddPCR system has been used to analyze H5N1 RNA strains of influenza and in the detection of bacterial infections.[68] In 2017, Lan et al.[69] demonstrated ultra-high throughput sequencing of more than 50,000 cells per run of Gram negative/positive bacteria and fungi.

In 2015, Hayes et al.[70] designed and implemented a qPCR droplet based microfluidic system. The study team utilized active droplet generation for the deployment of 300 nL droplets. Post thermal cycling, a bio-fluorescence detection platform was used to interrogate each droplet, and TaqMan assays were used to evaluate the performance of their system. It was found that the designed instrument had the capabilities to perform 576 reactions per hour (at 25 s dips over 4 lines). This is equivalent to conducting analysis on 12,384 well plates in a day at significantly reduced costs.

4.1.3 Detection of protein-based biomarker

Conventional ELISA has TATs of over 2 h and requires about 100 μL of samples and reagents in each well. Moreover, the assay involves multiple washing steps to get rid of unbound antibodies or antigens. To circumvent the shortcomings of ELISA, Tang et al.[71] created a one-step droplet microfluidics immunoassay for detecting C-reactive protein (CRP). They reported a LoD of 0.01 μg/mL which is 10 times more sensitive than the

LoD of ELISA for CRP. Moreover, their TAT was 1 h, with a rapid throughput of 500 droplets analyzed each second.

In 2016, Tang et al.[72] achieved high throughput analysis of prostate cancer biomarker proteins (prostate specific antigen, prostate specific membrane antigen, interleukin-6, and platelet factor-4) in serum. Their system was able to conduct 256 measurements in less than an hour using eight 32-sensor immuno-arrays simultaneously.

4.2 Microfluidic flow cytometry

Flow cytometry is considered as fundamentally microfluidic technology since it entails conducting single cell analysis.[73] Microfluidic-based flow cytometry platforms provide ultra-high-throughput (>50,000 cells/s) manipulation and analysis of cells.[74] As a result, these platforms have been employed in clinical applications, including the diagnosis of blood cancers[75] and CD4+ T-cell monitoring in HIV infected patients.[76]

Brower et al.[77] made use of flow cytometry to phenotypically sort double emulsion droplets based on their targets of interest. The study team achieved high sorting frequencies ranging between 12 and 14 kHz. qPCR was used to establish that all droplet-encapsulated DNA can be completely recovered post sorting, with minimal well-to-well cross contamination.

5. Micro and nanofluidics for high throughput drug screening

5.1 Microfluidics

Over the past decade, the application of fabricated micro-engineered systems, such as microfluidics has allowed researchers to develop innovative methods that have high efficiency for the processing of ultra-small analytical volumes of sampling.[38] The fabrication and architecture of microfluidics systems are mainly driven by their ability to integrate complex functional analytical structures that can potentially handle sample volumes, reduce turnaround time for sample processing, operational flexibility and can be fully automated.[38] The ability to process ultra-small analytical volumes of sampling in tens or hundreds of micro-meter channels is said to be a key characteristic feature of microfluidics systems. This ability also allows for the integration of the key functional components responsible for sampling, biochemical reactions, analytes separation, product isolation and detection.[38] Previously there has also been very limited reporting on the applications of digital microfluidics technology for drug development.[78,79] In this report it is evident that digital microfluidics technology can provide the ease of integration with conventional high-throughput screening systems, however, this technology is relatively at its infancy stages of development.[78,79] In essence, microfluidic systems can enable efficient serial processing of high throughput systems and analysis during drug development processes.[80]

5.2 Micro-reactors

Micro-reactors are specialized devices that consist of confined lateral micro-channels with dimensions below 1 mm, where chemical reactions can take place.[81] Micro-reactors are known for generating higher yields of pure chemical products at reduced time frames.[81] This ability to generating chemical products at reduced time frames makes micro-reactors highly suitable for their application in drug development, where the production of compounds, either with multi chemical reagents or under adaptable conditions, is essential.[81] Previously, micro-reactors have also been shown to function more efficiently when integrated via a purification step with a highly sensitive microfluidics biological assay system.[81] This ability to conduct reactions within an integrated multi-channel biological system, such as the microfluidics system and micro-reactor holds great potential for the development of a highly sensitive drug discovery tool. Furthermore, it is envisioned that such a system will allow scientists the opportunity to perform real-time separation of chemical compounds. It is also hoped that this level of integrated functionality within a single device will address the rapid compound production and screening, a fundamental requirement in the pharmaceutical industry.[81] Also, both of these systems offer reduced reaction times, thus assay and chemical reagent costs will be greatly reduced. Nevertheless, the ability of micro-reactors, segmented flow microfluidics systems holds the suitability potential in the processing of chemical and biological compounds and can be efficiently adapted for applications in high-throughput screening systems.[79,80]

5.3 Micro-plate based high throughput screening (HTS)

The development for an HTS system in drug development is currently and fundamentally driven by the miniaturization of Micro and Nanofluidics.[37] A current system that is largely driven by this miniaturization process, is the microplate based HTS technology.[82] Micro-plate based HTS involves the shift from the conventional 96-μL well plate to 1563 higher density microplate formats.[83] And uses the incorporation of miniaturized general methods, which includes robotics, Nano-liter dispensing devices (used for liquid handling), high-density wells on assay plates, sensitive detectors, and data processing and control software tools. It is also worth noting that even though miniaturized processes consist of several benefits, so are the number of challenges that do exist. These challenges range from but are not limited to, poor reliability of data and the standardization of data generated.[83] Moreover, some of the observed challenges with the use of higher density micro-well plates is the inaccuracy of dispensing Nano-liter volumes of liquids and that of uncontrolled evaporation of analytes.[37] Nevertheless, converting assays from traditional well plates to higher density well plates provides the benefits of increased high throughput screening of analytes and the minimization of used biological samples.[82,84] The combination of varying components of which many are existing is important because it will allow the pharmaceutical industries to effectively conduct thousands of pharmacological tests in parallel to biochemical and proteomics screening and to analyze physical properties and mechanisms that are not obvious on a macroscopic scale.[80,85]

5.4 Microarrays

Microarray is a microchip printed with chemical reaction areas containing immobilized DNA sequence information encoding RNA for specific genes.[86] Traditionally, microarrays technology are used to measure the expression of genes and proteins, thus allowing researchers to expand their knowledge about gene regulation and function.[86] Over the years, this technology has revolutionized its application in life sciences research, and it is for this reason that microarrays are proposed that they will have a significant impact in the pharmaceutical industry. According to ref.[87] microarrays techniques have become increasingly and widely used systems in the drug development and life sciences research. It is envisioned that in drug development processes, microarrays will allow for a synchronized analysis of thousands of chemical entities and biological analytes within a single experimental step.[88,89] Thus, allowing scientist to conduct chemical compound profiling, safety analysis and the identification of novel drug discoveries.[87–89]

5.5 Three dimensional (3D) human organs in high throughput systems

The use of engineered human multi-organs holds great promises for the pharmaceutical industries. Human engineered 3D models employ the use of some currently available technologies such as self-organized spherical 3D human organoids, micro-fabricated organ on chip technology, and human 3D bio-printed organ constructs that can mimic the crucial structural human organs and vital functional properties.[90] These technologies allow for the precise control of multicellular activities such as the composition and organization of the ECM (extracellular matrix), the spatial distributions of cells and their micro-environmental cues.[90] Such a system holds the potential to replace the currently existing multi-scaled drug screening platforms to allow for the development and substitution of the current in vitro cellular molecular assays and in vivo models.[91] It is envisioned that the bio-printing of human organs for drug screening will allow the pharmaceutical industries the ability to test specific chemical compounds while developing effective drugs, with reliable efficacy and accuracy.[90] Thereby, allowing the pharmaceutical industries the ability to predict the toxicity of chemical compounds early in the drug discovery process, before clinical trials and thus reducing the cost, time, and failure rates of drugs in clinical trials.[88,91]

6. Micro and nanofluidics for tuberculosis diagnostics

Tuberculosis (TB) is the leading cause of death among all infectious diseases. According to the World Health Organization (WHO) 2019 Report, more than 1.5 million people globally lost their lives to TB in 2018, while approximately 10 million people were infected.[92] TB is caused by the *Mycobacterium Tuberculosis* bacillus (MTb) and is spread when aerosols are expelled (e.g., coughing, sneezing, etc.) by those who are infected with the disease. There are two variations of TB, depending on the site of infection. Pulmonary TB, which affects the lungs, is the most common type of TB. While Extra-Pulmonary TB (EPTB) affects other organs besides the lungs. Furthermore, multi-drug resistant (MDR) TB is caused by MTb that is resistant to at least two of the first-line anti-TB drugs, primarily Isoniazid and Rifampicin.[92]

The WHO has set the goal to eradicate the global TB epidemic by 2030. This is also a Sustainable Development Goal set by the United Nations under Target 3.3. Since timely diagnosis and initiation of treatment is imperative in reducing the on-going transmission of TB, the End TB Strategy includes scaling up TB testing, treatment and care in the top 30 countries with the world's highest TB burden, which accounted for 87% of new TB cases in 2018. The bacteriological confirmation of TB (by means of the identification of MTb) is essential in ensuring that patients are correctly diagnosed and started on the appropriate regimen of anti-TB drugs. However, only 55% of Pulmonary TB cases were bacteriologically confirmed in 2018. Thus, making under-diagnosis a major contributor to the on-going TB transmission.

6.1 The role of micro and nanofluidics in scaling up TB testing

Highly sensitive TB diagnostic tools at Point of Care (PoC) settings are required to close the gap between TB diagnosis and treatment. To achieve this, the WHO has come up with criteria that is meant to guide innovators working on new diagnostic tests for infectious diseases in resource-constrained settings. This is known as the ASSURED criteria: affordable, sensitive, specific, user-friendly, rapid and robust, equipment free and deliverable to end-users.[92,93] In recent years, micro and nanofluidics technology have gained traction in medical research. The miniaturization, low cost and increased sensitivity in the detection of pathogens makes micro and nanofluidics devices suitable for use at PoC settings.[94,95] In addition, features such as visualization by means of colorimetric detection are attractive particularly in markets where advanced laboratory equipment is scare and trained technicians are few. Micro and nanofluidics devices require small amounts of samples and reagents, have a high surface area to mass ratios and yield high-speed analysis compared to conventional analytical assays.[95] Here-in we review currently available TB diagnostics, assessing them based on the ASSURED criteria, and investigating the role of micro and nanofluidics in the diagnostics of TB.

6.2 Culture

Culturing MTb remains the gold standard for TB testing. Conventional TB culture is *flat*, two-dimensional as it entails growing MTb in-vitro using solid medium. It typically takes 3–4 weeks for MTb to grow on Lowenstein-Jenson (LJ) medium.[96] Drug Susceptibility Testing (DST) is done by growing MTb in the presence of anti-TB drugs and resistance is indicated by MTb showing metabolic activity, despite exposure to the drug. The Minimum Inhibitory Concentration (MIC) of drugs needed to kill MTb is also determined in this manner.

2D cell culture has several disadvantages. The method is highly susceptible to contamination, with contamination rates of up to 31% being reported using LJ media.[96] This leads to incorrect results and delayed diagnosis. Moreover, MTb culturing lacks an appreciation of the complex molecular and physiological mechanisms that underlie cellular behavior in-vivo. For example, the absence of an extra-cellular matrix and lack of pharmacokinetics modeling varying drug concentration levels as time goes by limit the extent to which cellular processes can be studied.[97]

Bielecka et al.[98] developed a system which enables the growth of mycobacterium to be investigated in host cells that are grown in 3D microspheres. A bioelectrosprayer was used to generate microspheres from host blood samples infected with MTb. To mimic fluctuating drug concentrations as seen in patients, the microspheres were integrated with a microfluidic system, equipped with two inlets and a single outlet to allow the flow of medium through the microspheres. Specifically, the authors studied the effectiveness of Pyrazinamide (a first-line drug used in the treatment of TB) in killing MTb using the 3D microsphere system they developed compared with 2D culture. They found that Pyrazinamide was able to effectively kill MTb in 3D culture, but not in 2D culture or liquid medium (Middlebrook 7H9). Moreover, increases in administered drug concentrations resulted in rapid killing of MTb. This is the same outcome that is observed in TB patients. A downfall of this technique is that it requires approximately 21 days of cell cultivation, which is perhaps inherent in MTb being a slow-growing organism.

6.3 Imaging

TB diagnosis through imaging is largely done by Sputum Smear Microscopy (SSM). This is the most widely used diagnostic tool, with over 77.6 million SSM tests performed each year globally.[99] Despite being the most widely used test, it has the lowest sensitivity (approximately 50%), among all currently available TB tests.

To the best of our knowledge, micro and nanofluidics have not yet been employed to improve the way SSM is done (i.e., heat-fixing, Auramine or Ziehl Neelsen staining and counting MTb under a microscope). However, the technology has been used to improve TB diagnosis by automating microscopic imaging and immobilizing bacteria using cell-fixing materials, instead of heat-fixation as is often done in SSM. This makes micro and nanofluidics useful in exploring DST applications since the MTb remains viable after chemical cell-fixation, compared to heat-fixing which destroys the cell wall, thereby killing the mycobacterium.

Golchin et al.[100] used a confocal fluorescence microscope to study the growth of *Mycobacterium smegmatis* cells, expressing green fluorescent protein. Each cell was immobilized individually in an alginate hydrogel. A time-lapse series was conducted for over 12 h. Their work showed the feasibility of using real-time analysis and long-term culture of mycobacterium for phenotypic testing. Similar work was done by Choi et al.[101] using the mycobacterium H37Rv strain immobilized in an agarose matrix. An automated microscopic imaging system with digital signal processing was used to analyze MTb growth based on acquired images. Moreover, it took only 9 days to determine the MIC of drugs.

6.4 Immunoassays

Immunoassays are currently not extensively used in the diagnosis of MTb as they are unable to differentiate between Latent TB Infection (LTBI) and active TB disease. The Tuberculin Skin Test (TST) works by injecting a purified protein derivative into the intradermal skin on the forearm. This test is known to have high false-positive rates because it reacts to recent Bacillus Calmette Guerin (BCG) vaccination. The performance of TST is further reduced by factors such as patient age, immune system status and exposure to non-tuberculosis mycobacterium (NTM). On the other hand, the Interferon Gamma Release Assay (IGRA) is a serological immunoassay that is more specific than TST. IGRAs indicate MTb exposure by detecting interferon-gamma (IFN-γ) that is released when CD4 T-lymphocytes harvested from the patient's blood, are in contact with specific TB antigens (e.g., ESAT-6 and CFP-10). Since IGRA tests are more sophisticated, requiring Enzyme-Linked Immunosorbent Assay (ELISA) for detecting IFN-γ, they are not typically used at PoC.

Lipoarabinomannan (LAM) is a lipopolysaccharide that is found on the cell wall of mycobacterium (including NTM). Commercially available LAM tests detect the LAM antigen in urine samples using either ELISA or lateral flow assays. These assays have a very specific indication for use among HIV infected patients only. This is because LAM assays have poor performance in identifying the LAM antigen among non-HIV, TB infected patients. Furthermore, this assay detects the presence of other NTM expelled from the body (i.e., it is not specific for MTb). Using microfluidic technology, Ramirez-Priego et al.[102] sought out to increase the utility of LAM assays in diagnosing TB in people living without HIV co-infection. They made use of a microfluidic cartridge integrated with a photonic sensor chip that is based on the Mach-Zehnder Interferometer configuration. Their system was able to correctly identify LAM-antigens in the undiluted urine samples of five patients with and five patients without HIV co-infection. While reporting negative test results for the 10 patients without confirmed TB disease. These results are promising, despite the small sample size of only 20 patients used in their work.

In another study, Jing et al.[103] created a microfluidic platform that directly captures and conducts immunoreaction analysis of airborne MTb using the TB antigen Ag85B. The device comprised of two subsystems, namely bacterial enrichment, and microfluidic chip immunoassay. Capture efficiency of 100% was achieved in the bacterial enrichment subsystem, which comprised of a Staggered Herringbone Mixer. Their device was reported to have successfully performed bacterial capture, enrichment, lysis, and diagnosis of Ag85B, all within 50 min.

6.5 Nucleic acid detection

Polymerase Chain Reaction (PCR) requires three main steps to create DNA copies. These are denaturation (≈95 °C), annealing (≈56 °C) and extension (≈72 °C). Since nano/microfluidic platforms use small volumes, they are able to provide low thermal mass and rapid heat exchange, making them useful in the detection of nucleic components of MTb for diagnostic purposes.[104]

6.5.1 Digital droplet PCR

Digital Droplet PCR (ddPCR) is known as the third generation of quantitative PCR. It works by segmenting samples using water-in-oil emulsions to create droplets from which their genetic material can be identified and quantified.[105] Droplet digital polymerase chain reaction (ddPCR) is a third generation of polymerase chain reaction (PCR) that enables the exact quantification of nucleic acid targets within a sample. The capability of ddPCR to accurately detect and quantify low abundant targets has led to its fast-growing applications in detection of different pathogens. This review summarizes the ddPCR technology and its applications in tuberculosis diagnosis. From current studies including a total of nine publications on the applications of ddPCR in tuberculosis research, it is clear that ddPCR technology offers enormous advantages, such as unparalleled sensitivity, high precision, and absolute quantification without a standard curve, over common molecular diagnostic platforms like the real-time quantification PCR. The latest study also showed that rapid drug susceptibility test of *Mycobacterium tuberculosis* in sputa could be achieved within 4 days. However, the high cost is the main limitation for its wide applications, especially in developing countries. As we near the vision 2030 goal for sustainable development and ending the tuberculosis epidemic by 2030, ddPCR techniques may help achieve this objective and many more discussed in the UNGA-HLM-TB.[105] This is beneficial over conventional real-time (RT) PCR methods such as the GeneXpert (Cepheid, CA, USA) for the reasons that ddPCR is able to detect MTb DNA at limits of detection as low as 1.2 copies/μL.[106]

In 2017 Yang et al.[107] was among the first to use the QX100 ddPCR (Bio-Rad, CA, USA) to diagnose TB in clinical samples. In their work, Yang et al. used whole blood samples from 84 patients (28 with pulmonary TB, 28 with EPTB and 28 healthy participants). They found that ddPCR could be used to diagnose TB in clinical samples with low levels of MTb DNA.

Since then, the use of ddPCR for MTb detection has grown in popularity. Several studies have explored the use of this technique, in detecting different MTb gene targets. Fig. 4 shows the performance of ddPCR reported in literature against RT-PCR and NCBA methods.

6.5.2 Nanoparticle-based colorimetric bio sensing assay (NCBA)

The Surface Plasmon Resonance (SPR) of gold-nanoparticles is the underlying mechanism by which gold nanoprobes (Au-NP) are used as biosensors for molecular analysis.[118] When Au-nanoparticles are placed in a salt solution with a complementary target, binding occurs causing the nanoparticles to appear red. However, if no binding occurs, the Au-nanoparticles

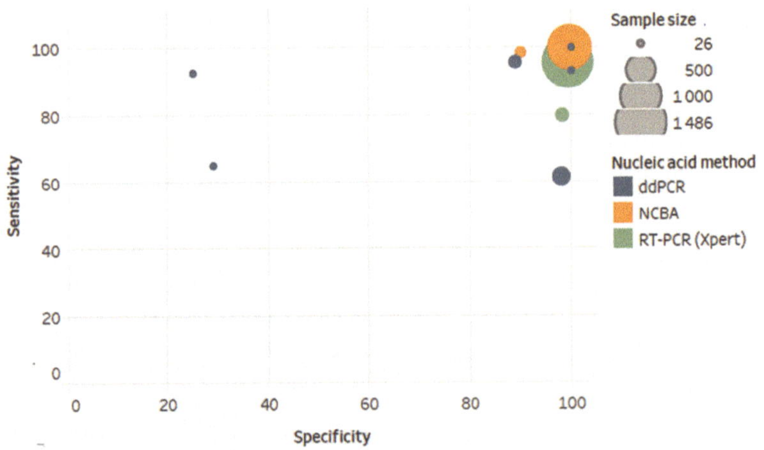

Fig. 4 The performance of different nucleic acid detection methods used to analyze clinical MTb samples. The size of the circle corresponds to the sample size of the study.[106–117]

aggregate.[119] This results in a shift in the SPR band toward higher wavelengths, thus exhibiting a blue color. It is this colorimetric detection scheme that Bernacka-Wojcik et al.[120] made use of in the DNA detection of MTb. A microfluidic channel was fabricated in PDMS using SU-8 photoresist. In addition, the system was equipped with optic fibers for transporting light from the source to the photodetector. The microfluidic channel and optic fibers are disposable, to avoid cross-contamination between the examination of different samples. While the remaining sub-systems, including the photodetector and LEDs are fixed components of the biosensor. A PCR-amplified 395bf segment of MTb RNA polymerase was used as the target molecule for the Au-NP. While the complementary and non-complementary targets are derived from MTb culture positive and NTM samples respectively. Only 3 μL of sample solution was required for DNA detection using this assay. The output voltage at the end of the photodetector was recorded for two intensities. This was done to obtain an objective assessment of the colorimetric change that occurs when successful binding with the complementary targets occur.

In 2017, Mansour et al.[110] distinguished between 72 negative and positive MTb sputum samples using Au-NP based Fluorescence Resonance Energy Transfer (FRET) assay. In the absence of the MTb DNA target, aggregation occurs on the AU-NP quenching its fluorescence. The authors reported achieving a sensitivity and specificity of more than 90%.

6.6 Discussion

Micro and nanofluidics have the potential to positively impact the lives of those affected by TB, by bringing highly sensitive and specific TB diagnostic tools closer to patients. Although there is currently no WHO endorsed microfluidic-based TB assays, this is likely to change in the near future due to recent advancements in the field. NCBA methods have been shown to have the capacity to colorimetrically detect MTb, lending the technology useful in Point-of-care applications where TB diagnosis could be done by detecting color changes through the naked eye. One area of potential concern in the use of small volumes required by microfluidic devices is that the sputum samples used for TB testing commonly comprise of constituents other than MTb (e.g., mucin, saliva, blood, oral flora, etc.). Hence, making the selection of the most purulent part of the specimen (which contains MTb) challenging. Currently, this is overcome by using larger volumes of the raw sputum sample (e.g., at least 1 mL and 3 mL for Gene

Xpert[121] and SSM[122] respectively). Alternatively, mycobacterium concentrating techniques such as bleach sedimentation and centrifugation are employed prior to testing. Perhaps these methods could be used in micro and nanofluidics devices, to ensure that the small volume of liquid selected for testing does indeed contain MTb if ever there is any at all in the sample. ddPCR has been tested using clinical samples on small-scale studies. In addition, it is noted that the performance of the test tends to vary based on the gene target used. That being the case, future work should include a comprehensive study investigating the gene target which results in the best performance using this technique.

7. Future perspectives

The success of drug development traditionally resides with the pharmaceutical industry's ability to produce novel chemical entities (NCEs) and the ability to optimize the development of marketable drugs.[81] However, drug development is inherently a lengthy process and often involves high development costs. Therefore, it is important to develop innovative technologies that will enable the rapid synthesis and screening of novel chemical entities. Moreover, the benefits of using HTS that are highly integrated, consisting of automated robotics, sensitive detectors, micro and or Nanofluidics going forward will allow the pharmaceutical industry to efficiently perform a large number of assays that can determine the drug-like properties of massive data sets.[85] It is also becoming increasingly important to develop technologies capable of screening thousands of drugs on cells or tissue isolated from human patients.[90,91] Although this type of research is still at its early stages, it is hoped that the isolation of human patient cells for drug screening will allow scientists to identify drugs that are most likely to be effective against specific human diseases. It is also hoped that this method of drug screening using patient cells will provide a more accurate and reliable method than the traditional drug-screening approaches that use laboratory cell models that are either old or are not from the origin of the patients.[90]

For organ on a chip microfluidic devices, future work should include the integration of various organs and their vasculature systems to provide a more wholistic simulation of the drug metabolism process.[123] Moreover, these systems should be coupled with high efficiency real-time biosensors to monitor the physiological effects of varying drug concentrations within each organ. The upcoming fabrication of organ on a chip includes 4D bio-printing where biometric tissues change their shapes and functionalities in response to external stimuli.[124]

Ultra-high throughput systems used in drug screening unavoidably generate high volumes of data. Artificial intelligence could be leveraged to analyze the generated performance data of therapeutic agents against a specific disease. Moreover, predictive models may possibly be employed to determine the likelihood of a given drug making it through the pre-clinical phase, based on metrics tracked from biosensors.[125] In doing so, the duration of the drug development process may be shortened.

Acknowledgments

We thank all members of the Deepak laboratory for their contributions. DBTGR is supported by ICGEB Early Career Grant, CSIR Parliamentary Grant and is recipient of NRF competitive unrated Grant. ZN is supported by CSIR Young Researcher Establishment Fund.

References

1. *What is Nanotechnology?*. https://ec.europa.eu/health/scientific_committees/opinions_layman/en/nanotechnologies/l-3/1-introduction.htm.
2. Bayda S, Adeel M, Tuccinardi T, Cordani M, Rizzolio F. The history of nanoscience and nanotechnology: from chemical—physical applications to nanomedicine. *Molecules*. 2020;25:112.
3. Castelletto S, Rosa L, Blackledge J, Al Abri MZ, Boretti A. Advances in diamond nanofabrication for ultrasensitive devices. *Microsyst Nanoeng*. 2017;3:1–16.
4. Thimiri Govinda Raj DB, et al. A novel strategy for the comprehensive analysis of the biomolecular composition of isolated plasma membranes. *Mol Syst Biol*. 2011;7:541.
5. Raj DBTG, Khan NA. Surface functionalization dependent subcellular localization of superparamagnetic nanoparticle in plasma membrane and endosome. *Nano Converg*. 2018;5:4.
6. Raj DBTG, Khan NA. Designer nanoparticle: nanobiotechnology tool for cell biology. *Nano Converg*. 2016;3:22.
7. Raj DBTG, Khan NA, Venkatachalam S, Chu DT. Step-by-Step Protocol for Superparamagnetic Nanoparticle-Based Plasma Membrane Isolation From Eukaryotic Cell. *Methods in Molecular Biology*. Springer Nature; 2020:173–179:2125.
8. Raj DBTG, Khan NA, Venkatachalam S, Chu DT, Arumugam S. Step-by-Step Protocol for Superparamagnetic Nanoparticle-Based Endosome and Lysosome Isolation From Eukaryotic Cell. *Methods in Molecular Biology*. Springer Nature; 2020:167–172:2125.
9. Raj DBTG, Khan NA. Protocol for eukaryotic plasma membrane isolation using superparamagnetic nanoparticles. *J Magn Magn Mater*. 2019;476:628–631.
10. Thimiri Govinda Raj DB. *Superparamagnetic Nanoparticle Based (10nm) Isolation of Plasma Membrane for High Resolution Mass Spec. Analysis of the Proteome, Lipidome and Glycome.* KU Leuven; 2011:1–100. thesis.
11. Raj DBTG, Khan NA. Synthesis of hybrid AuNPs functionalised superparamagnetic NPs. *Micro Nano Lett*. 2018;13:292–296.
12. Raj DBTG, Khan NA. Fluorescent nano-switches for diagnostics applications. *Nat Protoc (Protocol Exchange)*. 2015;2015.
13. Kaur P, Aliru ML, Chadha AS, Asea A, Krishnan S. Hyperthermia using nanoparticles—promises and pitfalls. *Int J Hyperth*. 2016;32:76–88.

14. Loshitskiy PP, Nikolov NA. Magnetothermia utilization in the curing of malignancies. Part 1. *Radioelectron Commun Syst*. 2015;58:49–60.
15. Fornaguera C, García-Celma MJ. Personalized nanomedicine: a revolution at the nanoscale. *J Pers Med*. 2017;7:12.
16. Herrmann IK, Rösslein M. Personalized medicine: the enabling role of nanotechnology. *Nanomedicine*. 2015;11(1).
17. Wang Y, Yu L, Kong X, Sun L. Application of nanodiagnostics in point-of-care tests for infectious diseases. *Int J Nanomedicine*. 2017;12:4789.
18. Patra JK, et al. Nano based drug delivery systems: recent developments and future prospects. *J Nanobiotechnol*. 2018;16:71.
19. Shi J, Votruba AR, Farokhzad OC, Langer R. Nanotechnology in drug delivery and tissue engineering: from discovery to applications. *Nano Lett*. 2010;10:3223–3230.
20. Yeo LY, Chang H-C, Chan PPY, Friend JR. Microfluidic devices for bioapplications. *Small*. 2011;7:12–48.
21. Vogel HG. *Drug Discovery and Evaluation: Pharmacological Assays*. Springer Science & Business Media; 2002.
22. Regnault C, Dheeman DS, Hochstetter A. Microfluidic devices for drug assays. *High Throughput*. 2018;7:18.
23. Du G, Fang Q, den Toonder JMJ. Microfluidics for cell-based high throughput screening platforms—a review. *Anal Chim Acta*. 2016;903:36–50.
24. Ma X, et al. 3D bioprinting of functional tissue models for personalized drug screening and in vitro disease modeling. *Adv Drug Deliv Rev*. 2018;132:235–251.
25. Nie J, Gao Q, Fu J, He Y. Grafting of 3D bioprinting to in vitro drug screening: a review. *Adv Healthc Mater*. 2020;9:1901773.
26. Zhu W, et al. 3D printing of functional biomaterials for tissue engineering. *Curr Opin Biotechnol*. 2016;40:103–112.
27. Campbell RL, Wagner DB, O'connell JP. *Solid Phase Assay With Visual Readout*. US Patent; 1987:US4703017A.
28. Yamada K, Shibata H, Suzuki K, Citterio D. Toward practical application of paper-based microfluidics for medical diagnostics: state-of-the-art and challenges. *Lab Chip*. 2017;17:1206–1249.
29. Yang K, Peretz-Soroka H, Liu Y, Lin F. Novel developments in mobile sensing based on the integration of microfluidic devices and smartphones. *Lab Chip*. 2016;16: 943–958.
30. Zhang L, Yang W, Yang Y, Liu H, Gu Z. Smartphone-based point-of-care testing of salivary \$α\$-amylase for personal psychological measurement. *Analyst*. 2015;140: 7399–7406.
31. Xu D, Huang X, Guo J, Ma X. Automatic smartphone-based microfluidic biosensor system at the point of care. *Biosens Bioelectron*. 2018;110:78–88.
32. Kim SC, Jalal UM, Im SB, Ko S, Shim JS. A smartphone-based optical platform for colorimetric analysis of microfluidic device. *Sensors Actuators B Chem*. 2017;239:52–59.
33. Ephraim RKD, et al. Diagnosis of Schistosoma haematobium infection with a mobile phone-mounted foldscope and a reversed-lens CellScope in Ghana. *Am J Trop Med Hyg*. 2015;92:1253–1256.
34. Fletcher D, Douglas E, Sheng A, Lam W, Maamari R. *Cellscope Apparatus and Methods for Imaging*. US Patent; 2016:US20160248951A1.
35. Wang S, et al. A microfluidic biosensor for online and sensitive detection of Salmonella typhimurium using fluorescence labeling and smartphone video processing. *Biosens Bioelectron*. 2019;140:111333.
36. Entzeroth M, Flotow H, Condron P. Overview of high-throughput screening. *Curr Protoc Pharmacol*. 2009;44:4–9.

37. Hong J, Edel JB, Demello AJ. Micro-and nanofluidic systems for high-throughput biological screening. *Drug Discov Today*. 2009;14:134–146.
38. Dressler OJ, Maceiczyk RM, Chang S-I, deMello AJ. Droplet-based microfluidics: enabling impact on drug discovery. *J Biomol Screen*. 2014;19:483–496.
39. Sahoo H, Mandal C, Mishra S, Banerjee S. Burden of COVID-19 pandemic in India: perspectives from health infrastructure. *medRxiv*. 2020;98–112.
40. Glaziou P. Predicted impact of the COVID-19 pandemic on global tuberculosis deaths in 2020. *medRxiv*. 2020;1–10. https://doi.org/10.1101/2020.04.28.20079582v1.
41. Covid-19. *Behind SA Shortages of Test Materials*; 2020. https://www.spotlightnsp.co.za/2020/05/05/covid-19-behind-sas-shortages-of-test-materials/.
42. *GeneXpert Testing Platform for TB Repurposed to Accelerate Testing for Covid-19*; 2020. https://www.wits.ac.za/covid19/covid19-news/latest/genexpert-testing-platform-for-tb-repurposed-to-accelerate-testing-for-covid-19-.htm.
43. Su W, Gao X, Jiang L, Qin J. Microfluidic platform towards point-of-care diagnostics in infectious diseases. *J Chromatogr A*. 2015;1377:13–26.
44. Reece A, et al. Microfluidic techniques for high throughput single cell analysis. *Curr Opin Biotechnol*. 2016;40:90–96.
45. Kaushik AM, Hsieh K, Wang T-H. Droplet microfluidics for high-sensitivity and high-throughput detection and screening of disease biomarkers. *Wiley Interdiscip Rev Nanomed Nanobiotechnol*. 2018;10:e1522.
46. Zhu P, Wang L. Passive and active droplet generation with microfluidics: a review. *Lab Chip*. 2017;17:34–75.
47. Chong ZZ, et al. Active droplet generation in microfluidics. *Lab Chip*. 2016;16:35–58.
48. Moon H-S, et al. Inertial-ordering-assisted droplet microfluidics for high-throughput single-cell RNA-sequencing. *Lab Chip*. 2018;18:775–784.
49. Yobas L, Martens S, Ong W-L, Ranganathan N. High-performance flow-focusing geometry for spontaneous generation of monodispersed droplets. *Lab Chip*. 2006;6:1073–1079.
50. Mastiani M, Seo S, Mosavati B, Kim M. High-throughput aqueous two-phase system droplet generation by oil-free passive microfluidics. *ACS Omega*. 2018;3:9296–9302.
51. Mastiani M, Seo S, Riou B, Kim M. High inertial microfluidics for droplet generation in a flow-focusing geometry. *Biomed Microdevices*. 2019;21:50.
52. Sripadaraja K, Umesh G, Satyanarayan MN. Simulation studies on picolitre volume droplets generation and trapping in T-junction microchannels. *SN Appl Sci*. 2020;2:1–14.
53. *Digital Microfluidics: Microfluidic Droplets & Emulsion Science*; 2020. https://www.elveflow.com/microfluidic-reviews/droplet-digital-microfluidics/digital-microfluidics-microfluidic-droplet-emulsion-science.
54. Umbanhowar PB, Prasad V, Weitz DA. Monodisperse emulsion generation via drop break off in a coflowing stream. *Langmuir*. 2000;16:347–351.
55. Shams Khorrami A, Rezai P. Oscillating dispersed-phase co-flow microfluidic droplet generation: multi-droplet size effect. *Biomicrofluidics*. 2018;12:34113.
56. Jiao-Yuan L, ZHENG S-X, Zhong-Bin XU, Xiao-Dong R. Study on co-flow effect on Janus droplet generation based on step emulsification. *Chin J Anal Chem*. 2020;48:57–65.
57. Postek W, Kaminski TS, Garstecki P. A passive microfluidic system based on step emulsification allows the generation of libraries of nanoliter-sized droplets from microliter droplets of varying and known concentrations of a sample. *Lab Chip*. 2017;17:1323–1331.
58. Opalski AS, Makuch K, Lai Y-K, Derzsi L, Garstecki P. Grooved step emulsification systems optimize the throughput of passive generation of monodisperse emulsions. *Lab Chip*. 2019;19:1183–1192.

59. Chen C-T, Lee G-B. Formation of microdroplets in liquids utilizing active pneumatic choppers on a microfluidic chip. *J Microelectromech Syst.* 2006;15:1492–1498.
60. Hsiung S-K, Chen C-T, Lee G-B. Micro-droplet formation utilizing microfluidic flow focusing and controllable moving-wall chopping techniques. *J Micromech Microeng.* 2006;16:2403.
61. Park S-Y, Wu T-H, Chen Y, Teitell MA, Chiou P-Y. High-speed droplet generation on demand driven by pulse laser-induced cavitation. *Lab Chip.* 2011;11: 1010–1012.
62. Yin Z, Huang Z, Lin X, Gao X, Bao F. Droplet generation in a flow-focusing microfluidic device with external mechanical vibration. *Micromachines.* 2020;11:743.
63. Gong J. Others. All-electronic droplet generation on-chip with real-time feedback control for EWOD digital microfluidics. *Lab Chip.* 2008;8:898–906.
64. van Ginkel JH, Huibers MMH, van Es RJJ, de Bree R, Willems SM. Droplet digital PCR for detection and quantification of circulating tumor DNA in plasma of head and neck cancer patients. *BMC Cancer.* 2017;17:1–8.
65. Zonta E, et al. Multiplex detection of rare mutations by picoliter droplet based digital PCR: sensitivity and specificity considerations. *PLoS One.* 2016;11:e0159094.
66. Heritier S, et al. BRAF mutation correlates with high-risk Langerhans cell histiocytosis and increased resistance to first-line therapy. *J Clin Oncol.* 2016;34:3023.
67. Sacher AG, et al. Prospective validation of rapid plasma genotyping as a sensitive and specific tool for guiding lung cancer care. *JAMA Oncol.* 2016;2:1014.
68. Hu Y, Xu P, Luo J, He H, Du W. Absolute quantification of H5-subtype avian influenza viruses using droplet digital loop-mediated isothermal amplification. *Anal Chem.* 2017;89:745–750.
69. Lan F, Demaree B, Ahmed N, Abate AR. Single-cell genome sequencing at ultra-high-throughput with microfluidic droplet barcoding. *Nat Biotechnol.* 2017;35: 640–646.
70. Hayes CJ, Dalton TM. Microfluidic droplet-based PCR instrumentation for high-throughput gene expression profiling and biomarker discovery. *Biomol Detect Quantif.* 2015;4:22–32.
71. Tang MYH, Shum HC. One-step immunoassay of C-reactive protein using droplet microfluidics. *Lab Chip.* 2016;16:4359–4365.
72. Tang CK, Vaze A, Shen M, Rusling JF. High-throughput electrochemical microfluidic immunoarray for multiplexed detection of cancer biomarker proteins. *ACS Sens.* 2016;1:1036–1043.
73. Piyasena ME, Graves SW. The intersection of flow cytometry with microfluidics and microfabrication. *Lab Chip.* 2014;14:1044–1059.
74. Stavrakis S, Holzner G, Choo J, DeMello A. High-throughput microfluidic imaging flow cytometry. *Curr Opin Biotechnol.* 2019;55:36–43.
75. Hasegawa D, et al. Validation of flow cytometric phospho-STAT5 as a diagnostic tool for juvenile myelomonocytic leukemia. *Blood Cancer J.* 2013;3:e160.
76. Liu Q, et al. The ARTμS: a novel microfluidic CD4+ T-cell enumeration system for monitoring antiretroviral therapy in HIV patients. *Lab Chip.* 2016;16:506–514.
77. Brower KK, et al. Double emulsion flow cytometry with high-throughput single droplet isolation and nucleic acid recovery. *Lab Chip.* 2020;2062–2074.
78. Yang H, Luk VN, Abelgawad M, Barbulovic-Nad I, Wheeler AR. A world-to-chip interface for digital microfluidics. *Anal Chem.* 2009;81:1061–1067.
79. Wolff A, et al. Integrating advanced functionality in a microfabricated high-throughput fluorescent-activated cell sorter. *Lab Chip.* 2003;3:22–27.
80. Sekhon BS, Kamboj S. Microfluidics technology for drug discovery and development-an overview. *Int J Pharm Tech Res.* 2010;2:804–809.
81. Watts P, Haswell SJ. Continuous flow reactors for drug discovery. *Drug Discov Today.* 2003;8:586–593.

82. Wunder F, Kalthof B, Muller T, Huser J. Functional cell-based assays in microliter volumes for ultra-high throughput screening. *Comb Chem High Throughput Screen.* 2008;11:495–504.
83. Hajare AA, et al. Review on: high-throughput screening is an approach to drug discovery. *Am J PharmTech Res.* 2013;4:112–129.
84. Berg M, Undisz K, Thiericke R, Moore T, Posten C. Miniaturization of an enzyme assay ($β$-galactosidase) in the 384-and 1536-well plate format. *J Assoc Lab Autom.* 1999;4:64–67.
85. Mere L, et al. Miniaturized FRET assays and microfluidics: key components for ultra-high-throughput screening. *Drug Discov Today.* 1999;4:363–369.
86. Wang Z, et al. Combinatorial patterns of histone acetylations and methylations in the human genome. *Nat Genet.* 2008;40:897–903.
87. Afshari CA, Nuwaysir EF, Barrett JC. Application of complementary DNA microarray technology to carcinogen identification, toxicology, and drug safety evaluation. *Cancer Res.* 1999;59:4759–4760.
88. Szymański Pawełand Markowicz M, Mikiciuk-Olasik E. Adaptation of high-throughput screening in drug discovery—toxicological screening tests. *Int J Mol Sci.* 2012;13:427–452.
89. Frotschl R, Kasper P. Microarrays in drug development: regulatory perspective. In: *Microarrays in Inflammation.* Springer Nature; 2009:199–209.
90. Park J, Wetzel I, Dréau D, Cho H. 3D miniaturization of human organs for drug discovery. *Adv Healthc Mater.* 2018;7:1700551.
91. Peng W, et al. 3D bioprinting for drug discovery and development in pharmaceutics. *Acta Biomater.* 2017;57:26–46.
92. World Health Organization. *Global Tuberculosis Report 2019.* Geneva: World Health Organization; 2019.
93. Kosack CS, Page A-L, Klatser PR. A guide to aid the selection of diagnostic tests. *Bull World Health Organ.* 2017;95:639.
94. Chen H, Liu K, Li Z, Wang P. Point of care testing for infectious diseases. *Clin Chim Acta.* 2019;493:138–147.
95. Wang S, Inci F, De Libero G, Singhal A, Demirci U. Point-of-care assays for tuberculosis: role of nanotechnology/microfluidics. *Biotechnol Adv.* 2013;31:438–449.
96. Kassaza K, et al. Lowenstein-Jensen selective medium for reducing contamination in Mycobacterium tuberculosis culture. *J Clin Microbiol.* 2014;52:2671–2673.
97. Duval K, et al. Modeling physiological events in 2D vs. 3D cell culture. *Phys Ther.* 2017;32:266–277.
98. Bielecka MK, et al. A bioengineered three-dimensional cell culture platform integrated with microfluidics to address antimicrobial resistance in tuberculosis. *MBio.* 2017;8.
99. Kik SV, Denkinger CM, Chedore P, Pai M. Replacing smear microscopy for the diagnosis of tuberculosis: what is the market potential? *Eur Respir J.* 2014;43:1793–1796.
100. Golchin SA, Stratford J, Curry RJ, McFadden J. A microfluidic system for long-term time-lapse microscopy studies of mycobacteria. *Tuberculosis.* 2012;92:489–496.
101. Choi J, et al. Rapid drug susceptibility test of Mycobacterium tuberculosis using microscopic time-lapse imaging in an agarose matrix. *Appl Microbiol Biotechnol.* 2016;100: 2355–2365.
102. Ramirez-Priego P, et al. Label-free and real-time detection of tuberculosis in human urine samples using a nanophotonic point-of-care platform. *ACS Sens.* 2018;3:2079–2086.
103. Jing W, et al. Microfluidic platform for direct capture and analysis of airborne Mycobacterium tuberculosis. *Anal Chem.* 2014;86:5815–5821.
104. Ahrberg CD, Manz A, Chung BG. Polymerase chain reaction in microfluidic devices. *Lab Chip.* 2016;16:3866–3884.
105. Nyaruaba R, Mwaliko C, Kering KK, Wei H. Droplet digital PCR applications in the tuberculosis world. *Tuberculosis.* 2019;117:85–92.

106. Song N, et al. Detection of circulating Mycobacterium tuberculosis-specific DNA by droplet digital PCR for vaccine evaluation in challenged monkeys and TB diagnosis. *Emerg Microbes Infec.* 2018;7:1–9.
107. Yang J, et al. Use of digital droplet PCR to detect Mycobacterium tuberculosis DNA in whole blood-derived DNA samples from patients with pulmonary and extrapulmonary tuberculosis. *Front Cell Infect Microbiol.* 2017;7:369.
108. Cho S-M, et al. A novel approach for tuberculosis diagnosis using exosomal DNA and droplet digital PCR. *Clin Microbiol Infect.* 2019;942.e1–942.e5.
109. Bhusal N, Shrestha S, Pote N, Alocilja EC. Nanoparticle-based biosensing of tuberculosis, an affordable and practical alternative to current methods. *Biosensors.* 2019;9:1.
110. Mansour A, Tammam S, Althani A, Azzazy HME. A single tube system for the detection of Mycobacterium tuberculosis DNA using gold nanoparticles based FRET assay. *J Microbiol Methods.* 2017;139:165–167.
111. Briceno RK, Sergent SR, Benites SM, Alocilja EC. Nanoparticle-based biosensing assay for universally accessible low-cost TB detection with comparable sensitivity as culture. *Diagnostics.* 2019;9:222.
112. Luo J, et al. Rapid direct drug susceptibility testing of Mycobacterium tuberculosis based on culture droplet digital polymerase chain reaction. *Int J Tuberc Lung Dis.* 2019;23:219–225.
113. Patterson B, et al. Detection of Mycobacterium tuberculosis bacilli in bio-aerosols from untreated TB patients. *Gates Open Res.* 2017;1:11.
114. Ushio R, et al. Digital PCR assay detection of circulating Mycobacterium tuberculosis DNA in pulmonary tuberculosis patient plasma. *Tuberculosis.* 2016;99:47–53.
115. Pandey P, et al. Diagnostic accuracy of GeneXpert MTB/RIF assay in comparison to conventional drug susceptibility testing method for the diagnosis of multidrug-resistant tuberculosis. *PLoS One.* 2017;12:e0169798.
116. Zeka AN, Tasbakan S, Cavusoglu C. Evaluation of the GeneXpert MTB/RIF assay for rapid diagnosis of tuberculosis and detection of rifampin resistance in pulmonary and extrapulmonary specimens. *J Clin Microbiol.* 2011;49:4138–4141.
117. Sharma SK, et al. Evaluating the diagnostic accuracy of Xpert MTB/RIF assay in pulmonary tuberculosis. *PLoS One.* 2015;10, e0141011.
118. Li Z, et al. Plasmonic-based platforms for diagnosis of infectious diseases at the point-of-care. *Biotechnol Adv.* 2019;37, 107440.
119. Costa P, Amaro A, Botelho A, Inácio J, Baptista PV. Gold nanoprobe assay for the identification of mycobacteria of the Mycobacterium tuberculosis complex. *Clin Microbiol Infect.* 2010;16:1464–1469.
120. Bernacka-Wojcik I, et al. Bio-microfluidic platform for gold nanoprobe based DNA detection—application to Mycobacterium tuberculosis. *Biosens Bioelectron.* 2013; 48:87–93.
121. Badal-Faesen S, et al. Impact of larger sputum volume on Xpert® MTB/RIF assay detection of Mycobacterium tuberculosis in smear-negative individuals with suspected tuberculosis. *J Clin Med.* 2017;6:78.
122. Rashid I, Mabuza LH, Govender I, Pretorius D. Volume of sputum to detect acid-fast bacilli as a measure of quality for the diagnosis of pulmonary tuberculosis at the Dr George Mukhari Hospital, South Africa. *Afr J Prim Health Care Fam Med.* 2011;3:240.
123. Wang Z, Samanipour R, Kim K. Organ-on-a-chip platforms for drug screening and tissue engineering. In: *Biomedical Engineering: Frontier Research and Converging Technologies.* Springer; 2016:209–233.
124. Li Y-C, Zhang YS, Akpek A, Shin SR, Khademhosseini A. 4D bioprinting: the next-generation technology for biofabrication enabled by stimuli-responsive materials. *Biofabrication.* 2016;9:12001.
125. Fetah KL, et al. Cancer modeling-on-a-chip with future artificial intelligence integration. *Small.* 2019;15:1901985.

CHAPTER FIVE

Design, fabrication and assembly of lab-on-a-chip and its uses

Aarathi Pradeep[a,b], Jeethu Raveendran[a], and T.G. Satheesh Babu[a,b,*]
[a]Amrita Biosensor Research Lab, Amrita School of Engineering, Amrita Vishwa Vidyapeetham, Coimbatore, India
[b]Department of Sciences, Amrita School of Engineering, Amrita Vishwa Vidyapeetham, Coimbatore, India
*Corresponding author: e-mail address: tg_satheesh@cb.amrita.edu

Contents

Abstract

Lab-on-a-chip diagnostic devices can be used as quick tools to identify the onset of diseases at an early stage. An integrated LoC platform usually consists of a set of microfluidic elements, each of which has dedicated functions like fluid mixing, fluid manipulation, and flow control, sample preparation, detection, and a read-out that can perform the conventional laboratory procedures on a miniaturized chip. The lab-on-a-chip device can be developed on a paper or polymeric platform and is usually fabricated

Progress in Molecular Biology and Translational Science, Volume 187
ISSN 1877-1173
https://doi.org/10.1016/bs.pmbts.2021.07.021
Copyright © 2022 Elsevier Inc.
All rights reserved.

using pattern transfer techniques or additive and subtractive manufacturing processes. Thorough knowledge of the physics involved in microfluidic technology is essential for developing miniaturized components required for a stand-alone Point-of-Care LoC device. This chapter discusses different types of lab-on-a-chip devices, the essential principles governing the design of these systems, and different fabrication techniques. The chapter concludes with some of the prominent applications of lab-on-a-chip devices.

1. Point of care diagnostics

The point of care technology (PoCT) aims at providing cost-effective health care at every doorstep. The spatial parameters are very significant as the word "point" indicates in PoCT systems. Effective prevention, diagnosis, and management of NCDs or any other diseases alike in rural and urban areas requires healthcare to be more patient-centered rather than centralized. The centralized system does not represent a convenient process for many patients hence the testing process is often being disconnected from the consultation process and more than one visit to the doctor is required to complete the assessment process. Numerous advantages of PoCT systems such as minimal manual input, portability, low cost, immediate clinical decision-making and "ease-of-use" formats make them a good alternative to enhance the health in rural areas. The world health organization has provided guidelines known as "ASSURED" for the development of PoCT devices.[1] The guidelines are shown in Table 1.

Table 1 Guidelines for the development of PoCT devices.[2]

Characteristic	Target specification
A = affordable	Less than 500 US$ per machine, Less than 10 US$ per test
S = sensitive	The lower limit of detection, 500 HIV RNA copies per mL, 350 CD4+ T cells per μL
S = specific	
U = user-friendly	1–2 days of training, easy to use
R = rapid and robust	Less than 30 min for diagnosis, less 1.5 h for HIV load monitoring, minimal consumables, shelf life greater than 1 year at room temperature, high throughput
E = equipment-free	Compact, battery powered, on-site data analysis, easy disposal, easy sample handling
D = delivered	Portable, hand-held

1.1 Types of PoCT technology

The PoCT systems are classified into small handheld devices and much bigger but still portable benchtop devices.[3]

1.1.1 Benchtop devices

Design of these kinds of instruments is similar to those present in the centralized laboratories. Over the years, laboratory instruments have evolved to be benchtop PoCT devices to meet constraints, space, and complexity. These systems emerged as a successive product of large laboratory analyzers. DCA vantage analyzer by Siemens is a bench top point of care system which can monitor glycemic index and kidney functioning.[4] It allows cost-effective diagnosis in decentralized settings with a minimal sample volume. Similarly, Cobas b 101 PoC system allows analysis of diabetes and dyslipidemia from capillary blood volumes.[5] Comparatively bigger, biochemical analyzer like Piccolo Xpress also belongs to the benchtop analyzer category.[6] They contain completely self-contained reagent discs and carry out dilutions by which tests can be configured. But the high capital cost of these systems discouraged them from becoming popular. Samsung lab product line up with an immunoassay, clinical chemistry and hematology and Nanogen's nexus Dx are popular examples of PoCT benchtop systems.[7]

The most common benchtop PoCT device is the blood gas analyzer.[8] A panel of analytes such as blood gases, metabolites, and electrolytes is tested with a smart card or cartridge which is loaded with reagents. These devices were designed such that the complexities associated with the operation are very minimal and a person with limited technical training will be able to operate them.

1.1.2 Handheld technologies

Handheld technology involves completely portable and mobile devices that allow the decentralization of healthcare in its complete sense. Testing with these devices can be performed by the patient individually at the comfort of their home or point of requirement. The well-known globally adopted example of handheld technology is the clinical thermometer. Similarly, handheld glucometers and pregnancy test strips have become inevitable providing fast and reliable results. These devices can be generally categorized into one of the four subclasses given below.[3] The schematic representations of these technologies are shown in Fig. 1.

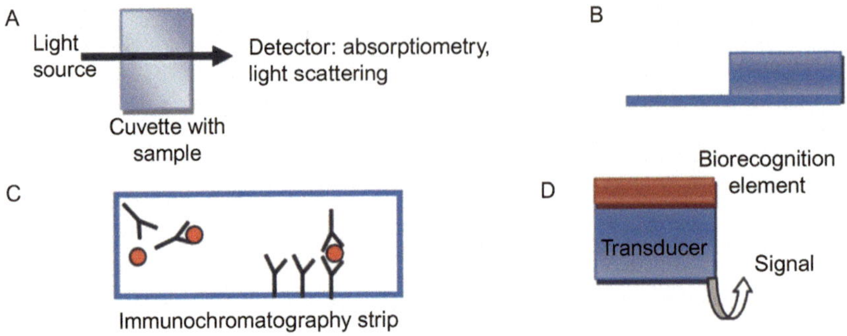

Fig. 1 Schematic representations of handheld PoCT.[3]

(i) *Cuvettes containing reagents*

Traditional colorimetric tests in the test tubes are replaced with reagents stored in cuvettes or cassettes. On the addition of samples such as serum or urine, a color is produced which can be monitored visually or with photometers. Blood glucose and proteins were commonly analyzed with this method.

(ii) *Dipsticks*

Dipsticks have reagent impregnated zones and on the addition of sample, a color is produced. Depending on the analyte, different colorimetric assays are incorporated in the reagent pads. The reactions are read visually or with the help of a reflectance meter.

(iii) *Immunochromatographic strip*

The principle of the immunochromatographic strip or lateral flow assay (LFA) is a sandwich or competitive immunoassay. Antigen (primary antibody-chromogen conjugate) moves through the chromatographic test strip and the movement is rendered by the complexation with the immobilized secondary antibodies. The color produced in the test line, i.e., secondary antibody immobilized zone indicates the presence of the antigen. Colored nanoparticles were extensively used as labels. These tests are intended to operate on the qualitative basis. The most common example of the immunochromatographic strip is the pregnancy test strip.

(iv) *Biosensors*

Biosensors that use the electrochemical principles have been globally commercialized (e.g., glucometers with test strips). The electrodes with non-enzymatic catalyst/enzyme/antibody/whole cell/aptamer are used to identify the biorecognition event. The electrochemical signal produced by the reaction is monitored and quantified with a detector.

1.2 Multiplexed PoCT

Multiplexed PoCT (xPoCT) is the simultaneous on-site detection of multiple analytes from a single sample. The approach of xPoCT has gained much attention in the field of clinical diagnosis with applications in developing countries and resource-limited settings. Early diagnosis and detection of biomarkers help in the identification of diseases from initial manifestations and this ensures timely treatment, which in turn, reduces the morbidity levels associated with the diseases. In most cases, monitoring of multiple biomarkers is essential to analyze the entire spectrum of the disease.[9] Monitoring of multiple biomarkers has potential to identify early events in the disease and helps in timely treatment. Therefore, the concept of multiplexing has become an important concept in PoCT. There has been a great demand for xPoCT devices since the complete analysis could be performed in a short span of time by non-experts. The ideal characteristics of xPoCT devices are consumption of lower sample volumes, measurement from non-invasive samples, faster results, long shelf life, automatic or semiautomatic system with minimum user interventions, cheap and portable readouts and test strips and cartridges.[10] Based on the spatial and regional separation of reagents and electrodes, multiplexing technologies can be classified into paper-based, array-based, bead-based and microfluidic multiplexed systems.[11]

1.2.1 Paper-based systems

Lateral flow assays allow the detection of target analytes from the solution without the use of sophisticated instruments. Simplicity, affordability, and ease of use make paper-based systems the most popular candidate to xPoCT. The technology uses capillary beds such as porous paper, microstructured or sintered polymers that have the capacity to transport fluids.

Generally the paper platform consists of four zones (Fig. 2) namely: (i) sample zone where the blood sample will be placed, (ii) conjugation zone where the analyte selectively binds to conjugated labeled antibodies and these clusters moves to the detection zone, (iii) detection zone which selectively capture analyte-conjugated labeled antibodies clusters.[12] Most of the tests use one more zone known as the control line. The control line consists of polyclonal antibodies which catch free latex/gold nanoparticle and the optical signal produced ensures that the strip is working.

The conjugate zone of the paper platform consists of nanomaterial-biomolecule conjugates like metallic, carbon-based and layered double

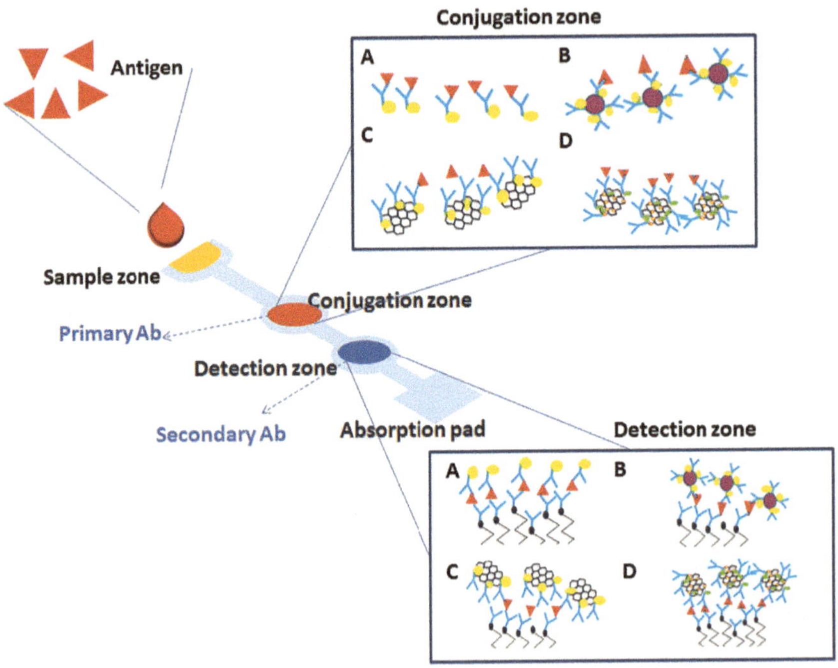

Fig. 2 Schematic of paper-based immunosystems.

hydroxide nanomaterials.[13] The biofunctionalized nanomaterial can be used as recognition probes or to obtain the amplified detection signal. The change of spectral or surface plasmon resonance signal occurring with the nanoparticle aggregation in the detection zone can be used to quantify the concentration of the analyte (Fig. 2. inset A and B). Metal nanoparticles (e.g., AuNP and AgNP), carbon-based nanostructures (e.g., graphene and its derivatives, multiwall carbon nanotubes) or magnetic nanoparticles (e.g., Fe_2O_3) can be used for the signal amplification.[13]

Primary antibody functionalized graphene and its derivatives or layered double hydroxides were used for the designing of chemiluminescence systems (Fig. 2. inset D).[14] Composites of graphene and metal nanoparticles such as PtNP, AgNP or AuNP can also be used (Fig. 2. inset C). Noble metal nanoparticles with graphene enhance the catalytic activity in chemiluminescence systems.

The concept of paper-based analytical devices (μPADs) has been introduced by Whitesides et al.[15] They have demonstrated the quantitative detection of liver biomarkers on paper. In the last decade, the application of

3D-μPADs has expanded tremendously resulting in easy performance of different fluidic operations like mixing, splitting, filtration, etc.[16]

1.2.2 Array-based systems

Array-based systems are one of the most common technologies found in xPoCT in which the sensors are arranged in a particular geometrical pattern. The sensors work on electrochemical and optical principles. Optical microarrays often rely on colorimetric, fluorometric and luminescence-based technology for signal detection. They generally use an array of chemical indicators of different reactivity embedded in hydrophobic membranes or nanoporous sol-gel matrix to produce the required optical change.[17] The selective and distinct color change of each indicator in the array creates a specific fingerprint allowing detection of a single or multiple species. A very broad range of chemical species can be selectively identified with this technique and hence the reactive indicators are chemically different. The detection of the analyte can be performed by using printed dots that change color based on the sample which can be recorded with an imaging device. Typically "before" and "after" images of the colorimetric sensor array is sufficient to determine the uniqueness of the sample. A color difference map is obtained by superimposing and subtracting the two images. This can be quantified as a high dimensional vector of RGB values. These vectors can be used to discriminate different analyte while retaining their chemical resemblances. When compared with the conventional techniques like gas chromatography this method was found less sophisticated and inexpensive.

Array-based electrochemical systems also have been developed and one of the most popular examples is electronic tongue systems.[18] Electronic tongue uses a group of sensors with cross responses. In the presence of a sample, a complex response is generated and analyzed with an appropriate chemometric tool. Similarly, odor quantification tools or electronic noses have also been developed. An array of nonspecific sensors which mimics mammalian olfactory system was used for the detection of vapor.[19]

1.2.3 Bead-based systems

Beads have been used as one of the substrate material for analytical biotechnology. Microbeads are solid plastic particles of less than 1 mm in their largest dimension. Beads of different color, size, functionalization, and materials are available. For the point of care multiplexing, three concepts are mainly used. First is the distinction of beads based on their difference in shape, size, and color. The second approach is by changing the labels used in the beads.

Enzymes, metal nanoparticles, redox materials, quantum dots, etc. have been used as the tags. The third is based on the spatial separation of beads in the microchannels. Flow cytometry systems are the most commonly adopted technique for the bead-based analysis.[20] A typical flow cytometer consists of a microfluidic flow system, optical excitation, and detection units. The beads coupled with immunoagents are incubated with the sample. It may be treated with antibodies specific to the analyte and/or chromogenic agents. The bead-based fluidic stream is introduced through the system with hydrodynamic focusing. The stream is focused through a micro path where one or more lasers passes through the fluid stream resulting in scattering of light. The fluorophore labeled analytes were excited and detected with a series of photomultiplier tubes.[21,22]

The beads allow simultaneous detection of different analytes from a single complex sample solution. The bead-based flow cytometry enables detection of an analyte in very lower sample volume that is too low to be analyzed with traditional ELISA. The most common markers include apoptotic mediators, adhesion molecules, chemokines, immunoglobulin isotypes, chemokines, inflammatory mediators, intracellular signaling molecules and antibodies.[23]

1.2.4 Lab-on-a chip-based systems

Lab-on-a-chip (LoC) systems are an attractive approach which can revolutionize the point of care (PoC) device market dramatically by converging complex laboratory functions into a small chip that is of a few centimeters in size. Development of LoCs for clinical analysis is highly challenging, as it requires expertise in diverse areas such as computational fluid dynamics, microfluidics, biosensors and microelectronics. Regardless of their interdisciplinary nature, LoCs have attracted tremendous attention due to their enormous advantages.

LoCs are a subset of microelectromechanical systems (MEMS), also called "micro total analysis systems" (μTAS).[24] The history of μTAS started in the 1960s with the invention of microtechnologies for the use in semiconductor industries and pressure sensors. This field has four parents: molecular analysis, biodefence, molecular biology, and microelectronics. With the outburst of genomics in the 1980s, the development of LoCs for molecular analysis has progressed drastically with several applications in DNA sequencing. Most of the LoC systems developed till date use inputs from two interdisciplinary areas namely (i) microfluidics and (ii) biosensors. A good balance between two disciplines is necessary for the successful fabrication of LoCs (Fig. 3).

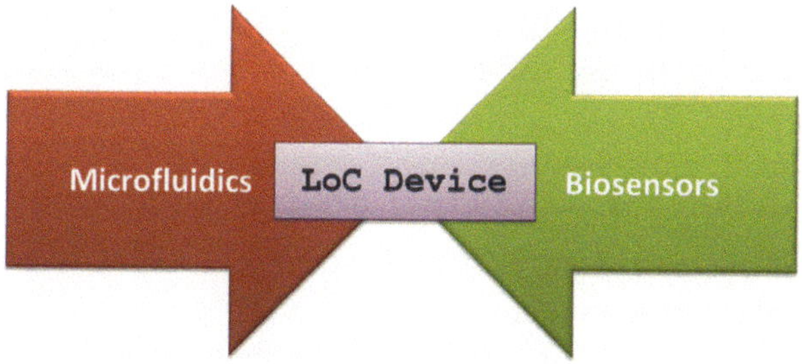

Fig. 3 LoC systems as the interplay between biosensors and microfluidics.

The recent advances in microfluidic technology have paved way for numerous microfabrication techniques.[25] A wide range of materials including polymers, hard substrates like silicon, glass and paper with unique qualities have been employed in microfabrication. The materials for microfabrication should possess good mechanical properties, thermal properties and chemical resistance. In the case of polymeric microstructures, they should have dimensional stability and accuracy. Depending on the type of application, features like permeability and biocompatibility must also be considered. Early works involved microfabrication using substrates like silicon but with the advent of biocompatible polymers like PDMS, soft lithography using such polymers dominated microfabrication. More recently, devices fabricated using 3D printing and paper-based diagnostic devices have been used for the development of lab-on-a-chip devices. The following section discusses on some of the common materials employed for the fabrication of microfluidic devices.

(i) *Silicon*

Silicon was the initial material used for microfluidics and has advantages high resistance to organic solvents, high thermal conductivity and easy deposition of metals in the device. However, the material is hard and opaque and limits its application in microfluidic devices requiring moving components and optical detection techniques.

(ii) *Glass*

Glass is another popular material for the development of microfluidic devices. Although it has advantages like optical transparency and compatibility with biological samples, it has limitations like hardness and requires clean room facilities for fabrication.

(iii) *Elastomeric polymers*

The most commonly used elastomeric polymer for microfluidic device fabrication is polydimethylsiloxane (PDMS). Advantages including biocompatibility, high optical transparency, good mechanical durability, stability against variation in temperature and humidity made PDMS the most suitable choice for techniques like soft lithography.[26]

The empirical formula of the PDMS is $(C_2H_6OSi)_n$ and its structure is shown in Fig. 4. PDMS exists in liquid or semi-solid state and is dependent on the size of the monomer chain. Dow Corning Sylgard 184 elastomer kit is the most popular choice for the stamp fabrication. The silicon encapsulates is a two-part liquid component (base and curing agent).

PDMS is cured by an organometallic crosslinking reaction (Fig. 5). The siloxane base oligomers contain vinyl groups. The cross-linking oligomers

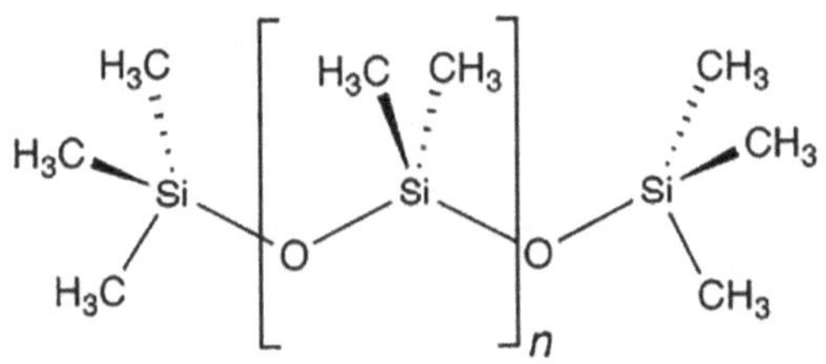

Fig. 4 Structure of PDMS.

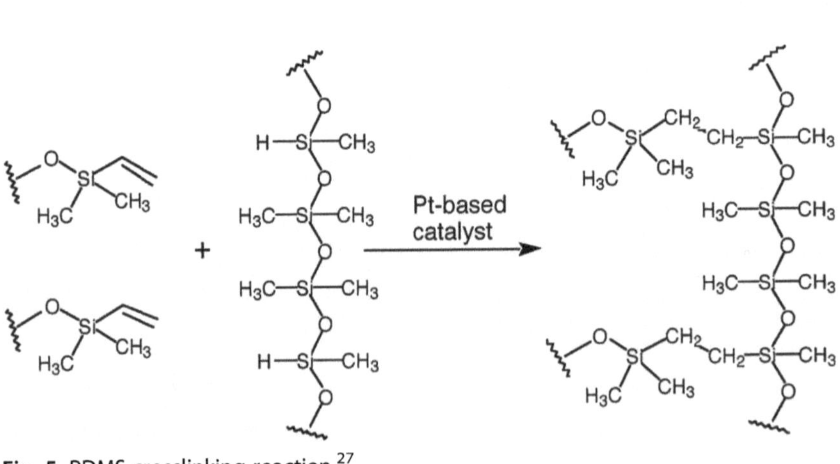

Fig. 5 PDMS crosslinking reaction.[27]

contain at least three silicon hydride bonds each. The curing agent contains a proprietary platinum-based catalyst that catalyzes the addition of the SiH bond across the vinyl groups, forming Si-CH_2-CH_2-Si linkages.[28] The multiple reaction sites on both the base and crosslinking oligomers allow three-dimensional crosslinking. One advantage of this type of addition reaction is that no by-product such as water is generated. If the ratio of curing agent to the base is increased, a harder, more cross-linked elastomer is formed. The crosslinking reaction is further accelerated by heating.

Some of the properties of PDMS may be disadvantageous in certain situations. For example, PDMS is incompatible with many organic solvents and therefore has been limited to aqueous solutions.[29] PDMS is natively hydrophobic in nature; hence nonspecific adsorption of biological samples may occur. Methods to control the surface chemistry of PDMS are being developed to overcome these problems.

(iv) *Thermoplastic polymers*

Thermoplastics are highly cross-linked polymers that can be that remolded by heating the polymer to glass transition temperature and set to a desired shape by cooling. These include polystyrene (PS), polycarbonate (PC) and polymethylmethacrylate (PMMA) polymeric materials. PS has advantages like optical transparency, biocompatibility, inertness and easy surface modification. PC is made by polymerization of bisphenol A and phosgene that results repeating carbonate groups. Advantages like low cost, structural stability and optically transparency, makes it well suited for DNA thermal cycling applications. PMMA based microfluidic chips can be easily fabricated by CO_2 laser micromachining and the chips have high mechanical rigidity, optical transparency. The low cost of these polymers enables the production of disposable microfluidic chips.

(v) *Paper*

Paper substrates are promising for the development of microfluidic devices due to several advantages including low cost, biocompatibility, easy surface functionalization and can be safely disposed by burning or natural degradation. These substrate scan be easily patterned by simple printing processes and mostly rely on the passive mechanism for fluid flow.

1.3 Principles of microfluidics in designing LoC

Microfluidics is a subclass of fluid mechanics dealing with fluid flow in channels in the micrometer range. If any of the geometric dimensions of the flow channel is in the micrometer scale, then the flow can be considered a microfluidic flow.[30] Because of the small size of the flow channel, the flow behavior is different from that of the conventional bulk flow of fluids. Some of the critical aspects of microfluidic flows are briefed below.

1.3.1 Validity of continuum assumption

The majority of the equations and laws of fluid mechanics are based on the continuum assumption. According to this assumption, fluid is considered as a continuous system, with its properties defined and varying continuously in the space. The assumption is valid if the molecules of the fluid are packed so closely when compared to the characteristic length of flow. The validity of the continuum assumption can be found using the Knudsen number, which compares the molecular free path to the physical size of the flow channel, and the assumption is valid for Knudsen numbers less than 1.[31] Continuum assumption is usually valid in liquids and becomes important in microfluidic gas flows when the mean free path of molecules is comparable to the channel dimensions.

1.3.2 Scaling effect

The reduced dimension in a microfluidic environment introduces new dominant forces that are insignificant in macro flows. The important effects which become dominant in the microfluidic environment include laminar flow behavior over turbulent flow, diffusive mass transport over convective transport, viscous forces dominated flow over inertial forces, and increased surface area to volume ratio. Dimensionless numbers are used to signify the relative importance of various competing forces acting on a fluidic system.

1.3.3 Dimensionless numbers

Dimensionless numbers are essential in fluid mechanics to give real insights about the fluid flow.[32] They are very useful in comparing different systems, identifying the dominating forces, and predicting the system behavior. Some of the essential dimensionless numbers in analyzing microfluidic systems are discussed below.

1.3.3.1 Reynolds number

Reynolds number is significant in characterizing the flow as either laminar or turbulent.[33] It is given as the ratio of inertial forces to viscous forces and depends on the characteristic dimension of the channel, velocity of flow, density, and viscosity of the fluid. Mathematically, it is given as

$$Re = \frac{Dv\rho}{\mu} = \frac{\text{inertial forces}}{\text{viscous forces}} \tag{1}$$

where D is the characteristic diameter (m), v is the velocity of the fluid ($m\,s^{-1}$), ρ is the fluid density ($kg\,m^{-3}$), and μ is the viscosity of the fluid (Pa s). Due to the small size involved in microfluidics, the Reynolds number is often less than 100, and the flow is primarily laminar ($Re < 2000$).

1.3.3.2 Knudsen number

Knudsen number is essential in validating the continuum assumption in microfluidics.[34] Often, when the Knudsen number is less than 1, the continuum assumption for fluid flow is valid. Mathematically, it is given as,

$$Kn = \frac{\lambda}{L} = \frac{\text{length of mean free path}}{\text{characteristic dimension}} \tag{2}$$

1.3.3.3 Peclet number

Peclet number is often used in identifying the dominant mode of mass transport in microfluidic channels.[35] It is the ratio of mass transport by convection to that by diffusion. Mathematically, it is given as,

$$Pe = \frac{uL}{D} = \frac{\text{mass transport by convection}}{\text{mass transport by diffusion}} \tag{3}$$

where u is the local fluid velocity ($m\,s^{-1}$), L is the characteristic dimension (m), and D is the diffusion coefficient ($m^2\,s^{-1}$).

1.3.3.4 Fourier number

The Fourier number represents the dimensionless time and is often used in analyzing mass transport phenomena in microfluidic channels.[35] As the mass transport in microchannels is primarily by diffusion, the Fourier number is used to calculate the average time for diffusion over a mixing length. Mathematically, it is given as,

$$Fo = \frac{D\tau}{L^2} \tag{4}$$

where D is the diffusion coefficient ($m^2\,s^{-1}$), τ is the diffusion time (s), and L is the characteristic length (m).

1.3.3.5 Capillary number

Due to the high surface area to volume ratio in microfluidic devices, surface tension forces are significant in many applications. The capillary number compares the magnitude of surface tension and viscous effects.[36] Mathematically, it is given as,

$$Ca = \frac{u\mu}{\sigma} = \frac{viscous\ drag\ force}{surface\ tension\ force} \quad (5)$$

where u is the characteristic velocity (m s^{-1}), μ is the viscosity of the fluid (Pa s), and σ is the surface tension between two phases (N m).

1.3.4 Governing equations of fluid flow

The fundamental governing equations of fluid flow are the conservation of mass, momentum, and energy.[33] They are partial differential equations from which suitable approximations and boundary conditions can deduce various parameters of interest.

1.3.4.1 Conservation of mass

The conservation of mass states that the mass of fluid entering a control volume is equal to the mass of fluid, leaving the control volume.[33] This relation is also known as the continuity equation, which is given as,

$$\frac{\partial \rho}{\partial t} = \nabla.\left(\rho \vec{V}\right) \quad (6)$$

For an incompressible fluid, the continuity equation can be written as,

$$\nabla.\vec{V} = 0 \quad (7)$$

1.3.4.2 Conservation of linear momentum

The conservation of linear momentum states that a change in momentum occurring within a control volume is equal to the total sum of forces acting upon the control volume.[33]

$$\rho\left[\frac{\partial \vec{V}}{\partial t} + \left(\vec{V}.\nabla\right)\vec{V}\right] = -\nabla p + \nabla.\bar{\bar{\tau}} + \rho\vec{f} \quad (8)$$

Where, $\vec{f}$ is the body force per unit mass, and $\bar{\bar{\tau}}$ is the viscous stress tensor.

For incompressible flow, the viscous stress tensor becomes zero, and dividing Eq. (8) by density yields the Navier-Stokes equation.

(a) *Navier-Stokes equation*

The Navier-Stokes (N-S) equation solves the velocity profile of flow in the microfluidic channels that can be used to evaluate various critical parameters like flow rate, pressure drop, and hydraulic resistance. The N-S equation for an incompressible, uniform-viscous, Newtonian fluid is given as,[33]

$$\rho\frac{\partial\vec{u}}{\partial t} = -\rho\vec{u}.\nabla\vec{u} - \nabla p + \eta\nabla^2\vec{u} \tag{9}$$

i.e., (Rate of change of momentum) = (convective force) + (pressure force) + (viscous force).

where $\vec{u}$ is the velocity field (m s^{-1}), ρ is the density of the fluid (kg m^{-3}), and η is the apparent viscosity of the fluid (Pa s).[21]

(b) *Poiseuille flow*

The Poiseuille flow is a particular case of the N-S equation, which describes the balance between net viscous force and net pressure force.[33] By assuming a steady, unidirectional and laminar flow with no acceleration in the fluid, it can be mathematically deduced from N-S equation as:

$$\nabla p = \eta\nabla^2\vec{u} \tag{10}$$

For a circular channel, the velocity profile of a Poiseuille flow is parabolic with maximum velocity in the center of the channel, given as

$$u = \frac{R^2 - r^2}{4\eta}\left(-\frac{dp}{dx}\right) = u_{max}\left(1 - \frac{r^2}{R^2}\right) \tag{11}$$

where R is the diameter of the channel (m).

The flow rate through the channel can be obtained by spatially integrating the velocity contributions, which yields,

$$Q = \frac{\pi R^4}{8\eta}\left(-\frac{dp}{dx}\right) \tag{12}$$

1.3.4.3 Conservation of energy

The conservation of energy equation is the statement of the first law of thermodynamics, which states that the total energy due to work and heat transfer within a control volume is constant.[33] Mathematically, it is written as,

$$\rho C_p\left[\frac{\partial T}{\partial t} + \left(\vec{V}.\nabla\right)T\right] = k\nabla^2 T + \varphi \tag{13}$$

where C_p represents the specific heat of the fluid at constant pressure ($J\ K^{-1}$), k is the thermal conductivity of the fluid ($W\ m^{-1}\ K^{-1}$), and ϕ is the dissipation function representing the work done against viscous forces.

1.3.5 Fluid flow in microchannels

Several methods can achieve fluid manipulation in lab-on-a-chip devices. Some of the commonly used driving forces for fluid flow are discussed below.

1.3.5.1 Pressure driven flow

Fluid actuation by creating a pressure difference between the inlet and outlet is one of the frequently used techniques for driving the fluid in the microchannel. The required pressure difference is achieved by using a syringe or a syringe pump at the inlet or outlet and has a parabolic flow profile (Fig. 6). Alternatively, pressure or vacuum can be generated by pressing a finger on an elastomeric surface on the LoC device.[37]

1.3.5.2 Electric-field driven flow

Electrokinetic fluid flow is the method of driving fluid using an electrical field, and it depends on the type of liquid, the interface of the liquid and surface, and the nature of the electric field.[38] The basic phenomenon in the electrokinetic flow is the formation of an electrical double layer resulting from the interaction of a charged surface and electrolyte. A few of the typical electrically-manipulated fluid flow is explained below.

(a) *Electroosmosis*

Electroosmosis is a method of electrokinetic fluid flow observed in ionized liquids due to interaction between an applied electric field and a double layer formed between the electrode surface and the electrolyte.[39] The applied electric field induces a slip velocity on the ions in the diffuse layer adjacent

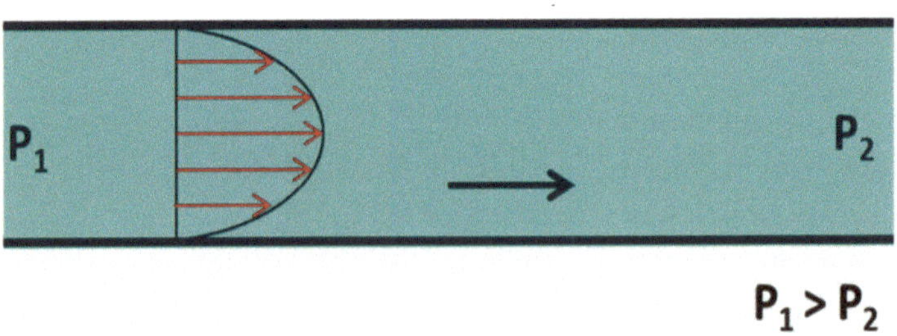

Fig. 6 The typical fluid flow of a pressure-driven flow.

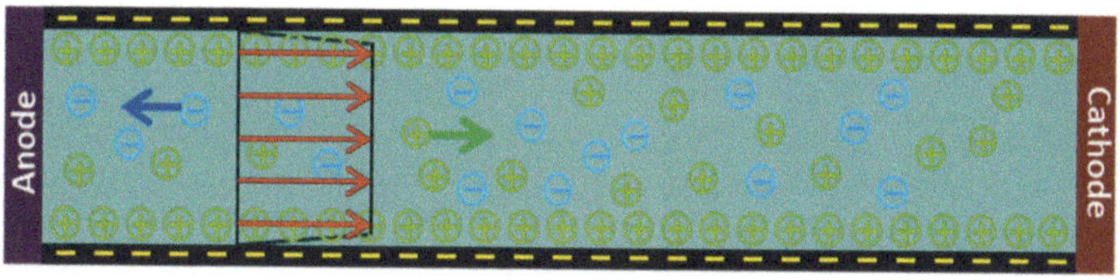

Fig. 7 Typical electroosmosis in a microfluidic channel.

to the interface, thus causing the fluid flow. A pictorial representation of the electroosmotic flow is shown in Fig. 7. They are further classified as DC, AC, time-periodic, and induced charge electroosmosis depending on the nature of the applied electric field. AC electroosmosis has higher charge build-up, and hence a more significant driving force for fluid flow as the surface charges are induced by the applied electric fields.

(b) *Electrophoresis*

Electrophoresis relies on inducing differences in the migration nature of charged species in the presence of an applied electric field. This technique is mainly used for separations and sample pre-concentration. They are further classified as zone electrophoresis, moving boundary electrophoresis, isotachophoresis, isoelectric focusing depending on the electrolyte used.

(c) *Dielectrophoresis*

Dielectrophoresis is the technique used to manipulate neutral or uncharged particles by inducing polarization using a non-uniform electric field. The resulting dielectrophoretic force depends on the applied electric field and the dielectric properties of the particle and the surrounding medium. The polarized particles are then moved to the respective electric-field zone depending on the polarizing ability of the particle.[40] This technique can handle particle sizes from 1 nm to mm and is widely used for particle manipulation, including particle rotation, separation, translation, trapping, and orientation.

1.3.5.3 Capillary-driven flow

In capillary-driven flows, the fluids are displaced by using capillary pumps eliminating the requirement of external pumping source. Capillary pumps are specially designed microstructures that can provide a desired capillary pressure for the fluid to flow. Shen et al. have demonstrated an autonomous capillary-driven based microfluidic device to carry out an immunoassay for the detection of C-reactive protein.[41]

1.3.5.4 Magnetic-field driven flow

In magnetic-field driven fluid, fluid flow is achieved using magnetic particles in a fluid, like ferrofluids, that can be manipulated using an external magnetic field. Chang et al. had demonstrated fluid manipulation using ferrofluid for the development of an immunoassay chip.[42]

1.3.5.5 Rotational forces driven flow

Fluid flow in rotating platforms is achieved using centrifugal forces due to the rotation of the substrate. In this method, complex microfluidic assays are carried out on a rotating disc, and wherein centrifugal forces are used for fluid displacement, metering, and mixing. Czilwik et al. had demonstrated a LabDisk system for bacterial detection that uses centrifugal forces for a fully automated assay.[43]

1.3.6 Numerical methods

With the development of powerful simulation software and faster computation, most scientists rely on simulation as the first step of the design process. Simulation studies are critical in evaluating flow parameters that are difficult to be measured experimentally. They are inevitable to analyze the design feasibility and predict experimental results at a low cost. Most of the commercial CFD tools allow Multiphysics simulation studies, i.e., couple various physics together to model real-world problems and get solutions with a high level of accuracy. Simulation studies can act as a validation tool for experiments that are hypothesis-driven and vice versa. Computational modeling and numerical analysis are critical for a thorough understanding and optimization of parameters for efficient performance.[44] The finite element method is a numerical analysis technique that discretises the domain of analysis into several mesh elements, and the governing partial differential equations of the physics are solved in each of these mesh elements.

1.4 Classification of microfluidic devices

Microfluidic devices are generally classified into (a) continuous flow microfluidic systems, (b) droplet-based microfluidic systems and (c) digital microfluidic systems.[45] Conventional microfluidic systems use continuous flow regimes of microsize for the fluid flow. For the fabrication of these microchannels, soft lithography is employed as the most common technique. The fluidic actuation through these channels are performed via active and passive pumping principles. These techniques are found to be more suitable when the volume of the sample is comparatively higher. To create isolated reaction areas and to reduce the sample volumes, droplet-based systems have been developed. Droplet-based systems use a continuous stream of two immiscible fluids at the input of T or Y mixers.[46] At the point of intersection of the streams, independent droplets separated by an immiscible carrier phase are generated. By increasing the number of miscible input streams, fluidic manipulations such as mixing and chemical reactions like synthesis are performed within the droplets. A new era of droplet-based microfluidics namely, digital microfluidics was introduced in the 2000s.[47] These systems completely removed the requirement of channels for fluid flow. The systems use an array of electrodes to control microdroplets based on the principle of electrowetting on the dielectric surface. Operation methods, flow actuation, advantages and disadvantages of these microfluidic techniques are compared in Table 2.

2. Fabrication of lab-on-a-chip devices

In microfluidic devices, it is not the overall device size but rather the length scale that determines flow behavior. It may be reasonable to say that the device size should be smaller than a centimeter with microchannels in the order of 10^{-6} m. Microfluidic chips are the devices used in microfluidics in which a microchannels network has been molded or patterned. For the fabrication of the microchannels or other microcomponents within the chip different fabrication strategies are used. They involve building down from bigger chunks or building up from smaller materials. The basic microfabrication techniques are divided into four categories namely, additive techniques, subtractive techniques, pattern transfer techniques, and lithography[49] (Fig. 8).

2.1 Additive techniques

Additive techniques are bottom-up approaches. The most common additive techniques are chemical vapor deposition, physical vapor deposition, thermal oxidation, and spin coating. Some of the main challenges of this technique include anisotropic properties in microstructures, stair-stepping and void formation.[50] As the additive technique builds the microstructure in a layer-by-layer method, there exists anisotropy in properties in different layers and in different directions. Stair-stepping is the appearance of layering effect in the fabricated microstructure and this can be reduced with methods like sand sintering. Void formation results due to poor bonding between different layers of fabricated material. Paul et al., had studied the effect of nozzle

Table 2 Comparison between different microfluidic systems.[48]

	Continuous-flow microfluidics	Droplet-based microfluidics	Digital microfluidics
Operating method	The motion of continuous fluid in microchannels	The motion of droplets in microchannels using streams of immiscible fluids	Motion of discrete droplets on an array of planar electrodes
Flow actuation	Mechanical (syringe) pumps, pneumatic pressure, electrokinetic, capillarity	Mechanical (syringe) pumps, pneumatic pressure	Electrowetting on dielectric, dielectrophoresis
Advantages	Ease of fabrication and operation, suitable for applications that require a continuous flow with relatively high sampling volume, and being compatible with most of the current screening and sensing mechanisms	Ease of fabrication and operation, suitable for a applications that require isolated reaction sites to avoid cross-contamination	Lower sample consumption, scalability, better localization, reconfigurability, and portability
Disadvantages	High sample volume consumption compared to other microfluidic systems, possible contamination, and not being scalable due to fabrication and physical limitations	No control over individual droplets, challenging to create droplets of different sizes using the same setup, and challenging to implement stable gas-liquid systems	Complicated fabrication procedure, and bio-adsorption and evaporation

geometry on void formation and had found that rectangular nozzles led to reduced voids than cylindrical nozzles.[51]

In chemical vapor deposition (CVD), the precursor of the material to be deposited is taken in the gaseous form. After the reaction between the precursor gases and catalysts is completed, the material of interest is formed in

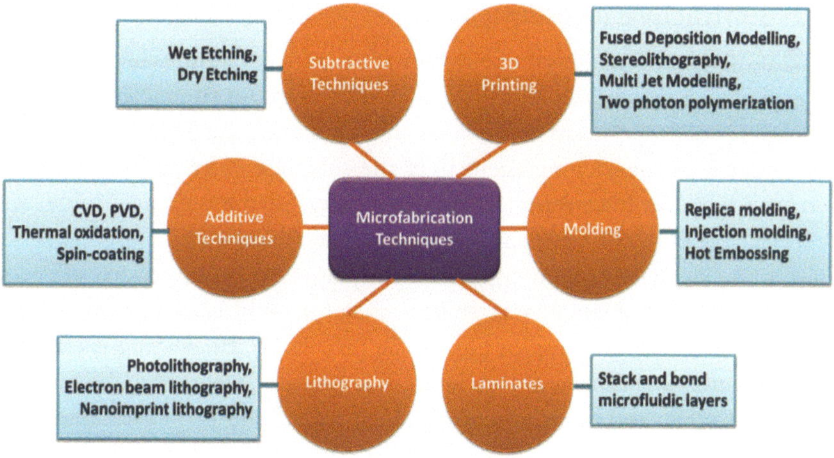

Fig. 8 Classification of basic microfabrication techniques.

solid form and deposited over the surface. CVD involves the dissociation and/or chemical reactions of gaseous reactants in an activated (heat, light, plasma) environment, followed by the formation of a stable solid product.[52]

In physical vapor deposition (PVD), the material to be deposited is taken in the solid state, vaporized by physical means and deposited on the wafer without any chemical reaction. PVD technique has been used for the deposition of electrically conducting layers such as metals and silicides and is done by sputtering or evaporation.[53]

Thermal oxidation is the most common and simplest method employed for the development of silicon dioxide on the silicon surface. This method is used for the fabrication of silicon-based microfluidics where submicron accuracy in the channel and filter pores dimension is required. The technique can be wet or dry thermal oxidation.[54] In dry thermal oxidation pure oxygen is made to react with the silicon surface at very high temperatures typically in the range, 8,001,200 °C. In the wet oxidation process, the water vapor reacts with the silicon at higher temperatures.

Spin coating is the simplest method to fabricate a film on a substrate. This method is used for coating resists for photolithography.[55] It is also used for coating of polymers or chemical precursors onto a polymer called prepolymers. Spin coating results in a relatively planar surface, even on a non-planar substrate. Therefore, this technique can be used for planarization purposes.

2.2 Subtractive techniques

The most common subtractive technique used is etching, which is further classified as dry or wet etching.[56] Wet etching employs chemical solutions to selectively etch the substrate material and is generally carried out by dipping or spraying of the etching agent over the substrate. The advantage of wet etching includes high repeatability, specificity, planar etching profile, and controllable etch rate with concentration. Mostly wet etching process is isotropic and it can lead to under-etching and this can cause an adverse effect on lateral structures. Hence, wet etching is commonly used to remove thin layers or used for thinning of films. The process of dry etching is classified as physical and chemical dry etching. The physical etching process utilizes ions, electrons, and protons to etch the surface. The main challenges that must be addressed in this manufacturing process included under-etching, difficulty to form enclosed microstructures and material wastage.

2.3 Microfluidic devices using laminates

Laminates based microfluidic devices are fabricated by stacking independent cut layers that can be bonded together to form microfluidic channels.[57] A wide range of materials can be used for microfabrication using this technique with glass, polymers like PMMA and polycarbonate and adhesive tapes as the commonly used materials. Each layer of the microfluidic device is cut individually using precise cutting techniques like a knife plotter or laser cutting tools. A knife plotter uses a sharp blade to cut the material to the required microfluidic geometry but limits the fabrication less than 500 μm. Laser cutting tools have better resolution as it employs a focused beam to micropattern the substrate material. They can be used to pattern microstructures up to 25 μm. After removing the inner material of the substrate, the patterned layers are bonded with other layers to form closed microfluidic channels. The different layers are carefully aligned using an alignment tool with alignment holes as guiding points. Thermal bonding and bonding using adhesives are commonly used to bond the different layers of the laminate. Although this technique can enable mass fabrication of microfluidic device, this technique is limited by the depth of the microchannels that can be achieved, which is typically 50–200 μm. Also, careful aligning and leak-proof of the different layers is important for efficient performance of the device.[58]

2.4 Lithography

2.4.1 Photolithography

Photolithography is used to create geometric patterns on a photoresist (light-sensitive polymer) coated substrate. The photoresist can be positive or negative.

Fig. 9 shows the pattern differences generated from the use of a positive and negative resist. Light changes the chemical structure of the positive resist so that it becomes more soluble in the developer. The master formed is an exact copy of the pattern which is to remain on the substrate as a stencil for subsequent processing such as etching. Examples of positive photoresist are PMMA, S1800, ma-P1200 series and SP2-220.

Negative tone resists behave in the opposite manner as compared to positive resists. When exposed to UV light, the negative resist undergoes crosslinking or polymerization and becomes more difficult to dissolve in the developer. Therefore, the exposed regions become hard and difficult to be removed by the developer. Masks used for negative photoresists, therefore, contain the inverse or photographic "negative" of the pattern to be transferred. Examples of negative photoresist are SU-8, KMPR, ma-N1400 series and UVN-30.

SU-8 is sensitive to near the UV range and is also the most commonly used resist for permanent applications.[59] SU-8 photoresist consists of three basic components, namely: (i) epoxy resin (EPON SU-8), (ii) solvent (gamma-butyrolactone), (iii) photo-initiator (triarylium-sulfonium salt). A single molecule of SU-8 contains eight epoxy groups in a bisphenol—a novolac glycidyl ether. The chemistry of SU-8 polymerization is shown in Fig. 10. When the SU-8 photoresist is irradiated with light, the protolysis

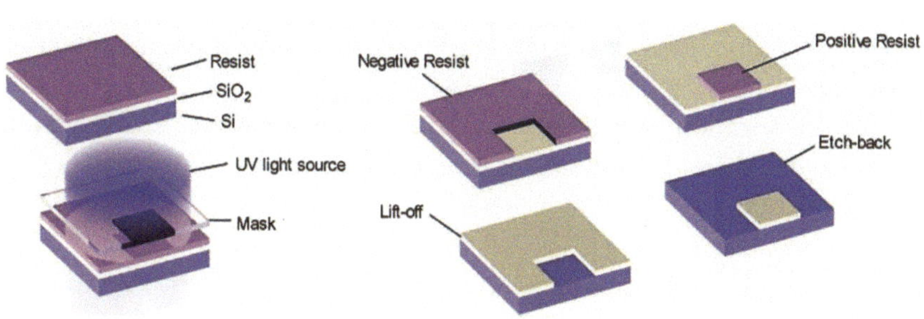

Fig. 9 Comparison between positive and negative photoresists.

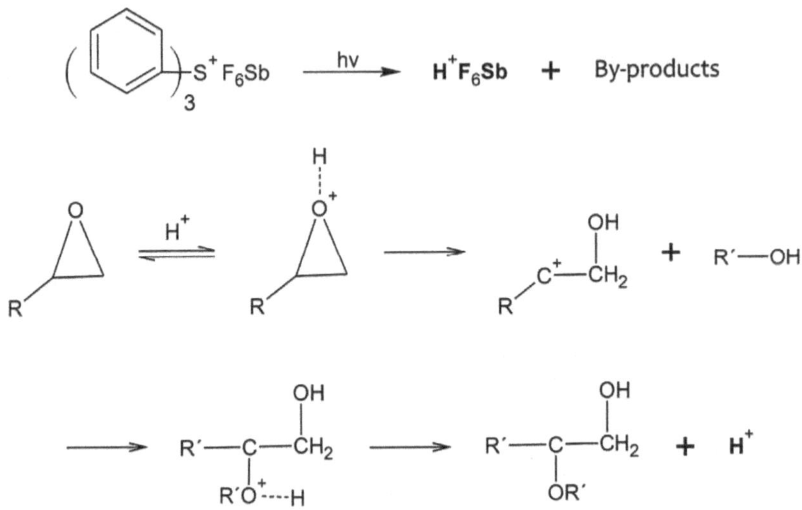

Fig. 10 Chemistry of SU-8 polymerization.

of photoacid generator occurs and the resulting hexafluoroantimonic acid protonates the epoxides on the oligomer.

After heating, oxonium ions are available to react with neutral epoxides in a series of crosslinking reactions. Each monomer molecule contains eight reactive epoxy sites, and therefore a high degree of cross-linking can be obtained after photothermal activation. This results in high mechanical and thermal stability of the lithographic structures. In fact, fully cross-linked SU-8 has a glass transition temperature (Tg) around 200 °C and the degradation temperature of 380 °C.

The processing of SU-8 is much similar to other negative photoresist and it includes several stages. Fig. 11 depicts the different stages of SU-8 based photolithography and each stage of the process is briefly discussed below.

(a) *Substrate preparation*

Prior to fabrication, it is required to clean the glass substrate with piranha solution to ensure the removal of impurities on the substrate. This is done for better adhesion of the photoresist on the substrate.

(b) *Spin coating*

Spin coating is the process by which a thin layer of photoresist is applied to the substrate. Spin coating is characterized by the spread and spin cycles respectively for uniform spreading and excess photoresist removal. The spinning of the glass plate with highly viscous photoresist causes the fluid to

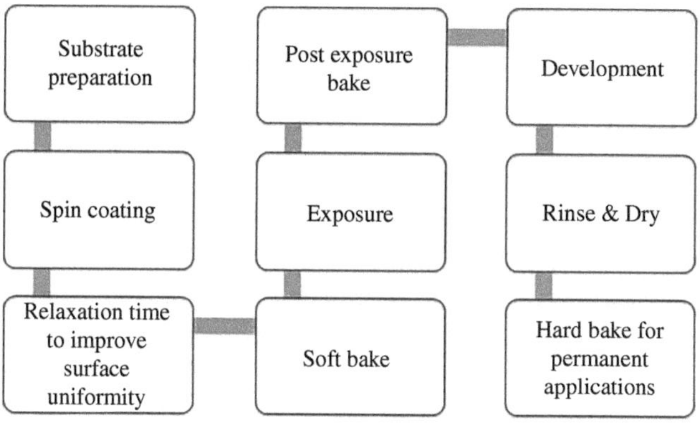

Fig. 11 Various steps involved in the photolithography using SU-8 photoresist.

spread uniformly due to centrifugal force. The thickness of the photoresist film depends on the speed and time of spin. These parameters should be studied and optimized for desired thickness of the photoresist.

(c) *Pre-bake*

After spin coating, prebaking is carried out to remove the excess solvents. This leads to shrinkage of the film and helps improve the adhesion between the film and resist.

(d) *Exposure*

The laser head writes on the loaded substrate with a UV laser and this initializes the polymerization of photoresist. There are two parameters that affect the energy of the UV laser. One is the output power of the laser and the other is the pixel pulse duration. The edge quality of the patterns could be improved by adjusting the exposure speed.

(e) *Post-exposure*

Post-exposure treatment is carried out to enhance polymerization process by hardening the UV exposed areas.

(f) *Development*

This process aids the removal of excess photoresist. The substrate is immersed in a suitable developer solution for a period of time.

The technique is not well-suited to pattern non-planar surfaces and requires the use of photo sensitive polymers.[26] Microsystems for mass spectroscopy, capillary electrophoresis-electrospray ionization microchips, microprobes for drug delivery, microneedles for neural applications, polymerase chain reaction biochips are a few of the examples wherein SU-8 was used for fabrication of microfluidic devices.

2.4.2 Electron beam lithography

In this class of high resolution lithography, a focused beam of electrons is used to pattern an electron reactive resist. The substrate to be patterned is kept on a stage which is placed in a high vacuum chamber. The electron beam is raster scanned on a thin layer organic resist coated substrate according to the desired pattern. The substrate is then developed in a suitable developer solution to obtain the pattern.[60] The steps involved in the fabrication of a master using this technique are shown in Fig. 12A.

The essential feature of this system is a source for the electron beam from which electrons will be generated and will ultimately be used for the patterning of the substrate. An ideal electron beam generator must be capable of generating high intensity electron beam, have a small spot size, long life and good stability. The commonly used electron sources include thermionic emitter, photo emitter or a field emitter. In a thermionic emitter, the electron beam is released due to the thermal energy, while in photo emitter, it is released due to photons or incident radiation. In a field emitter, electron bean is generated due to the applied electric current and the quantum

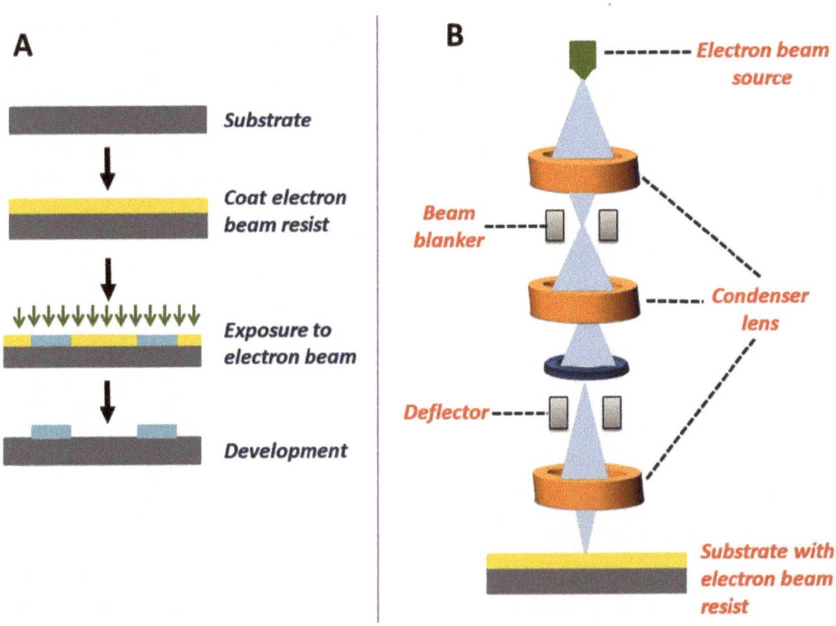

Fig. 12 (A) Fabrication steps involved in electron beam lithography and (B) schematic representation of the working of an electron beam lithography unit.

mechanical properties of the electrons. A combination of thermionic and field emitter, called as thermal field-aided emission is also sometimes used.

A pair of electromagnetic lenses and deflection coils focuses the electron beam. Beam deflectors are then used to raster scan the electron beam on the surface of the target. A beam blanker is used to turn the electron bean on and off. A special type of lens, called stigmator, is used to compensate for the imperfections in the creation and alignment of the electron beam. The schematic illustration of an electron beam lithography is shown in Fig. 12B.

The stigmators remove the astigmatism in the assembly and ensures an ideal circular sot of the beam. Also, the instrument makes use of a vacuum to remove interferences from the electron beam. The resist on interaction with the high energy electron beam either polymerizes or increases the solubility of the resist, depending on the chemical properties of the resist. Due to the short wavelength of electrons, this technique is very useful to fabricate extremely small feature size is required, even less than 10 nm. But this technique is expensive and slower compared to other optical techniques and may have interferences from secondary and back-scattered electrodes.

One of the main limitations of this technique lies in the reduced precision of lenses and deflectors due to aberrations and limited field size and convergence angle. Another challenge encountered is the due to the small deflection of the beams, which causes a defect called stitching and stems. An inaccurate stage shift while patterning can cause stitching inaccuracies, in which the writing fields are not exactly titled against one another. EBL finds vast applications in the field of photonics, electronics and micro electro mechanical systems. Some of the examples include fabrication of transistors for high frequency applications, surface plasmonic devices and highly efficient solar cells. Also, nanofluidic channels for biological studies in lab-on-a-chip devices can be fabricated using this technique.

2.4.3 Nanoimprint lithography

Nanoimprint lithography depends on the direct mechanical deformation of a thin polymer film or a resist coated on a substrate using a template. The desired pattern is imprinted onto the resist by pressing a mold of the desired pattern on the resist and the feature size is typically in the nanoscale. Therefore, there is a direct contact between the mold and the polymer. A surface treatment is done on the template to create a low surface energy release layer. This layer prevents the adhesion of the resist materials onto the template. The minimum resolution that can be fabricated depends on the smallest feature size of the template. Therefore the technique often has higher throughput and lower costs compared to the other optical based lithographic techniques.[61] Also, this technique allows for the easy fabrication of 3D patterns in an affordable manner. After pressing the mold, further polymerization of the resist is achieved using optical/thermal methods.

There are two commonly used NIL processes: (i) Thermal NIL and UV-NIL. In both the techniques, a thin layer of the resist is initially coated onto the substrate (Fig. 13). In a thermal NIL, the resist is a thermoplastic polymer that can soften on raising its temperature.[62] The mold is brought into direct contact with the resist and subjected to a pressure. When the substrate is heated above the glass transition temperature of the polymer, it softens and takes the shape of the mold thereby imprinting the pattern onto its surface. This is followed by cooling and separation of the mold from the substrate. In a UV-NIL, the resist is typically an UV-curable photoresist that is initially in a liquid state. The mold is brought into direct contact with the photoresist and complete filling of the template with the resist is ensured. This is followed by UV exposure resulting in the solidification of the photoresist.

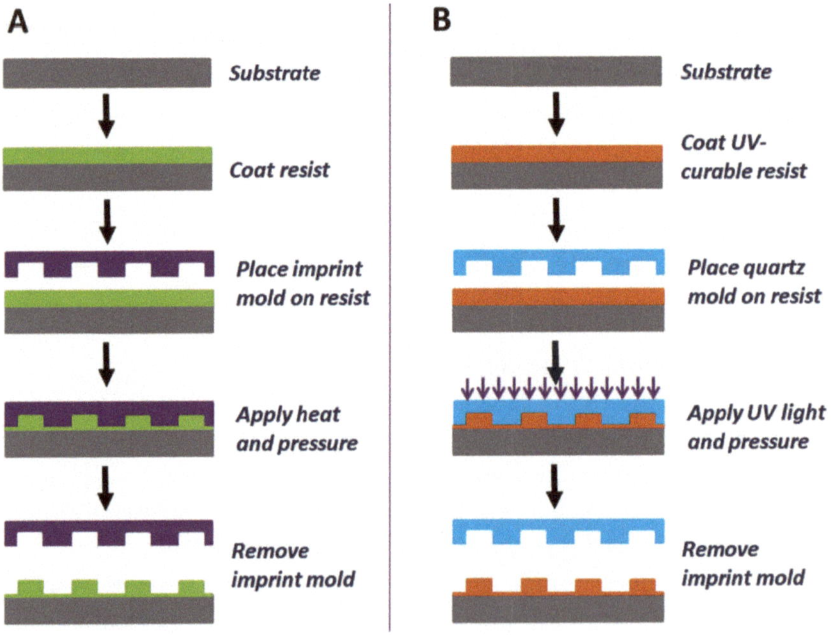

Fig. 13 Fabrication of a microfluidic patterns using nanoimprint lithography (A) thermal NIL and (B) UV-NIL.

This technique is extensively used in a wide range of applications including nanoelectronics, nano-biology, nano-optoelectronics and have been used in the development of biological devices, hard disk drives, flexible displays and OLED. Current research works focus on overcoming different limitations in NIL technique like template fabrication, defect control, alignment accuracy and better nanoimprint resist materials with better characteristics like better etching resistance.[63]

2.5 Molding

Molding is a technique to produce microfluidic devices from a template. In microfabrication, this technique is classified into three types: (i) Replica molding, (ii) Injection molding and (iii) hot embossing. The following section deals with each of the following microfabrication technique.

2.5.1 *Replica molding*

Replica molding, also referred to as "Soft lithography" was introduced by Whitesides group.[64] This technique requires the pre-fabrication of a mold, typically on glass or silicon, using photolithography. It includes a family of techniques based on printing, molding, and embossing with an elastomeric stamp. The biocompatible polymers are allowed to cure on the master and on peeling off the microfluidic channels stamps are obtained.

The process of soft lithography is shown in Fig. 14. Master-stamp technique is the most extensively used technology in microfabrication for

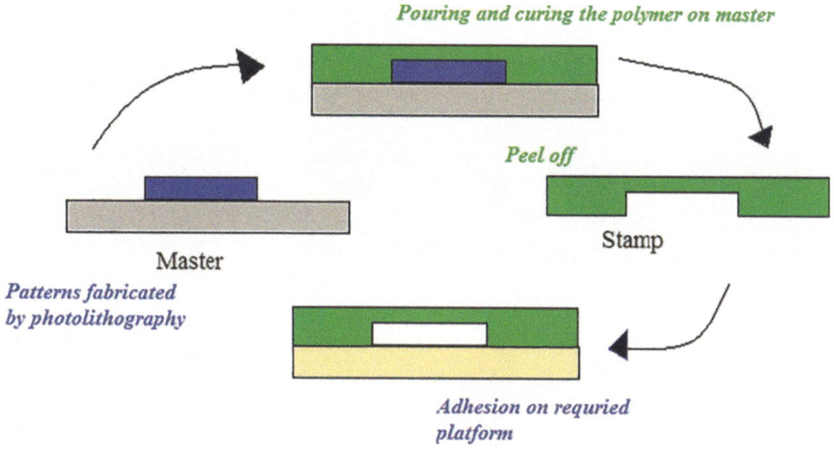

Fig. 14 Steps involved in soft lithography.

low-cost parallel replication or pattern transfer of microfluidic platforms. Generally, the master development is carried out using permanent application photoresists.

2.5.2 Injection molding

Injection molding is the process of injecting molten material into a mold and allows it to set, thereby taking the shape of the mold. Thermoplastic material is fed into a heated barrel where the thermoplastic material becomes soft and molten. This molten material is then subjected to pressure inside a mold cavity. The temperature is then decreased below the glass transition temperature of the thermoplastic, thereby producing a solid material. This technique allows microfabrication using thermoplastics enabling cost-effective and mass production. This technique basically consists of four steps. In the first step, the thermoplastic material is heated to molten liquid state. This is followed by creation of a mold cavity on to which the molten material is injected at a specific rate.[65] Fig. 15 shows the sequence of steps employed in microfabrication using injection molding.

2.5.3 Hot embossing

Microfabrication by hot embossing technique involves stamping of a microfluidic pattern on to a polymer that has been softened by raising the temperature of the polymer to above glass transition temperature. It makes use of a molding tool, also called as the mold insert that is pressed at a high pressure on to the softened polymer. The mold insert contains the pattern of the microstructure that must be transferred on to the polymeric substrate. The mold insert is then pressed carefully on to the softened polymer at a low velocity to reduce the lateral forces and the embossed polymer is then cooled slowly.[65] This technique can be used to create delicate plastic

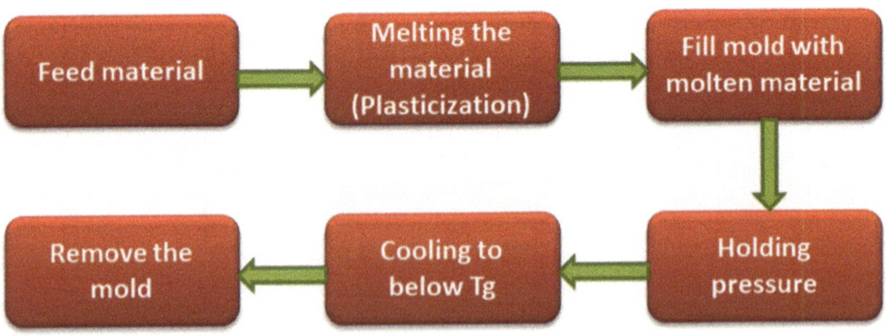

Fig. 15 Steps in microfabrication using injection molding.

microstructures that have a high precision. This technique is widely used for optical applications where high quality and highly precise microdevices are required. One main limitation of this technique is the requirement of expensive molds for microfabrication. To overcome this challenge, Debono et al. had demonstrated a single step process for microfabrication enabling rapid template mold fabrication using thermoplastics.[66]

2.6 3D printing

3D printing is the fabrication of a three dimensional material from a 3D drawing in a single process. The geometry is fed into a computer that is able to slice the geometry into layers and fabricated them using a 3D printer.[67] This technology enables rapid and low cost fabrication of microfluidic devices. The most commonly used technologies for 3D printing includes Fused Deposition Molding (FDMM), Stereolithography (SL), Multi Jet Modeling (MDM) and two photon polymerization. The main challenge associated with this technique is the lower resolution attainable and much of the research is done to overcome this disadvantage.

2.6.1 Fused deposition modeling (FDM)

FDM is the most commonly used 3D printing technology which is based on extrusion printing. The material to be printed is melted in a nozzle and printing is done in a layer-by-layer method. The deposited layer usually has a round texture as the filament layer is extruded through a circular hole in the nozzle.[68] The typical set up of an FDM based 3D printing set up is shown in Fig. 16. A wide range of biocompatible thermoplastics including acrylonitrile butadiene styrene (ABS), PLA, polycarbonate, polyamide, and polystyrene and also reinforced composite materials are employed with this technique. The main parameters that influence this technique include temperature, nozzle diameter, raster width and raster angle. By varying these parameters, intricate microstructures like honeycomb pattern can be developed with varying properties like roughness, strength and accuracy. Although this is a low-cost technique, it suffers from drawbacks like compressive stress fracture, high surface roughness, large filament size and has a higher probability of producing leak-borne materials.

2.6.2 Stereolithography

SL uses optical technology to polymerize a photopolymer to build the structure by layer by layer.[69] A vat of a resin is polymerized as per requirement by exposing it to a structured light source, typically UV light. There are two

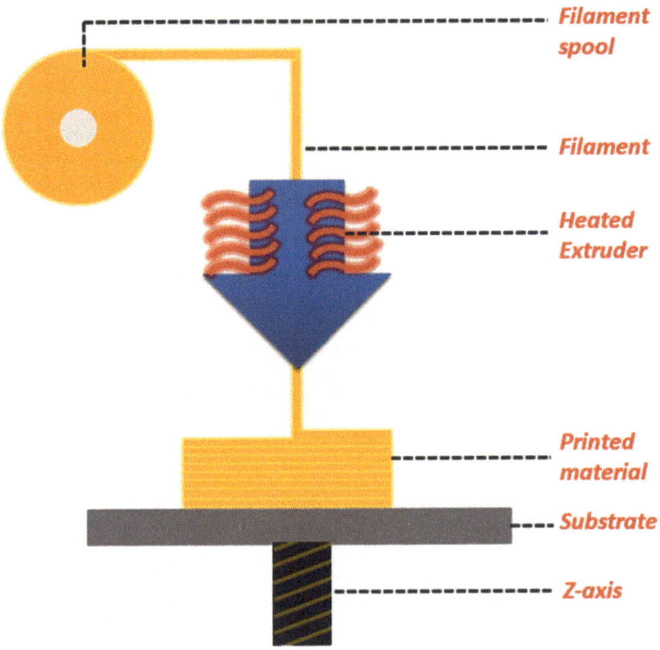

Fig. 16 Working principle of FDM based 3D printing.

common ways of exposing light on to the photopolymer: scanning laser and digital light projector. SLA apparatuses use a galvano-mirror and focused light source for spot polymerization of the photopolymer. In DLP based technique, the entire surface is exposed to light through a mask to obtain the desired pattern. This approach makes use of digital micro-mirror display technology. The layering process is done either in free surface or constrained surface configurations. In free surface configuration, also known as bath configuration, the substrate is submerged into a tank containing photoactive resin that polymerizes when exposed to a UV laser. A blade that is loaded with resin ensures uniformity of layer. This is followed by polymerization of the next layer on the previously polymerized layer. In constrained surface configuration, also referred to as bat configuration, the object is created on a movable substrate in a hanging manner, similar to how a bat hangs on a ceiling. The movable substrate is above the resin containing tank and the light source is positioned below the tank. The major factors affecting this technique are in the resolution of the light source and properties of the resin.

2.6.3 Multi jet modeling

Multi Jet Modeling is used for fabricating complex geometries using hard plastics with high resolution. Like most of the other 3D printing techniques, a 3D geometry of the design is sliced into several layers. Then a photosensitive resin is ejected from an inkjet onto the substrate. This is an additive technique and the subsequent layers are printed on the previously printed layer. The resin is then cured by a light source that is attached to the print head. This printing technique has high accuracy when compared to SL technique. Extra finishing processes are not required as this technique can produce a smooth finish for the products.

2.6.4 Two photon polymerization

This is an optical technique which uses direct near-infrared femtosecond laser writing to pattern arbitrary spots within the material. Three dimensional structures can be created directly with the transparent photoresist. The laser pulse scans the photoresist from bottom to top slice. While in a single photon polymerization process the photochemical reaction is initiated by the photo-initiator absorbing only a single photon, in a multi photon polymerization process, two photons are absorbed by the photo-initiator species thereby initiating the photochemical reaction. Positive and negative photoresists are used for the fabrication. Microstructures are built from unit volume of a material (voxel) that can overlap each other. The major parameters influencing the fabrication include the properties of laser, resin and voxel.[70] The major limitations of employing this technique in the microfabrication are the extensively long time requirement and high costs.

3. Assembly of a lab-on-a-chip device

An important aspect in LoC fabrication is the integration of multiple sensors. Dry etching techniques such as plasma, chemical, physical and deep reactive ion etching and additive techniques such as chemical vapor deposition, physical vapor deposition, and screen printing have been commonly employed for the integration of sensors into the microfluidic systems.[71] Of these, screen printing technique has been widely employed for the fabrication of disposable biosensors in the microfluidic platform as it is relatively simple inexpensive industrially acclaimed process.

The development of a fully functional lab-on-a-chip requires the bonding of the different layers of the device, fluidic connections and electrical connections. A good bonding is crucial for the performance of the device and serves different functions including to hold reagents for the analysis and inhibit its evaporation, preventing unwanted spread of the reagents and to protect the microchannels against dust and contaminants. Numerous methods are available for bonding of the PDMS, such as partial curing, surface oxidation with wet chemicals, corona or plasma treatment, adhesives and adhesive tapes. As discussed earlier, PDMS is widely used for the fabrication of microfluidic devices and it can be easily bonded either reversibly or irreversibly to another substrate. The reversible bonding is achieved by applying pressure on the PDMS stamp after placing it on top of the substrate. Irreversible bonding of PDMS is achieved by using various surface treatments. The most commonly reported method employs the use of O_2 plasma or corona to activate the surface groups in PDMS followed by exposure to heat. This helps in the irreversible or permanent bonding of the PDMS based microfluidic devices.[72] Edding et al. studied different PDMS bonding techniques and found that partially cured and uncured PDMS adhesive techniques show high bond strength than those obtained with oxygen plasma.[73] The fabrication of microfluidic structures by patterned adhesive bonding using PDMS and the bonding between polyethylene terephthalate (PET), the common substrate used for the screen printing, and PDMS have also been studied. Different organofunctional silanes like aminosilanes and epoxysilanes have also been used for the permanent bonding of PDMS based devices on different plastic substrates like PMMA, PET and polycarbonate. Other than PDMS, thermoplastics have been widely used to develop microfluidic devices. Thermal fusion bonding has been used to permanently bond such devices.[74] In this method, the substrate is heated to high temperatures, often near its glass transition temperature and then a force is applied to bond the substrates. One of the limitations of thermal bonding is the possibility of microfluidic structure warping due to the high temperature. Other methods of bonding thermoplastic based devices include solvent bonding and ultrasonic welding. In solvent bonding technique, a solvent is applied between the substrates to the bonded.[75] The substrates are then carefully aligned as per requirements and are subjected to a high pressure to achieve the bonding. Bonding of therrmoplastics are also done by using ultrasonic welding, which ultrasonic energy is used to locally weld the interface of the substrates.[76] In the case of bonding of hard substrates like silicon and glass, techniques like anodic bonding and fusion bonding are employed. The most economical way of bonding has been demonstrated using adhesive tapes. Also UV-curable adhesives, parylene and SU-8 have also been used to bond the different layers of a microfluidic device.

Another important consideration for the development of a fully assembled lab-on-a-chip device is the fluidic interconnections. These interconnections play a major role in interfacing the microfluidic components to the sample. They typically have low dead volumes, reliable at extreme operating conditions, reusable, chemically inert and must be carefully integrated with the device to avoid sample contamination. Press-fit connectors are commonly used to interface PDMS based microfluidic devices. These connectors were based on compression sealing between the inserted tubing and the PDMS microchannel.[77] One of the recent works have developed fit-to-flow interconnects, which enables self-aligned interconnects to a microfluidic network. It consisted of tubing inserted into a PDMS socket that could be interfaced with a microfluidic chip using interfacial adhesion.[78] O-rings are also extensively used for fluidic interconnections. Although reversible interconnects offers advantages in terms of reusability, they are usually not capable of withstanding high pressures. Therefore, in high pressure applications, permanent fluidic interconnects are preferred.

4. Applications

The following section discusses some of the prominent applications of lab-on-a-chip devices in different domains.

4.1 LoC in healthcare management

LoC devices are being increasingly developed for healthcare management. The most successful commercial Point-of-Care LoC device is the blood glucose monitors and the pregnancy test strips.[79] Other commercial PoC devices include Abbott i-STAT hand analyzers for quantification of a wide range of biomolecules including sodium, potassium, creatinine, hematocrit, hemoglobin, blood glucose, and blood gases[80]; Robert Bosch Vivalytic and Vivascope for molecular diagnostics and non-invasive cellular imaging respectively[81,82]; CLINITEK Status + analyzers, Stratus CS cardiac analyzers, blood gas analyzers, DCA Vantage analyzer for diabetes monitoring, all from Siemens Healthcare[83]; Cobas by Roche[83]; all of these devices are capable of a wide range of biomarker analysis. Most of these devices are used in clinical settings, and hence there is a rising demand for the development of microfluidic chips for point-of-care health management. LoC devices capable of quantifying several biomolecules like urea, creatinine, cholesterol, uric acid, dopamine, bilirubin and performing immunoassays for the quantification of various cancer biomarkers like CEA, PSA, HER2, C-reaction protein, HPV have been extensively studied in microfluidic platforms.[79,84,85]

4.2 LoC in defense applications

LoC devices are highly useful in defense applications to detect explosives like nitroaromatic compounds, chemical and biological warfare agents such as sarin, organophosphates, toxins, and pathogens.[86] Tan et al. had reported a lab-on-a-chip device for the early detection of the nerve agent, sarin, in blood.[87] Such devices can prove crucial in the border regions where there may be limited access to laboratory.

4.3 Environmental testing

LoC devices used to monitor water quality by detecting trace amounts of heavy metals like arsenic, copper, cadmium, chromium, lead, and mercury present in water samples.[88] López et al. had developed a LoC device to estimate the amount of nitrates and potassium in the water sample.[89] LoC devices have developed to detect pathogens like *E. coli*, Hepatitis B Virus, Salmonella typhimurium and other pathogens in water and food samples.[90] By developing such LoCs that are affordable, the quality of living can be greatly enhanced.

4.4 LoC in drug delivery

LoC devices allow high throughput analyses, thus allowing significant contributions in drug discovery and drug delivery.[91,92] The drug delivery applications ofLoC devices are at the cellular level, tissue level, or at the organism level. At the cellular level, they are mainly used for drug testing and screening. They are also employed to study the cell response to various chemical and biochemical stimuli, including cell toxicity studies. A single microfluidic chip can be used to study the effect of varying doses of drugs administered to several cells using techniques like microfluidic gradient generators. At the tissue level, they are mainly used for tissue engineering and regenerative medicine studies. Also, engineered tissue-on-a-chip studies can mimic in-vivo analysis of tissue response to a range of drug concentrations. They are used for making microneedles and implantable microfluidic drug delivery devices at the organism level.[93]

4.5 LoC in biochemical assays

LoC devices are widely being researched for the point-of-care diagnosis of bacterial and RNA or DNA viral infections.[94] The detection is usually performed by on-chip immunoassays or on-chip nucleic acid amplification techniques. For detection based on nucleic acid amplification, the LoC devices must be capable of performing sample preparation procedures like cell lysis, RNA isolation or DNA extraction and purification, nucleic acid amplification, and controlled thermal cycling. Polymer-based lab-on-chip devices for carrying out automated RT-PCR have been widely researched and developed.[95] These studies paved the way for the on-chip detection of different viruses, including Zika, Ebola, Hendra, Influenza, and Dengue and bacterial infections like sepsis.[96,97] Recent advances employ paper-based LoC for the detection of such diseases.[98] These devices are crucial for the rapid testing of conditions at the site-of-need in situations like a pandemic.

4.6 Organ on a chip

LoC devices have been utilized to develop microfluidic cell cultures that can replicate the function of different organs of the human body. This concept is called "Organ-on-a-Chip." It has led to revolutionary research to study tissue development, physiology of organs, the genetic, biochemical, and metabolic activities of cells in a functional environment. Microfluidic has paved way for the development of lung-on-a-chip, heart-on-a-chip, brain-on-a-chip, liver-on-a-chip, gut-on-a-chip, kidney-on-a-chip, intestine-on-a-chip, muscle-on-a-chip, tumor-on-a-chip and bone-marrow-on-a-chip.[99,100] Multiple organs can also be modeled on a single device to analyze the inter-organ responses to different drugs.

5. Conclusion

A lab-on-a-chip diagnostic device enables point of care testing of a wide range of biomolecules and is developed keeping in mind the requirements of low cost, ease of use and instantaneous diagnosis at every door step. This chapter discusses different point-of-care technologies available for the development of such devices, its fabrication and concludes with a brief review on the applications of these devices in varied domains which signifies the impact of these devices in day-to-day living. The development of these devices is highly interdisciplinary and brings together basic sciences, engineering and technology with the common goal of improving the quality of living.

References

1. Wang Y, Yu L, Kong X, Sun L. Application of nanodiagnostics in point-of-care tests for infectious diseases. *Int J Nanomedicine*. 2017;12:4789–4803. https://doi.org/10.2147/IJN.S137338.
2. Wu G, Zaman MH. Low-cost tools for diagnosing and monitoring HIV infection in low-resource settings. *Bull World Health Organ*. 2012;90(12):914–920. https://doi.org/10.2471/BLT.12.102780.
3. John AS, Price CP. Existing and emerging technologies for point-of-care testing. *Clin Biochem Rev*. 2014;35(3):155–167.
4. Omoruyi FO, Mustafa GM, Okorodudu AO, Petersen JR. Evaluation of the performance of urine albumin, creatinine and albumin-creatinine ratio assay on two POCT analyzers relative to a central laboratory method. *Clin Chim Acta*. 2012;413(5–6): 625–629. https://doi.org/10.1016/j.cca.2011.12.013.
5. *Roche—Roche Launches New Cobas B 101 System for Management of Metabolic Syndrome*. [Online]. Available https://www.roche.com/media/releases/med-cor-2012-12-18.htm. [Accessed: 06-May-2021].
6. Park H, Ko DH, Kim JQ, Song SH. Performance evaluation of the piccolo xpress point-of-care chemistry analyzer. *Korean J Lab Med*. 2009;29(5):430–438. https://doi.org/10.3343/kjlm.2009.29.5.430.
7. Shin E, Jin C, Ku C, et al. Performance characteristics of cardiac biomarkers on the samsung LABGEOIB10 immunoassay analyzer for use in point of care (POC) and near patient settings. *Clin Chem*. 2013;59(10 SUPPL. 1):A194–A195.
8. Price CP. Regular review: point of care testing. *Br Med J*. 2001;322(7297): 1285–1288. https://doi.org/10.1136/bmj.322.7297.1285.
9. Nimse SB, Sonawane MD, Song KS, Kim T. Biomarker detection technologies and future directions. *Analyst*. 2016;141(3):740–755. https://doi.org/10.1039/c5an01790d.
10. Aguilera-Herrador E, Cruz-Vera M, Valcárcel M. Analytical connotations of point-of-care testing. *Analyst*. 2010;135(9):2220–2232. https://doi.org/10.1039/c0an00307g.
11. Dincer C, Bruch R, Kling A, Dittrich PS, Urban GA. Multiplexed point-of-care testing—xPOCT. *Trends Biotechnol*. 2017;35(8):728–742. https://doi.org/10.1016/j.tibtech.2017.03.013.
12. Nery EW, Kubota LT. Sensing approaches on paper-based devices: a review. *Anal Bioanal Chem*. 2013;405(24):7573–7595. https://doi.org/10.1007/s00216-013-6911-4.
13. Ge X, Asiri AM, Du D, Wen W, Wang S, Lin Y. Nanomaterial-enhanced paper-based biosensors. *TrAC Trend Anal Chem*. 2014;58:31–39. https://doi.org/10.1016/j.trac.2014.03.008.
14. Tiwari A, Dhoble SJ. Recent advances and developments on integrating nanotechnology with chemiluminescence assays. *Talanta*. 2018;180:1–11. https://doi.org/10.1016/j.talanta.2017.12.031.
15. Martinez AW, Phillips ST, Butte MJ, Whitesides GM. Patterned paper as a platform for inexpensive, low-volume, portable bioassays. *Angew Chem Int Ed Engl*. 2007;46(8):1318–1320. https://doi.org/10.1002/anie.200603817.
16. Martinez AW, Phillips ST, Whitesides GM, Carrilho E. Diagnostics for the developing world: Microfluidic paper-based analytical devices. *Anal Chem*. 2010;82(1):3–10. https://doi.org/10.1021/ac9013989.
17. Zhang C, Suslick KS. Colorimetric sensor array for soft drink analysis. *J Agric Food Chem*. 2007;55(2):237–242. https://doi.org/10.1021/jf0624695.
18. del Valle M. Sensor arrays and electronic tongue systems. *Int J Electrochem*. 2012; 2012:1–11. https://doi.org/10.1155/2012/986025.
19. Gardner JW, Shin HW, Hines EL. Electronic nose system to diagnose illness. *Sens Actuators B Chem*. 2000;70(1–3):19–24. https://doi.org/10.1016/S0925-4005(00)00548-7.

20. Holmes D, She JK, Roach PL, Morgan H. Bead-based immunoassays using a micro-chip flow cytometer. *Lab Chip.* 2007;7(8):1048–1056. https://doi.org/10.1039/b707507n.
21. McKinnon KM. Flow cytometry: an overview. *Curr Protoc Immunol.* 2018;2018: 5.1.1–5.1.11. https://doi.org/10.1002/cpim.40.
22. Alfonso AF, Al-Rubeai M. Flow cytometry. In: *Comprehensive Biotechnology.* Elsevier; 2019:541–560.
23. Lim CT, Zhang Y. Bead-based microfluidic immunoassays: the next generation. *Biosens Bioelectron.* 2007;22(7):1197–1204. https://doi.org/10.1016/j.bios.2006.06.005.
24. Castillo-León J. Microfluidics and lab-on-a-chip devices: history and challenges. In: *Lab-on-a-Chip Devices and Micro-Total Analysis Systems.* Springer; 2015:1–15.
25. Qin D, Xia Y, Rogers JA, Jackman RJ, Zhao X-M, Whitesides GM. *Microfabrication, Microstructures and Microsystems.* Springer Link; 1998:1–20.
26. Xia Y, Whitesides GM. Soft lithography. *Annu Rev Mater Sci.* 1998;28(1):153–184. https://doi.org/10.1146/annurev.matsci.28.1.153.
27. *GeSiM PDMS Flow Cell Overview of Casting Station.* [Online]. Available https://www.htsresources.com/overview_of_miccell_pdms_casting_station.php. [Accessed: 06-May-2021].
28. Gourley PL, et al. Biocompatible semiconductor optoelectronics. *J Biomed Opt.* 2002; 7(4):546. https://doi.org/10.1117/1.1506931.
29. Wagner M, Wolf BA. Interfacial tension between poly(isobutylene) and poly (dimethylsiloxane): influence of chain length, temperature, and solvents. *Macromolecules.* 1993;26(24):6498–6502. https://doi.org/10.1021/ma00076a029.
30. Whitesides GM. The origins and the future of microfluidics. *Nature.* 2006;442 (7101):368–373. https://doi.org/10.1038/nature05058.
31. Bruus H. Chapter 1. Governing equations in microfluidics. In: *Microscale Acoustofluidics.* Royal Society of Chemistry (RSC); 2014:1–28.
32. Berthier J, Silverzan P. Dimensionless numbers in microfluidics. *Microfluid Biotechnol.* 2009.
33. White FM. *Fluid Mechanichs.* 7th ed; 2011.
34. Mazare Y, Legeais G, Dufour P, Faure JC. Cholecystite Typhique Primitive Perfor'Ee Chez Un Enfant De Trois Ans., Pediatrie. 20;1965:209–212. http://doi.wiley.com/10.1002/1521-3773%252820010316%252940%253A6%253C9823%253A%253AAID-ANIE9823%253E3.3.CO%253B2-C.
35. Nguyen NT, Wu Z. Micromixers—a review. *J Micromech Microeng.* 2005;15(2). https://doi.org/10.1088/0960-1317/15/2/R01.
36. Tice JD, Song H, Lyon AD, Ismagilov RF. Formation of droplets and mixing in multiphase microfluidics at low values of the reynolds and the capillary numbers. *Langmuir.* 2003;19(22):9127–9133. https://doi.org/10.1021/la030090w.
37. Iwai K, Shih KC, Lin X, Brubaker TA, Sochol RD, Lin L. Finger-powered microfluidic systems using multilayer soft lithography and injection molding processes. *Lab Chip.* 2014;14(19):3790–3799. https://doi.org/10.1039/c4lc00500g.
38. Devasenathipathy S, Santiago JG. Electrokinetic flow diagnostics. In: *Microscale Diagnostic Techniques.* Springer; 2005:113–154.
39. Tadros T. Zeta potential in colloid science. Principles and application. *Colloids Surf.* 1982;5(1):79–80. https://doi.org/10.1016/0166-6622(82)80060-7.
40. R. Pethig, Dielectrophoresis: Status of the theory, technology, and applications. *Biomicrofluidics.* 2010;4(2). https://doi.org/10.1063/1.3456626.
41. Shen H, Khan R, Wang X, Li Z, Qu F. Capillary-based chemiluminescence immunoassay for C-reactive protein with portable imaging device. *Anal Bioanal Chem.* 2018;410(27):7177–7183. https://doi.org/10.1007/s00216-018-1321-2.

42. Chang YJ, Hu CY, Lin CH. A microchannel immunoassay chip with ferrofluid actuation to enhance the biochemical reaction. *Sens Actuators B Chem.* 2013;182: 584–591. https://doi.org/10.1016/j.snb.2013.03.078.
43. Czilwik G, et al. Rapid and fully automated bacterial pathogen detection on a centrifugal-microfluidic LabDisk using highly sensitive nested PCR with integrated sample preparation. *Lab Chip.* 2015;15(18):3749–3759. https://doi.org/10.1039/c5lc00591d.
44. Boy DA, Gibou F, Pennathur S. Simulation tools for lab on a chip research: Advantages, challenges, and thoughts for the future. *Lab Chip.* 2008;8(9):1424–1431. https://doi.org/10.1039/b812596c.
45. Nguyen NT, Hejazian M, Ooi CH, Kashaninejad N. Recent advances and future perspectives on microfluidic liquid handling. *Micromachines.* 2017;8(6). https://doi.org/10.3390/mi8060186.
46. Mashaghi S, Abbaspourrad A, Weitz DA, van Oijen AM. Droplet microfluidics: a tool for biology, chemistry and nanotechnology. *TrAC Trend Anal Chem.* 2016;82: 118–125. https://doi.org/10.1016/j.trac.2016.05.019.
47. Choi K, Ng AHC, Fobel R, Wheeler AR. Digital microfluidics. *Annu Rev Anal Chem.* 2012;5:413–440. https://doi.org/10.1146/annurev-anchem-062011-143028.
48. Luka G, et al. Microfluidics integrated biosensors: a leading technology towards lab-on-a-chip and sensing applications. *Sensors (Switzerland).* 2015;15(12):30011–30031. https://doi.org/10.3390/s151229783.
49. Wise KD, Najafi K. Microfabrication techniques for integrated sensors and microsystems. *Science.* 1991;254(5036):1335–1342. https://doi.org/10.1126/science.1962192.
50. Abdulhameed O, Al-Ahmari A, Ameen W, Mian SH. Additive manufacturing: challenges, trends, and applications. *Adv Mech Eng.* 2019;11(2). https://doi.org/10.1177/1687814018822880.
51. Paul SC, Tay YWD, Panda B, Tan MJ. Fresh and hardened properties of 3D printable cementitious materials for building and construction. *Arch Civ Mech Eng.* 2018; 18(1):311–319. https://doi.org/10.1016/j.acme.2017.02.008.
52. Zhao X, Wei C, Gai Z, Yu S, Ren X. Chemical vapor deposition and its application in surface modification of nanoparticles. *Chem Pap.* 2020;74(3):767–778. https://doi.org/10.1007/s11696-019-00963-y.
53. Harsha K, In: *Principles of Physical Vapor Deposition of Thin Films*, Elsevier, 2005; 6(s1):1155.
54. LeGoues FK, Rosenberg R, Meyerson BS. Kinetics and mechanism of oxidation of SiGe: dry versus wet oxidation. *Appl Phys Lett.* 1989;54(7):644–646. https://doi.org/10.1063/1.100905.
55. Hall DB, Underhill P, and Torkelson JM, Spin coating of thin and ultrathin polymer films, *Polym Eng Sci.* 1998;38(12):2039–2045. https://doi.org/10.1002/pen.10373.
56. Iliescu C, Tan KL, Tay FEH, Miao J. Deep wet and dry etching of Pyrex glass: a review. In: *ICMAT 2005 Symposium*; 2005:75–78. https://doi.org/10.1117/12.34127.
57. Martin PM, Matson DW, Bennett WD, Stewart DC, Lin Y. Laser-micromachined and laminated microfluidic components for miniaturized thermal, chemical, and biological systems. In: *Design, Test, and Microfabrication of MEMS and MOEMS.* 1999;3680:826. https://doi.org/10.1117/12.341279.
58. Gale BK, et al. A review of current methods in microfluidic device fabrication and future commercialization prospects. *Inventions.* 2018;3(3). https://doi.org/10.3390/inventions3030060.
59. Ceyssens F, Puers R, "SU-8 Photoresist," in Encyclopedia of Nanotechnology, 2016, pp. 3858–3873.
60. Munro E. Electron beam lithography. *Adv Electron Electron Phys.* 1980;73–131. https://doi.org/10.1201/b11626-4.

61. Sotomayor Torres CM, et al. Nanoimprint lithography: an alternative nanofabrication approach. *Mater Sci Eng C*. 2003;23(1–2):23–31. https://doi.org/10.1016/S0928-4931(02)00221-7.
62. Traub MC, Longsine W, Truskett VN. Advances in nanoimprint lithography. *Annu Rev Chem Biomol Eng*. 2016;7:583–604. https://doi.org/10.1146/annurev-chembioeng-080615-034635.
63. Schift H, Kristensen A. Nanoimprint lithography. In: *Springer Handbooks*, InTech; 2017:113–142.
64. Lipomi DJ, Martinez RV, Cademartiri L, Whitesides GM. Soft Lithographic Approaches to Nanofabrication. In: *Polymer Science: A Comprehensive Reference, 10 Volume Set*. vol. 7; 2012:211–231.
65. Becker H, Gärtner C. Polymer microfabrication technologies for microfluidic systems. *Anal Bioanal Chem*. 2008;390(1):89–111. https://doi.org/10.1007/s00216-007-1692-2.
66. Debono M, et al. One-step fabrication of microchannels with integrated three dimensional features by hot intrusion embossing. *Sensors (Switzerland)*. 2016;16(12). https://doi.org/10.3390/s16122023.
67. Nielsen AV, Beauchamp MJ, Nordin GP, Woolley AT. 3D printed microfluidics. *Annu Rev Anal Chem*. 2020;13:45–65. https://doi.org/10.1146/annurev-anchem-091619-102649.
68. Gaal G, et al. Simplified fabrication of integrated microfluidic devices using fused deposition modeling 3D printing. *Sens Actuators B Chem*. 2017;242:35–40. https://doi.org/10.1016/j.snb.2016.10.110.
69. Melchels FPW, Feijen J, Grijpma DW. A review on stereolithography and its applications in biomedical engineering. *Biomaterials*. 2010;31(24):6121–6130. https://doi.org/10.1016/j.biomaterials.2010.04.050.
70. Zhou X, Hou Y, Lin J. A review on the processing accuracy of two-photon polymerization. *AIP Adv*. 2015;5(3). https://doi.org/10.1063/1.4916886. In preparation.
71. Kauzlarick D, Fundamentals of microfabrication, the science of miniaturization, 2nd edition [Book Review], *IEEE Eng. Med. Biol. Mag.*, 2003;22(2):109–111. https://doi.org/10.1109/memb.2003.1195708.
72. Haubert K, Drier T, Beebe D. PDMS bonding by means of a portable, low-cost corona system. *Lab Chip*. 2006;6(12):1548–1549.
73. Eddings MA, Johnson MA, Gale BK. Determining the optimal PDMS-PDMS bonding technique for microfluidic devices. *J Micromech Microeng*. 2008;18(6).
74. Matellan C, Del Río Hernández AE. Cost-effective rapid prototyping and assembly of poly(methyl methacrylate) microfluidic devices. *Sci Rep*. 2018;8(1).
75. Ng SP, Wiria FE, Tay NB. Low distortion solvent bonding of microfluidic chips. In: *Procedia Engineering*. vol. 141. Elsevier; 2016:130–137.
76. Finkbeiner T, Soergel HL, Koschitzky MP, Ahrens R, Guber AE. Ultrasonic welding for the rapid integration of fluidic connectors into microfluidic chips. *J Micromech Microeng*. 2019;29(6).
77. Christensen AM, Chang-Yen DA, Gale BK. Characterization of interconnects used in PDMS microfluidic systems. *J Micromech Microeng*. 2005;15(5):928–934.
78. Chen A, Pan T. Fit-to-flow (F2F) interconnects: universal reversible adhesive-free microfluidic adaptors for lab-on-a-chip systems. *Lab Chip*. 2011;11(4):727–732.
79. Gubala V, Harris LF, Ricco AJ, Tan MX, Williams DE. Point of care diagnostics: status and future. *Anal Chem*. 2012;84(2):487–515.
80. *i-STAT Handheld | Abbott Point of Care*. [Online]. Available https://www.pointofcare.abbott/in/en/offerings/istat/istat-handheld. [Accessed: 04-May-2020].
81. *Vivascope Cell-Analysis Platform From Bosch—Bosch Media Service*. [Online]. Available https://www.bosch-presse.de/pressportal/de/en/vivascope-cell-analysis-platform-from-bosch-205290.html. [Accessed: 04-May-2020].

82. *Home | Vivalytic.* [Online]. Available https://www.bosch-vivalytic.com/en/. [Accessed: 04-May-2020].
83. *Point-of-Care Testing Equipment India—Siemens Healthineers India.* [Online]. Available https://www.siemens-healthineers.com/en-in/point-of-care-testing. [Accessed: 04-May-2020].
84. Gutiérrez-capitán M, Baldi A, Fernández-sánchez C. Electrochemical paper-based biosensor devices for rapid detection of biomarkers. *Sensors (Switzerland).* 2020;20(4).
85. Sharma S, Zapatero-Rodríguez J, Estrela P, O'Kennedy R. Point-of-care diagnostics in low resource settings: present status and future role of microfluidics. *Biosensors.* 2015; 5(3):577–601.
86. Diauudin FN, et al. A review of current advances in the detection of organophosphorus chemical warfare agents based biosensor approaches. *Sens Bio-Sens Res.* 2019;26.
87. Tan HY, et al. Lab-on-a-chip for rapid electrochemical detection of nerve agent sarin. *Biomed Microdevices.* 2014;16(2):269–275.
88. Jaywant SA, Mahmood Arif K. A comprehensive review of micro fluidic water quality monitoring sensors. *Sensors (Switzerland).* 2019;19(21).
89. Calvo-López A, Arasa-Puig E, Puyol M, Casalta JM, Alonso-Chamarro J. Biparametric potentiometric analytical microsystem for nitrate and potassium monitoring in water recycling processes for manned space missions. *Anal Chim Acta.* 2013;804:190–196.
90. Weng X, Neethirajan S. Ensuring food safety: quality monitoring using microfluidics. *Trend Food Sci Technol.* 2017;65:10–22.
91. Wu MH, Bin Huang S, Bin Lee G. Microfluidic cell culture systems for drug research. *Lab Chip.* 2010;10(8):939–956.
92. Dittrich PS, Manz A. Lab-on-a-chip: microfluidics in drug discovery. *Nat Rev Drug Discov.* 2006;5(3):210–218.
93. Kim YC, Park JH, Prausnitz MR. Microneedles for drug and vaccine delivery. *Adv Drug Deliv Rev.* 2012;64(14):1547–1568.
94. Zhu H, Fohlerová Z, Pekárek J, Basova E, Neužil P. Recent advances in lab-on-a-chip technologies for viral diagnosis. *Biosens Bioelectron.* 2020;153.
95. Houssin T, et al. Ultrafast, sensitive and large-volume on-chip real-time PCR for the molecular diagnosis of bacterial and viral infections. *Lab Chip.* 2016;16(8):1401–1411.
96. Ganguli A, et al. Hands-free smartphone-based diagnostics for simultaneous detection of Zika, chikungunya, and dengue at point-of-care. *Biomed Microdevices.* 2017;19(4).
97. Sinha M, Jupe J, Mack H, Coleman TP, Lawrence SM, Fraley SI. Emerging technologies for molecular diagnosis of sepsis. *Clin Microbiol Rev.* 2018;31(2).
98. Kaarj K, Akarapipad P, Yoon JY. Simpler, faster, and sensitive Zika virus assay using smartphone detection of loop-mediated isothermal amplification on paper microfluidic chips. *Sci Rep.* 2018;8(1).
99. Huh D, Torisawa YS, Hamilton GA, Kim HJ, Ingber DE. Microengineered physiological biomimicry: organs-on-chips. *Lab Chip.* 2012;12(12):2156–2164.
100. Bhatia SN, Ingber DE. Microfluidic organs-on-chips. *Nat Biotechnol.* 2014;32(8): 760–772.

CHAPTER SIX

Advances in microfluidic 3D cell culture for preclinical drug development

Maria Russo[a,*], **Cesare M. Cejas**[a,†], **and Gabriele Pitingolo**[b]
[a]Microfluidics, MEMS, Nanostructures (MMN), CNRS UMR 8231, Institut Pierre Gilles de Gennes (IPGG) ESPCI Paris, PSL Research University, Paris France
[b]Bioassays, Microsystems and Optical Engineering Unit, BIOASTER, Paris France
*Corresponding author: e-mail address: maria.russo@espci.fr

Contents

† Present address: MicrofluidX Ltd, Stevenage Bioscience Catalyst, Gunnels Wood Road, SG1 2FX Stevenage, United Kingdom.

Progress in Molecular Biology and Translational Science, Volume 187
ISSN 1877-1173
https://doi.org/10.1016/bs.pmbts.2021.07.022
Copyright © 2022 Elsevier Inc.
All rights reserved.

Abstract

Drug development is often a very long, costly, and risky process due to the lack of reliability in the preclinical studies. Traditional current preclinical models, mostly based on 2D cell culture and animal testing, are not full representatives of the complex in vivo microenvironments and often fail. In order to reduce the enormous costs, both financial and general well-being, a more predictive preclinical model is needed. In this chapter, we review recent advances in microfluidic 3D cell culture showing how its development has allowed the introduction of in vitro microphysiological systems, laying the foundation for organ-on-a-chip technology. These findings provide the basis for numerous preclinical drug discovery assays, which raise the possibility of using micro-engineered systems as emerging alternatives to traditional models, based on 2D cell culture and animals.

1. Introduction

In recent years, there have been significant advances in developing innovative drugs and personalized therapies for several diseases.[1] The drug development process goes through several phases: (I) research and development (R&D), (II) preclinical research, (III) clinical research, (IV) agencies (FDA, EMA, etc.) review, and (V) post-market safety monitoring. Typically, the process to develop one clinically approved drug is very long, risky, and costly, taking approximately more than 10–12 years and about 1–2 billion dollars[2] in total cost. During the drug discovery phase, biotechnology and pharmaceutical companies face many challenges, mainly due to the high costs of R&D and preclinical research, as well as the high attrition rate of developing compounds.[3-6] During the last 20 years, these costs continue to rise[7] despite less than five percent of these new drugs actually reaching the market.[8] The extremely low approval rate is due to failures that mainly arise from a lack of efficacy during the clinical trial phases,[9,10] insufficient therapeutic index, and poorly predictive preclinical models.[11,12] The lack of predictability of the preclinical models may also lead to the early failure of new compounds with a high potential healthy benefit because of toxic side effects observed only in animal models.[13] Indeed, costs are exacerbated by the element of risk. To reduce the costs and to increase the success of the clinical trials, ineffective drug candidates should be eliminated as early as possible (so-called fail early, fail cheaply).[7] This can be performed through high-throughput screening (HTS) in the preclinical stages.[14] This indicates a clear need for improvement of the preclinical studies that should be more predictive of human physiological responses[14] in order to perfectly select the good drug candidate.

Current preclinical models mainly rely on two-dimensional (2D) cell culture (in vitro)[15] and animal testing (in vivo). Although widely used, both are not fully representative of the human physiology and often fail at the human clinical trial stage, especially in the cancer area.[12,16,17] From one side, in traditional 2D cell culture systems, cells grow as 2D monolayer on plastic systems (e.g., petri dishes or flasks). These in vitro models are well-established, standardized and inexpensive. However, they are non-systemic and do not completely mimic the complexity of human tissues and organs, where cells are arranged in a three-dimensional (3D) microenvironment.[18] In 2D cell culture, cells are flattened and stretched, generating strong internal stresses and high cell sensitivity and efficacy to drugs.[19] Animal models more accurately represent this 3D complexity as well as provide systemic responses but nevertheless, these models remain complex, time-consuming, and very expensive.[20] Moreover, animals (mostly rodents) inherently possess low predictive capability for the drug responses due to the strong genomic differences between the different species.[21] Not least, ethical issues concerning sacrificing animals have been in the forefront leading to the banning of animal tests for the safety assessment of cosmetics, first in the European Union[22,23] with other countries following suit. For economic, ethic and scientific issues, biotechnology and pharmaceutical companies are pressured to explore alternative in vitro methods that are more predictive of clinical results, in order to advance the drug development process and to develop high effective drugs reducing costs (see Fig. 1).[24]

To overcome current limits, 3D cell culture in vitro is gaining increasing acceptance both in research and industry. This can preserve cell shape, growth, and interactions, thereby promoting cell differentiation[25] and tissue

Traditional models		Advanced model
2D cell culture	Animals	Microfluidic 3D cell culture
Advantages • Simple • Low cost • Well standardized	Advantages • Systemic • Functional 3D tissues • Well standardized	Advantages • Systemic and shear stress • Functional 3D tissues and ECM • Concentration gradient • Reduced volumes of sample • Improved imaging capabilities
Drawbacks • Static and nonsystemic • Difficult to reproduce ECM • Flat cell shape • High sensitivity to drugs • No concentration gradient	Drawbacks • Costly • Limited resolution of imaging • Genetic differences with human • Low visualization of single steps • Ethical issues	Drawbacks • Not standardized and automatized • Complex experimental set-up • Need for integrated platforms

Fig. 1 Comparison between traditional (2D cell culture and animals) and advanced (microfluidic 3D cell culture) models for preclinical drug development: advantages and drawbacks.

organization not possible in traditional 2D systems.[14,19,26-29] Broadly, 3D cell cultures are often classified either as scaffold-free or scaffold-based systems.[30,31] Scaffold-free systems allow the cells to self-assemble to form nonadherent multicellular aggregates or spheroids by using hanging-drop or forced-floating methods.[32] However, in many cases, in vivo tissue architectures do not present spheroids; therefore 3D cell culture methods need to force the cell assembly and the formation of their extracellular matrix (ECM).[33] For the case of 3D scaffold-based systems, cells grow in presence of a support mimicking the functions of ECM. These scaffolds use a variety of materials (hydrogels and natural or synthesized polymers, such as collagen, chitosan, and PEGDA) and a variety of geometries (sponges, micro- and nano-fibers, etc.) with different mechanical properties. Hybrid culture models in which spheroids are assembled into a 3D polymeric scaffold have also been investigated.[34,35] An ideal 3D cell culture system would mimic an in vivo microenvironment allowing cell-to-cell and cell-to-ECM interactions, oxygen exchanges, nutrient supply, and metabolic waste elimination.[14,36] Several studies have demonstrated the importance of the ECM for 3D cell cultures. Understanding the interaction between cells and extracellular environment is fundamental to guide cells for tissue engineering applications.[37-40] Various studies have demonstrated that elasticity, dimensionality, charge, adhesion, and topography of the matrix positively influence cell behavior.[41] Despite their advantages, most of these platforms are still limited due to their static culture maintenance. Under static control conditions, 3D cell cultures have shown differences in cell morphology, proliferation, and in drug responses, while under flow conditions similar to the in vivo ones, response and function are largely enhanced.[19] In this perspective, microfluidic technology can add flow to cell culture systems. Microfluidics[42] introduced in the 1990s has rapidly gained attention for several applications in the field of nanotechnology,[43] cell culture, and tissue engineering. Controlling the flow at the microscale[44] offers a unique opportunity to engineer highly complex, dynamic, well-controlled microenvironments that can also be automated.[45-48] A microfluidic chip permits visualization via microscopy of in vitro cellular microenvironments,[49] particularly the cell-to-cell interaction at a single level.[50] In microfluidics, the continuous perfusion/diffusion regime applied to the 3D cell culture allows for regular supply of nutrients, exchange of oxygen and removal of waste products (i.e., cellular debris).[45,51]

Microfluidics facilitates experimental modeling of in vitro conditions, for example, the shear stress,[52] which is the frictional force generated by blood flow that influences the vascular endothelial cells, can be achieved through controlled flow conditions inside the chip.[53] Moreover, in some cases, to protect cocultures of multiple cell types[54] from excessive shear rates, barriers between different compartments can be introduced in the microfluidic chip. Other components in addition to the flow control can be integrated into a microfluidic chip, e.g., hydrogels can guarantee desired and controlled delivery of nutrients[55] and oxygen gradients to cells.[56] This is especially useful for the optimization of cell culture and drug dosing studies.[57] A controlled injection and delivery of drugs with a dynamic perfusion/diffusion regime can be a powerful tool to better predict potential novel drugs in physiologically conditions.[58]

Different microfluidic platforms have been developed for 3D cell culture and several materials have been exploited as substrates in the fabrication of the microfluidic chip, which for cell culture purposes, is often referred to as a microfluidic biochip. Among glass, polymers, and paper-based microfluidic platforms, polydimethylsiloxane (PDMS)-based chips are the most widely used due to their low cost and ease of prototyping.[45] Especially for biological applications, the use of materials which allows the diffusion of small molecules into the microfluidic systems is highly convenient. Hydrogels have been extensively used in microfluidics[59] for free diffusion of small molecules and mimicry of ECM including biocompatibility, porosity, and representative stiffness. These are all key parameters that promote native-like tissue functions.[60] Microfluidics with advances in 3D cell culture, stem cell biology, and tissue engineering opened up the development of micro-engineered models of functional human organs, called organ-on-a-chip.[24,27] Organ-on-a-chip (OOC) technology[61,62] has been already investigated for a large variety of organs including skin, muscle, liver, and neural tissue.[38,63,64] Although animal cells can be used for in vitro assays, the development of 3D tissue structures on chip utilizing human cells or stem cells is mostly explored for drug discovery (Fig. 2).

The preclinical testing of a drug in these micro-engineered 3D human tissues is more accurate compared to 2D in vitro models, particularly in reflecting or predicting the outcomes in human clinical trials.[17,65] Moreover, primary cell culture and stem cell differentiation on chip offers the possibility of setting up preclinical drug screening platforms or disease models, potentially using cell sources from patients, going in the direction of personalized medicine.[66] Organs-on-a-chip is currently applied also for in vitro ADME-Tox (absorption, distribution, metabolism, excretion, and toxicity) studies.[64]

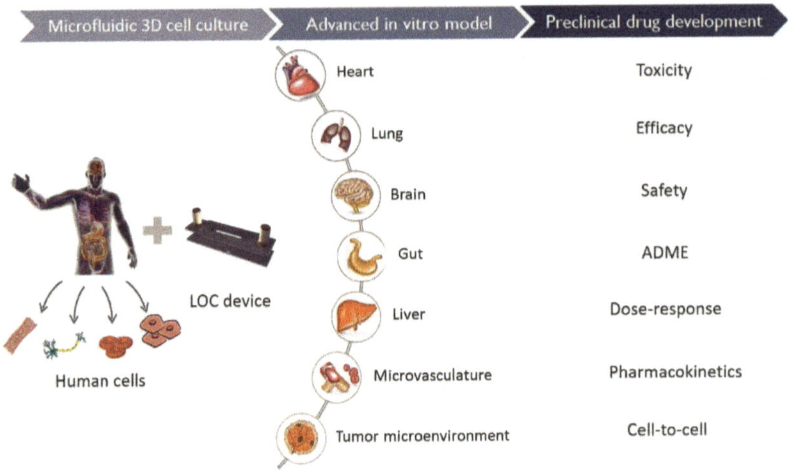

Fig. 2 Schematic illustration of microfluidic 3D cell culture, the main developed OOC models, and a list of applications for preclinical drug development.

This chapter explores the progress in microfluidics for 3D cell culture that is a precursor to the emerging organ-on-a-chip technology. This is a new route in the preclinical drug discovery and development research, obviating the gap between 2D cell culture and in vivo animal models and potentially curbing the cost of drug development.

2. Materials and microfabrication methods

Microfluidic-based biochips destined for cell culture have been designed to finely control parameters of the cell microenvironment in 2D space at microscale.[67–73] in vitro microfluidic models are relatively inexpensive and conducive for quantitative studies of cell physiology but remain far away from in vivo cell physiology. As advancement of the technology, microfluidics offers the possibility to control the cell culture also in 3D space,[45] particularly for drug discovery and development.[19,74] Within the last decade or so, there has been extensive research on moving toward 3D microfluidic cell culture[24,74–80] in large part to accurately mimic the in vivo extracellular microenvironment.[30] Moreover, there are benefits in increasing the use of these platforms is via parallelization and multiplexing[81] for increased throughput. The main materials and microfabrication techniques employed for microfluidic cell culture are outlined here.

2.1 Materials for microfluidic chip fabrication

Several microfluidic platforms exist for cell culture in 2D or 3D and are classified as a function of the type of material,[19,45,82] in which some examples are shown in Fig. 3.

Polymer-based materials (soft elastomers and thermoplastics) have been common in microfluidics. The most widely used are the soft elastomers, like thermoset polyesters (TPE),[95] and polydimethylsiloxane (PDMS) (e.g., Refs. 83, 96), such as in Fig. 3A. PDMS is more widely used due to its inert, nontoxic nature, as well as its permeability to oxygen and low cost, thereby permitting rapid prototyping.[19,45,97] PDMS is naturally hydrophobic and thus surface treatments, e.g., plasma oxidation or polymer grafting, are essential steps in rendering the surface hydrophilic. However, gas permeability of PDMS also risks media evaporation, thereby causing osmolality shifts that can have repercussions in cell culture.[98] For better control of gas diffusion inside the culture space, more rigid biocompatible thermoplastics (Fig. 3B) have also been used such as polycarbonate (PC),[86,99] poly(methyl methacrylate) (PMMA),[84,85,100] polystyrene (PS),[87] and cyclic olefin polymers or copolymers (COP/COC[88,101]). All possess the advantage of being transparent for optical visualization purposes, such as direct imaging of fluorescent tagged proteins.[19] In addition, thermoplastics are generally more resistant against changes in temperature and pressure and they also have low electrical conductivity and high chemical stability, thereby permitting mass production at a relatively lower cost.[102] In addition, thermoplastic surfaces can be easily tuned via surface modification. Gencturk, *et al.*[102] provide a summary of the different properties of thermoplastics. Some biochips have also been fabricated using other types of polymers, e.g., glycol acrylates, polyethylene therepthalates (PETs), and Teflon or perfluorinated polymers while other biochips demonstrate a hybrid combination of PDMS and thermoplastics (e.g., Refs. 93, 94, 103, 104) to take advantage of the properties of both components.

Glass or silicon-based materials are sometimes chosen because of their general ability to work with various organic solvents as well as at high temperatures (Fig. 3C). However, they are rarely used[45] on its their own due to expensive fabrication processes, often requiring specific equipment to properly manipulate and cut into desired geometries. Normally, glass or silicon-based materials are fabricated in conjunction with other materials or polymers to create hybrid microfluidic chips.[105,106] A combination of glass and Norland Optical Adhesive (NOA) and PDMS, has been fabricated

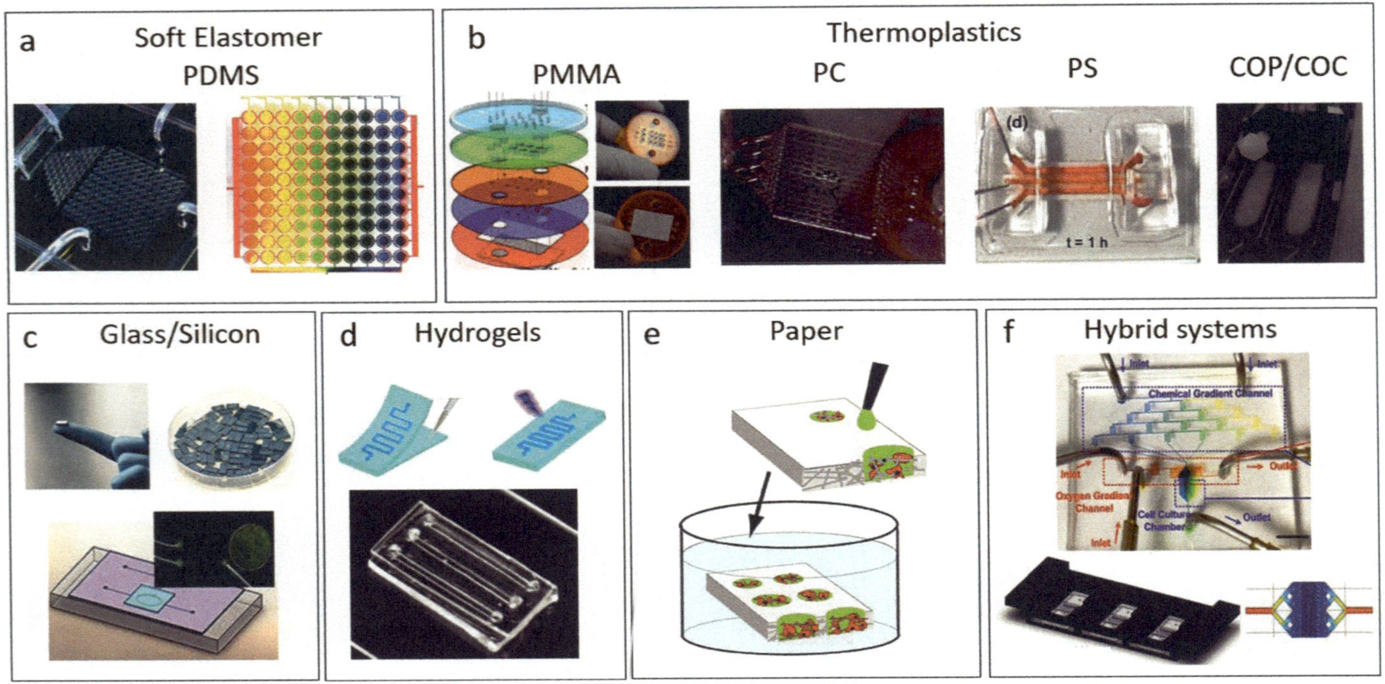

Fig. 3 See figure legend on opposite page.

as cell culture biochip.[89] Glass, which is amorphous silica, is chemically inert, naturally hydrophilic, and biocompatible. The inertness of glass makes it conducive for integration with other special components, such as integrating ITO (indium tin oxide) in a microheater chip destined for perfusion-based microscale cell culture.[45,107] Glass typically offers better optical properties than polymers because the latter often exhibit intrinsic fluorescence.[45] Glass also possess good surface stability appropriate for repeated experiments and solvent compatibility for reproducible electroosmotic flow. Thus, glass

Fig. 3 Types of materials used in the fabrication of microfluidic biochips for cell culture applications. (A) Soft elastomers, e.g., PDMS. (B) Rigid thermoplastics: PMMA, PC, PS, and COP/COC. (C) Glass/silicon. (D) Hydrogels. (E) Paper. (F) Hybrid systems, such as PDMS-PC. *(Panel A: Reprinted with permission from Hung PJ, Lee P, Sabounchi P, Lin R, Lee PL. Continuous perfusion microfluidic cell culture array for high-throughput cell-based assays. Biotechnol Bioeng. 2005; 89:1–8. © John Wiley & Sons, Inc; Panel B: Reprinted with permission from Coluccio ML, Perozziello G, Malara N, et al. Microfluidic platforms for cell cultures and investigations. Microelectron Eng. 2019; 208:14–28. © Elsevier and Meddens MBM, Lui S, Finnegan PS, Edwards TL, James CD, Lidke KA. Single objective light-sheet microscopy for high-speed whole-cell 3D superresolution. Biomed Opt Exp. 2016; 7:2219–2236. © The Optical Society; Tehranirokh M, Kouzani AZ, Francis PS, Kanwar JR. Microfluidic devices for cell cultivation and proliferation. Biomicrofluidics. 2013; https://doi.org/10.1063/1.4826935. © AIP Publishing and Baudoin R, Griscom L, Prot JM, Legallais C, Leclerc E. Behavior of HepG2/C3A cell cultures in a microfluidic bioreactor. Biochem Eng J. 2011; 53(2):172–181. © Elsevier, Reprinted from permission from Chan CY, Goral VN, DeRosa ME, Huang TJ, Yuen PK. A polystyrenebased microfluidic device with three-dimensional interconnected microporous walls for perfusion cell culture. Biomicrofluidics. 2014; 8:046505. © AIP Publishing; Becker H, Schulz I, Mosig A, Jahn T, Gartner C. Microfluidic devices for cell culture and handling in organ-on-a-chip applications. Proc SPIE. 2014; 8976:89760N. © SPIE The International Society for Optics and Photonics; Panel C: Reprinted from permission from Qi Z, Xu L, Xu Y, et al. Disposable silicon-glass microfluidic devices: precise, robust and cheap. Lab Chip. 2018; 18:3872–3880. © Royal Society of Chemistry and Li R, Lv X, Hasan M, et al. A rapidly fabricated microfluidic chip for cell culture. J Chromatogr Sci. 2015; 54:523–530. © Oxford University Press; Panel D: Reprinted from permission from Li R, Lv X, Hasan M, et al. A rapidly fabricated microfluidic chip for cell culture. J Chromatogr Sci. 2015; 54:523–530; Shen C, Li Y, Wang Y, Meng Q. Non-swelling hydrogel-based microfluidic chips. Lab Chip. 2019; 19:3962–3973. © Royal Society of Chemistry; Panel E: Reprinted from permission from Derda R, Laromaine A, Mammoto A, et al. Paper-supported 3D cell culture for tissue-based bioassays. PNAS. 2009; 106:18457–18462. © The National Academy of Sciences; Panel F: Reprinted from permission from Chiang HJ, Yeh SL, Peng CC, Liao WH, Tung YC. Polydimethylsiloxanepolycarbonate microfluidic devices for cell migration studies under perpendicular chemical and oxygen gradients. J Vis Exp. 2017; 120:55292. © JoVE and Macown RJ, Veraitch FS, Szita N. Robust, microfabricated culture devices with improved control over the soluble microenvironment for the culture of embryonic stem cells. Biotechnol J. 2014; 9:805–813. © John Wiley & Sons, Inc.).*

biochips are often reusable.[79] It is also gas impermeable, thereby creating a hypoxic environment.[108] Moreover, glass is often the substrate choice for droplet production due to its robustness in handling both multiphase fluids (oil and water). These droplet-based digital assays are often used as small bioreactors for 3D cell cultures.[109]

While polymers such as PDMS contain a silicon atom, silicon-based substrates often refer to the metalloid itself. In some cases, silicon wafers are etched to create microchannels for use as droplet generators for pharmaceutical microdroplet production,[110] though some works have demonstrated directly incorporating the silicon wafer into the biochip design.[51]

Hydrogel-based materials have offered the possibility to develop tissues/organs on chip, faithfully mimicking in vivo conditions, by integrating microtechnologies and cell biology (Fig. 3D). Hydrogels, mainly based on water and proteins, can mimic the natural microenvironment using biochemical and biophysical signals. As a pioneering method, gelatin, crosslinked with the transglutaminase enzyme, has been employed to fabricate microchannels suitable for cell culture and analysis of adherent cells.[91] Despite their advantages, hydrogels can be difficult to use for chip fabrication, due in large part to typical swelling that occur at typical cell culture conditions (in an incubator at 5% humidified CO_2 at 37°C) and a decrease in mechanical performance over repeated swelling/deswelling actions or deformation of the prepared slab. To overcome this limitation, a nonswelling hydrogel method has been proposed to make microfluidic chips via cross-linking of di-acrylated Pluronic F127 (F127-DA),[90] a biocompatible polymer. Gelatin thin films have been also integrated on chip using the GEL-D method and has been successfully applied to 3D microtissue formation.[111]

Paper-based materials are inexpensive and have been gaining attention by virtue of the fact that the fabrication of paper microfluidic chips does not require specialized engineering approaches (Fig. 3E). Paper, a cellulose-based material, is biocompatible, cheap, accessible, and be easily modified through surface chemistry.[19,95] This method uses paper typically used in paper chromatography to design hydrophobic barriers via wax printing.[58] The paper's natural cellulose matrix acts as a scaffold.[19,112] Derda, *et al.*[92] were one of the first who have demonstrated the use of paper in 3D cell culture by stacking several layers of paper containing cell suspension in a extracellular matrix hydrogel (e.g., Matrigel, Geltrex) to imitate in vivo nutrient and oxygen gradients in 3D space.

Hybrid systems refer to microfluidic devices where slabs, made of different materials, are combined and bonded (Fig. 3F).

2.2 Main microfluidic geometries for cell culture

When creating microfluidic biochips for cell culture, different structural features are incorporated for different functions, specifically to anchor the cells or serve substrate for cell adhesion and growth. These features, in which some examples are shown in Fig. 4, can be simple microchannels (e.g., Refs.

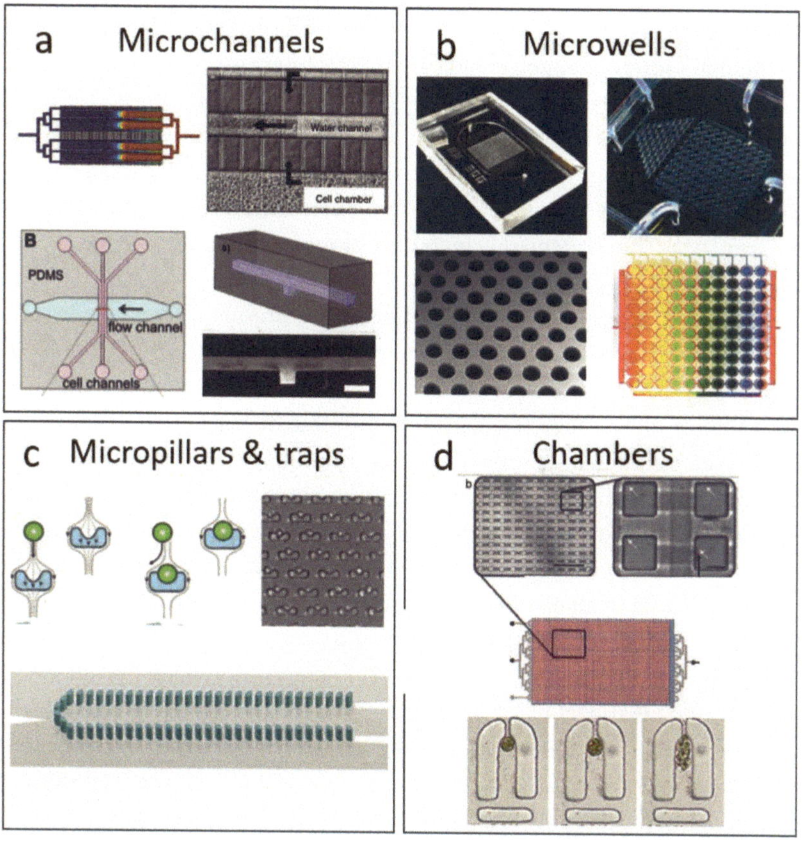

Fig. 4 Types of geometries used for microfluidic cell culture applications. (A) Microchannels, e.g., Refs. 113, 114, 125. (B) Micro-array wells,[83,116]. (C) Micropillars and traps,[63,117]. (D) Chambers, e.g., Ref. 122, 124. *(Panel A: Reprinted with permission © Springer Nature; © Springer Nature; © Oxford University Press; Panel B: Reprinted with permission © Royal Society of Chemistry; © John Wiley & Sons, Inc.; Panle C: Reprinted with permission © American Chemical Society; © Royal Society of Chemistry; Panel D: Reprinted with permission © Springer Nature; © Royal Society of Chemistry).*

51, 72, 79, 87, 113, 114), microwell arrays (e.g., Refs. 78, 83, 115, 116), micropillars or traps,[63,77,117-120] or cell retention chambers.[121-124]

Microchannels (Fig. 4A) are generally straight, linear rectangular channels where cells sediment along its length during seeding.[72,113] These simple geometries imply relatively nonintricate fabrication techniques but at the same time also permit straightforward continuous perfusion of cell culture medium. The unobstructed laminar flow ensures spatial delivery of nutrients and oxygen.[126] However, it also exposes the cells in direct contact with the convective fluid, which increases risk of shear stress, especially at higher flow rates. To mitigate this, smaller channels are fabricated adjacent to the main microchannel, which are used for perfusion purposes, such as in Lazar *et al.*[114] These small perfusion channels communicate with the main microchannels (where cells are cultured), ensuring delivery of nutrients without the damaging risk of shear stress. Microchannels can also serve as "flow strippers," when positioned just before the main culture chamber, by regulating/homogenizing incoming flow in order further reduce shear stress.[94]

To ensure that cells are protected from continuous convective flow from perfusion, **microwells or micro-arrays** (Fig. 4B) are also common features, which borrow its design from a typical microtiter plate, only with micrometer dimensions. Each well communicate with each other through connected microchannels that ensure continuous perfusion from one chamber to another. The microwell array design permits investigation of numerous factors (growth scales and time scales) that regulate proliferation and differentiation of cells.[45,127] It is also commonly used in screening as it allows simultaneous treatment of various conditions in different wells.

Other features such as **micropillars or traps** (Fig. 4C) act as physical micropatterned matrices ensuring that cells are immobilized during seeding, such as in DiCarlo *et al.*[117] In some cases, pillars serve as protection and shield the cells from the effects of direct convective flow.[126] Similar to small adjacent perfusion channels, the slits in between pillars permit continuous delivery of nutrients into the culture space.

Chambers (Fig. 4D) serve to confine the cells in a particular culture space and often are a combination of one or a few geometries or designs whose combination yields a high-performance biochip, such as in Lecault *et al.*, where microwells serve as retention chambers where cells sediment.[122] Often, these are integrated with valves and switches to control the entry or exit of matter, ensuring precise conditions within the chamber.[121,124]

2.3 Microfabrication methods

Different techniques have been employed for microfluidic chip fabrication and can be generally classified into laboratory scale and industrial methods, as shown in Fig. 5.

Laboratory methods include those that are typically present in any standard microfluidics laboratory. This includes photolithography equipment (e.g., mask aligner) by etching a design on a UV-exposed resin-coated silicon wafer. From this silicon wafer mold, PDMS polymer and crosslinker solution is poured into the mold via replica molding and then cured via heating at 70°C. Soft lithography is widely used in microfluidics, and thus, for this reason, general protocols can easily be found, e.g., Ref. 130.

Micromilling or micromachining is also an option[128,131] typically used for more rigid thermoplastics via cutting tools that perform subtractive manufacturing process, i.e., it removes bulk material[128] from the substrate, leaving only the intended design. Modern micromills are now computer numerical controlled (CNC) systems, which automate the process and improve reproducibility and precision.[128] Micromilling techniques either mill directly onto the polymer or indirectly onto a metal mold and then the mold is used to transfer the features onto the thermoplastic via hot embossing techniques.

Hot embossing is a technique where a mold is heated up to a certain temperature and pressed onto the thermoplastic at a certain pressure.[132] The temperature is usually above the glass transition temperature of the polymer to be embossed. The heat permits transfer of the micro-features from the mold onto the thermoplastic substrate. Hot embossing can be relatively low cost (depending on the quality of the heat press) and provides a method to mold thermoplastics scalable to industrial production.[133] The mold can be also a metal (e.g., stainless steel, aluminum, or brass), etched silicon, or electroplated nickel.[133]

Standard silicon microfabrication techniques were adapted for glass micromachining.[45] Fabrication of glass microchips usually requires special equipment, though it can be achieved in the laboratory but only following careful protocols, because it usually involves the use of hydrofluoric acid, a highly corrosive compound used to etch glass.[134]

Biochips made of polyether ether ketone (PEEK), a semicrystalline thermoplastic, have also been gaining attention due to their biocompatibility and excellent chemical and hydrolysis resistance as well as mechanical and thermal stability.[135] Biochips made of PEEK are often combined with other

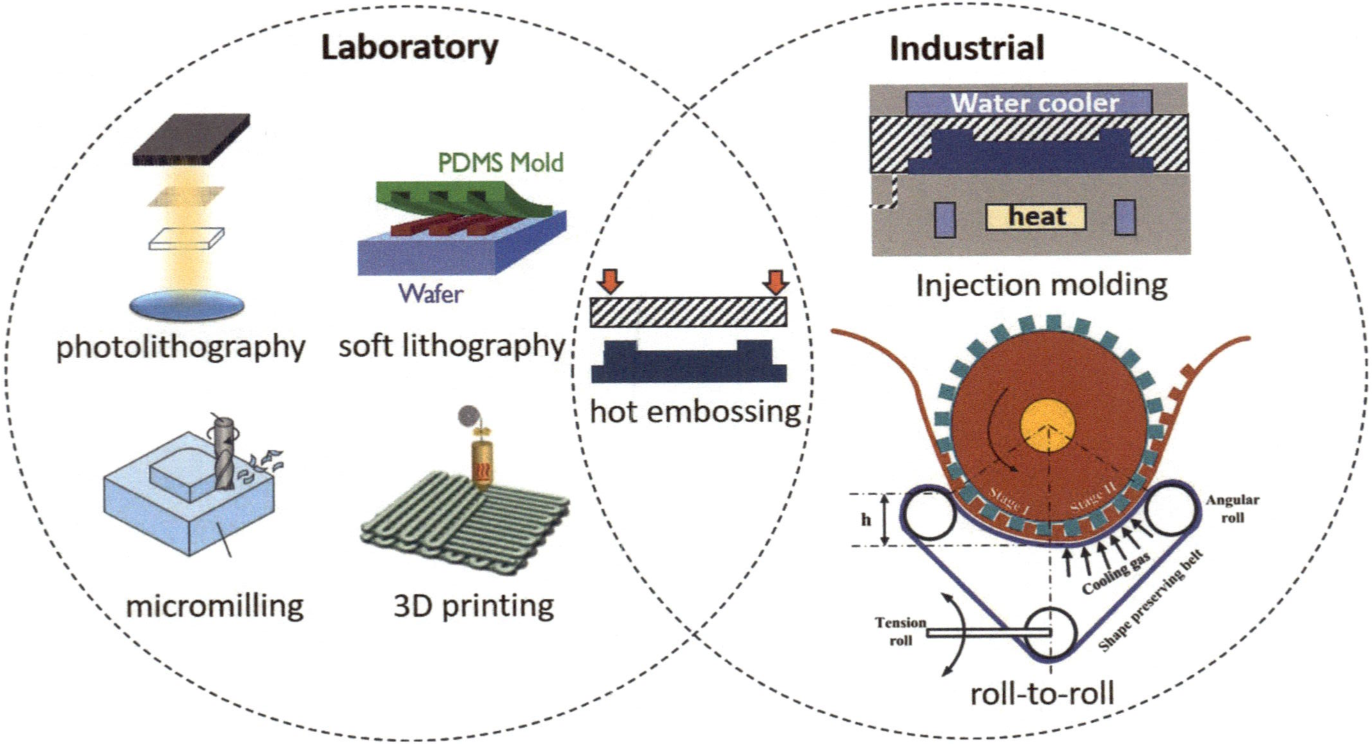

Fig. 5 See figure legend on opposite page.

thermoplastics, e.g., PS,[135] using nonconventional techniques such as autohesion, i.e., sealing without using adhesives upon application of pressure and mild temperature.

In recent years, 3D-printed biochips have been gaining attention due in large part to increasing accessibility to 3D printers in many laboratories. In some cases, 3D printers have been used to print foam-based scaffolds,[31,129] through the solidification of liquid foams.[136–138] This enables control of properties and can be incorporated into the biochip.[59]

Other laboratory scale methods used to fabricate the biochips also exist such as laser lithography and dry resist ion etching,[139] though they are not as popular because they require availability of certain equipment.

Industrial scale methods are those techniques which are more adapted for high-throughput manufacturing, thereby permitting scale-up at a relatively low cost. Often, laboratory scale methods make use of intricate procedures that naturally become costly to industrialize. Thus, these industrial methods optimize the trade-off between process technique and manufacturing cost.

Injection molding is a process in which hot liquid polymer is injected into a mold cavity that contains the template. The liquid polymer then conforms to its features prior to cooling and solidification. For fabricating cell culture devices, despite high efficiency and consistency, there are disadvantages, which include high initial tooling cost, expensive machinery, and long set-up lead times. Recent work has demonstrated injection-molding to integrate 3D cell culture patterns in microfluidic chips.[140]

Roll-to-roll embossing is another industrial technique, which is a relatively cheap process for large-scale production[141] and is capable of precise

Fig. 5 Main chip fabrication methods generally classified into laboratory or industrial scale. In laboratory scale: photolithography and soft lithography, micromilling, 3D Printing. In industrial scale: injection-molding and roll-to-roll embossing. Some techniques, e.g., hot embossing, can be used in both cases if the appropriate equipment is available. *(Reprinted with permission from Weibel DB, DiLuzio WR, Whitesides GM. Microfabrication meets microbiology. Nat Rev Microbiol. 2007; 5(3):209–218. © Springer Nature; Guckenberger DJ, de Groot TE, Wan AMD, Beebe DJ, Young EWK. Micromilling: a method for ultra-rapid prototyping of plastic microfluidic devices. Lab Chip. 2015; 15:2364. © Royal Society of Chemistry; Mohanty S, Larsen LB, Trifol J, et al. Fabrication of scalable and structured tissue engineering scaffolds using water dissolvable sacrificial 3D printed moulds. Mater Sci Eng C. 2015; 55:569–578. © Elsevier; Peng L, Wu H, Shu Y, Yi P, Deng Y, Lai X. Roll-to-roll hot embossing system with shape preserving mechanism for the large-area fabrication of microstructures. Rev Sci Instrum. 2016; 87:105120. © AIP Publishing.)*

replication of intricate microstructures via micropatterning[142] of microfluidic chips on large surface areas of polymer rolls. This technique, due to its relative cost-effectiveness, enables ease of prototyping.

2.4 Assembly and bonding methods

While multiple options exist on what type of substrate material to use, bonding or sealing is an important step to ensure that the entire microfluidic biochip remains a closed, leak-proof system. The sealing of microfluidic chips allows to: (1) confine biological specimens in reduced volume of liquid, (2) minimize contamination and (3) limit the evaporation rates of liquid samples and reagents while at the same time assuring the regular exchange of oxygen and nutrients. Some challenges exist because most thermoplastic materials do not possess intrinsic bonding compatibility to elastomers[143] and thus, the appropriate bonding mechanism must be employed to properly seal the biochip. Biochip bonding could either be irreversible or reversible.

Irreversible bonding methods generally imply a chemical or physical reaction between two surfaces. In the case of PDMS-PDMS or PDMS-glass bonding, one example is through plasma oxidation, by modifying the chemistry of both surfaces, albeit temporarily, thereby allowing the covalent bonding to take place at a prescribed temperature.[144] Several methods were proposed to irreversibly bond microfluidic slabs[145-154] but quite often, when using bonded thermoplastics, the resulting chips are not ideal for cell culture due to limited gas permeability, unless another feature is added specifically for this purpose such as a semi-permeable membrane. Adhesives or epoxy materials are sometimes also used for irreversible bonding. In certain cases of thermoplastic and elastomer bonding, additional surface treatments are needed using silanes.[150,155]

Reversible bonding methods refer to temporarily sealed and modular chips that can be selectively opened during cell harvest or tissue sampling for off-chip characterization. For this kind of bonding, there is a portion of the biochip design that can be adjusted, exchanged, or removed. These methods usually employ magnets, such as in Pitingolo *et al.*,[156] where magnets are incorporated into PMMA inserts fabricated from micromilling. The magnets have demonstrated bond sealing strength as is remains resistant to flow conditions, i.e., it remains more intact compared to conventional methods.

Other modular methods make use of standard screws or clamps[157] or more recently, microbuilding blocks that take their inspiration from the

Lego-like building blocks that interlock via tongue and groove joint connections[158,159] or even custom-designed and 3D-printed alignment mechanisms.[160] While such reversible sealing leads to devices that are easy to fabricate, cost-effective, functional, and relatively easy to scale-up for high throughput, major limitation still exists. In some cases, the poor sealing strength and occasional leakage and the difficulty to reproduce the in vivo cellular microenvironment remain to be a challenge. Recently, an innovative method has been proposed for the reversible bonding of a microfluidic chip using deposition of a gelatin layer. The resulting chip was found to have a pressure resistance above 0.70 MPa (7 bar). This high pressure limit implies that the said technique would be applicable to a majority of microfluidic applications.[111]

3. Microfluidic hydrogels and coating methods as extracellular matrix mimics

Extracellular matrix (ECM) is a complex cellular microenvironment that provides structural integrity as well as in vivo conditions that permit regulation and control of cell function, growth, and communication.[161] Substrate stiffness directly plays a role on cell migration, adhesion, and differentiation and it is precisely for this reason why ECM gels have lower stiffnesses to accommodate complex cell behavior. ECM consists of a complex hydrated protein matrix that allows adhesion of cells. The ECM also serves as a template for cell growth in 3D space.[19] It is a dynamic structure[161] because cells have shown to reconfigure ECM by fiber bundling or fiber degradation. Within the last decade, there has been engineering advancement in fabricating custom ECM as cell culture scaffolds for tissue engineering and regenerative medicine. Frantz *et al.*[162] and Yue[163] provide comprehensive information on the ECM physiology and biology.[162,163] In short, ECM contains high-molecular-weight proteins, e.g., fibronectin and collagen, and branched biomacromolecular structures that are organized in a fibrillar network. The pore spaces within the network are interspersed with adhesive binding sites that provide an active place for cells to reside.[164]

3.1 Importance to mimic ECM and use of hydrogels

The ECM mimics inherent physiological conditions, thereby increasing the chances of conducive cell growth as it substitutes natural cell microenvironments.[58] The importance of ECM in microfluidic biochips can be summarized as follows:

- ECM provides cell binding sites for cell adhesion, thereby immobilizing the cells[164] by serving as scaffolding to facilitate attachment.[97]
- ECM may provide as a natural barrier to cells to prevent abnormal migration and proliferation,[165] e.g., a recent work has demonstrated correlation of ECM stiffness and cell migration speed.[166]
- ECM maintains homeostasis of living organisms.[165]
- Depending on composition, ECM may regulate cell functions and activity of soluble factors[165] and have shown to improve gene expression in cells.[167]
- ECM may control cell phenotype and redirect mechanical signal transduction through cell-matrix interactions,[165,168] thereby regulating the shape, migration, proliferation, survival, and differentiation of cells.[169]

Synthetic or natural hydrogels are excellent examples of ECM[170,171] because they are hydrophilic and their stiffness can be tuned via precise polymerization control of crosslinking density. Some examples of hydrogels are based from synthetic polymers such as polyethylene glycol (PEG), poly-N-isopropylacrylamide (PNIPAM), polyvinyl alcohol (PVA), poly(2-hydroxyethyl methacrylate) (PHEMA),[164] or polyacrylic acid (PAA).[19] Some hydrogels are also made from natural polymers such as fibrin, collagen, and hyaluronic acid.[19] These natural polymers can take in large quantities of water or absorb biological fluid through the interstitial pore spaces. Seeding of cells into the hydrogel matrix can take place either during or after the polymerization process. This ultimately depends on which cell types or experimental variables are being studied.

Despite underlining the importance of ECM, certain limitations still exist:

- Hydrogel fabrication requires additional cost and expert skill for precise control of polymerization conditions.
- Thick hydrogel-based ECM may be opaque, difficult to image, and may limit access to apical and basal compartments.
- Collecting cells for biochemical assays is more challenging in some ECM gels, which require specific enzymes (e.g., protease) to dissolve and recuperate the embedded cells, though synthetic hydrogels can be designed to dissolve upon activation of a stimulus.[172]
- High costs of reagents for ECM fabrication.
- Difficulty to scale-up industrially because while it is entirely possible to fabricate hydrogels in a high throughput scale, there can be issues concerning consistency between hydrogel batches during polymerization.[19]

3.2 Microfluidic methods to mimic ECM

Chemically defined hydrogels are commonly used as biomimetic matrices or scaffolds for directing the formation of 3D microtissues in microfluidic chips. Hydrogels have been proven to facilitate the formation of organoids, thereby improving the maturity of advanced in vitro models via precise control of their shape, size, architecture, and mechanical characteristics. The different methods to mimic ECM are presented below.

Generation of shape-controlled hydrogels. Hydrogels can be fine-tuned to generate fibrous scaffolds using methods that produce microfibers (~10μm) and even nanofibers (~10nm) such as blow spinning and electrospinning.[19] Like all polymerization techniques, other components can be added to control the structural integrity of the hydrogel. Synthetic fibers generally degrade much slower than the natural ones. The elongated nature of the fibers has been shown to be suitable to certain types of cells. For example, fibers made from PLGA or poly(lactic-co-glycolic acid) have shown to permit neural progenitor cells to elongate along the length of the fibers.[173]

Hydrogel ECMs can also come in the form of microparticles using electrostatic spraying technique.[174] These uniformly sized particles can encapsulate and compartmentalize cells. Lu *et al.* have efficiently demonstrated the use of these hydrogel microparticles to culture size-controlled tumor microtissues.[174] Cell-encapsulating hydrogel microparticles have also been used as building blocks to assemble repeating patterns and mimic larger tissue constructs.[161]

Coating methods. Commercially available hydrogel-based ECMs can also be used, e.g., Matrigel, which is composed of basement membrane proteins derived from tumors as well as growth factors needed for cell differentiation.[161] Matrigel can be injected directly into the microfluidic biochip to coat the surface[175] or applied using replica molding and microcontact printing, where the protein solution is poured into the PDMS stamp mold before sealing.[24] Another commercial ECM, Geltrex, offers the same purpose as Matrigel but generally does not contain phenol red, which is useful for assays that require colorimetric detection or fluorescence. Both Matrigel and Geltrex come in different formulations for various cell culture applications and typically contain ECM proteins, which are charged such as fibronectin, laminin, or collagen. These ECM coatings also reduce PDMS hydrophobicity and promote cell adhesion.[97]

Micromolding. Micromolding coupled with photolithography has also been employed to fabricate 3D hydrogel microstructures with encapsulated cells.[161,176] In this method, the cells are first suspended in a gel precursor prior to being poured into a patterned mold to produce structures of various sizes. To form a reversible seal, a hydrophobic PDMS cover slip is then positioned on top of the precursor. The polymer precursor is then cross-linked via UV or heat application. After which, the PDMS cover slip is then removed and the hydrogels are removed from the pattern via washing.[176] Using materials with higher surface energies, micromolding permits fabrication of nanoparticles that have drug delivery applications.[161]

4. From microfluidic 3D cell culture to functional tissues and organs

The human body is highly complex and consists of designated biological subunits (cells) that form tissues and organs with specific functions. When human cells are cultured outside the body, they are often subjected to large exogenous contacts and uneven culture conditions.[177] The challenge then is to mimic and eventually rebuild the in vivo like tissue constructs for biomedical research and cutting-edge applications, including tissue engineering, drug screening and development.[178] 3D microfluidic cell culture has gained significant relevance to mimic the in vivo microenvironment[57] and to faithfully reproduce functional human tissues and organs.[24]

Recently, complex microfluidic chips have been designed for highly controlled microenvironments, 3D cell culture manipulation[179] and long-term differentiation of various stem cells.[45] In this section, the main human tissues and organs developed within microfabricated biochips are outlined.

4.1 Spheroids and tumor microenvironment (TME) on chip

From its discovery in the 1970s,[180] spheroids represent the most used 3D in vitro tumor models.[181] In general, spheroids are produced after aggregation and compaction of cells in nonadherent systems. Spheroids are employed as models for drug discovery and cell proliferation studies, and to investigate the interaction between tumor and immune cells.[182] A plethora of methods have been proposed for the production of spheroids: non-adherent surface, hanging drop, and microfluidic systems. The microfluidic chips, in particular, are advantageous for spheroids formation and analysis in terms of cost-effectiveness, high-throughput screening along with the capability of mimicking the tumor microenvironment and physiological

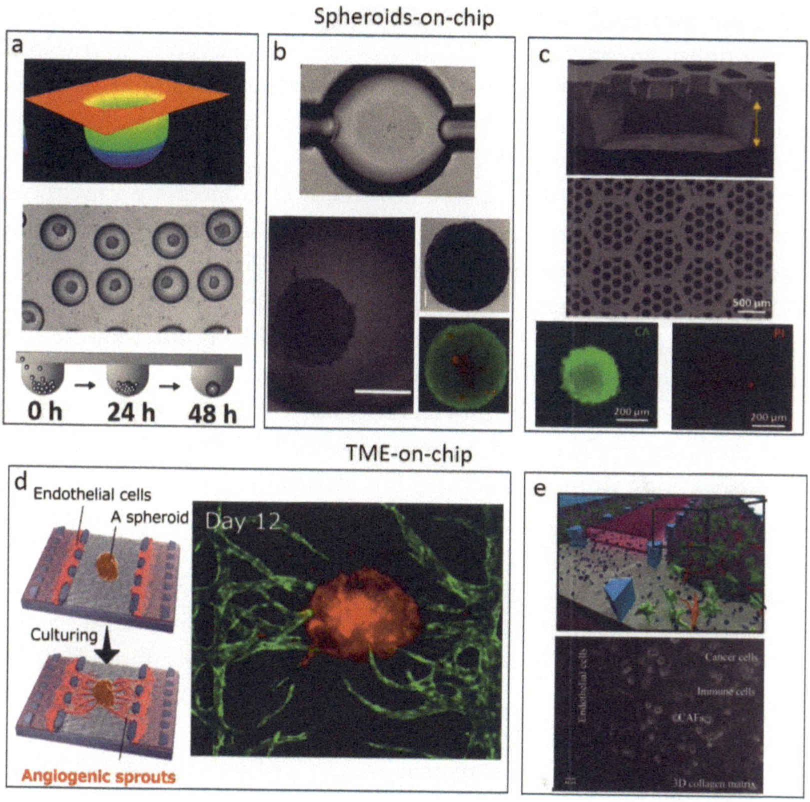

Fig. 6 Typical spheroids-on-chip (A, B, C) and tumor microenvironment (TME)-on-chip (D, E) platforms. (A) Formation of spheroids in microfluidic chips; (B) Spherical microwells connected via microhannels; (C) Micro-cage devices for tumor spheroids formation; (D) Culture patch microfluidic method[198]; (E) TME-on-chip.[197] *(Panel A: Reprinted with permission from Kwapiszewska K, Michalczuk A, Rybka M, Kwapiszewski R, Brzozka Z. A microfluidic-based platform for tumour spheroid culture, monitoring and drug screening. Lab Chip. 2014; 14(12):2096–2104. © Royal Society of Chemistry; Panel B: Reprinted with permission from Pitingolo G, Nizard P, Riaud A, Taly V. Beyond the on/off chip tradeoff: a reversibly sealed microfluidic platform for 3D tumor microtissue analysis. Sens Actuators B Chem. 2018; 274:393-401. © Elsevier; Panel C: Reprinted with permission from He Y, Huang B, Rofaani E, et al. Fabrication of micro-cages and caged tumor spheroids for microfluidic chip-based assays. Microelectron Eng. 2020; 225:111256. © Elsevier; Panel D: Reprinted with permission © Elsevier; Panel E: Reprinted with permission © Elsevier.)*

blood flow conditions.[183] The formation of spheroids in microfabricated chips has already been explored for several types of cells such as hepatocytes in a pioneering work,[184] cancer cells, osteoblasts, and endothelial cells[185] and other combinations as showed for example in Fig. 6A.[186,187] Modular and reversibly bonded biochips have also been proposed going beyond the conventional on/off chip trade-off analysis. For example, by exploiting magnetic techniques, a temporary bonded chip, constituted of PMMA and PDMS spherical microwells connected via microchannels, has been developed[188] for tumor analysis in Fig. 6B.

Recently, He *et al.* described a new method to integrate a microcaged patch into a modular microfluidic chip for tumor spheroids formation.[35] The microcage system is made up of an array of honeycomb compartments with a monolayer of cross-linked gelatin nanofibers at the bottom and a mesh of 200 μm hole size at the top as shown in Fig. 6C. Furthermore, in response to the need for increased throughput in the analysis of tumor cells, a droplet-based platform for high-density, 3D cell culture and visualization in a single device has been introduced.[189] Despite their numerous advantages, most of in vitro spheroid models fail to truly mimic the tumor vascularization that plays a key role in the development of a hostile tumor microenvironment (TME), thereby resulting in cancer progression and drug resistance.[190] In addition to the increased permeability of tumor vessels, the TME includes immune cells, fibroblasts, signaling molecules, and the ECM.[191] To better mimic the TME on biochip, several platforms have been recently introduced, taking into account this complex biological organization in Fig. 6D and E. In the cancer field, several works have reported 3D microarchitectures on-chip that mimic TME, building the basis for the technology of tumors-on-chip.[192–195] In particular, Nashimoto *et al.* presented a vascularized tumor on chip platform to mimic the influence of blood flow on tumor tissues.[196] In this work, the cultivation of fibroblasts in the tumor spheroid on chip induced angiogenic sprouts and the formation of a perfusable vascular network. To address the major challenge in the in vitro cancer research represented by the heterogeneity of the tumor microenvironment, including the host immune components, a 3D coculture (four cell types: cancer, immune, endothelial, and fibroblasts) reconstituting *ex vivo* a human tumor ecosystem (HER2+ breast cancer) has been proposed.[197]

4.2 Blood-brain barrier and heart on chip by stem cells differentiation

Human pluripotent stem cells (hPSCs), embryonic (hESCs),[199] and induced pluripotent (iPSCs)[200] stem cells hold great potential for disease modeling and preclinical drug development.[201,202] In particular, the differentiation of iPSCs can give rise to different functional tissues on biochips, such as nervous, cardiac, pulmonary alveolus, and blood vessel. The microfluidic modeling of the blood-brain barrier (BBB),[203-205] spinal cord and peripheral nerve tissues plays a leading role in new drug development for central nervous system disorders.[206] Moreno *et al.* obtained human neuroepithelial cells from iPSCs and have successfully differentiated them into dopaminergic neurons within phase-guided, three-dimensional microfluidic cell culture bioreactor[207] (Fig. 7A). Sharma, Moore and coworkers developed a microengineered microfluidic platform mimicking the human peripheral nerve (HNoaC) with iPSC-derived neurons (hNs) and primary human Schwann cells (hSCs)[208] (Fig. 7B). In 2020, a group from the Georgia Institute of Technology demonstrated a vascularized BBB on chip consisting of endothelial cells (ECs), pericytes, and astrocytes with aquaporin-4 (AQP4) expression[209] (Fig. 7C). Cardiac cells derived on chip from iPSCs are valuable for studying heart development, modeling the disease, and to test new drugs.[210,211] Several robust microfluidic-based approaches for the differentiation of iPSCs into cardiomyocytes directly on a chip have been reported.[212-214] However, these microfluidic approaches do not recapitulate the 3D organization of ECM proteins of stem cell niches. To respond to this need, the group of Yong Chen (ENS Paris) introduced a patch-based microfluidic system (Fig. 7D), based on gelatin nanofibers, for differentiation of human iPSCs into functional cardiomyocytes.[215] Marsano *et al.* proposed an innovative heart-on-chip platform for the culture of 3D cardiomyocytes, mimicking the mechanical stimulation in the native myocardium[216] (Fig. 7E).

4.3 Lung on chip

As recently highlighted for repurposing of already approved drugs,[217,218] it is fundamental to recreate lung/alveolar models on chip.[219] A pioneer in this field is the group of Donald E. Ingber (Wyss Institute) introducing a biomimetic microsystem that reconstitutes the critical functional alveolar–capillary interface of the human lung.[220] After such early work, other research groups introduced novel microfluidic platforms to mimic the functional area of the human lung[221-223] (Fig. 7F). In particular, the group headed by Olivier Guenat (University of Bern) designed and developed a biochip with stretchable biological membrane to mimic the complex architecture of the lung parenchyma and the air-blood barrier.[224] This work proposed a substrate made of collagen and elastin with less adsorption of biomarkers in

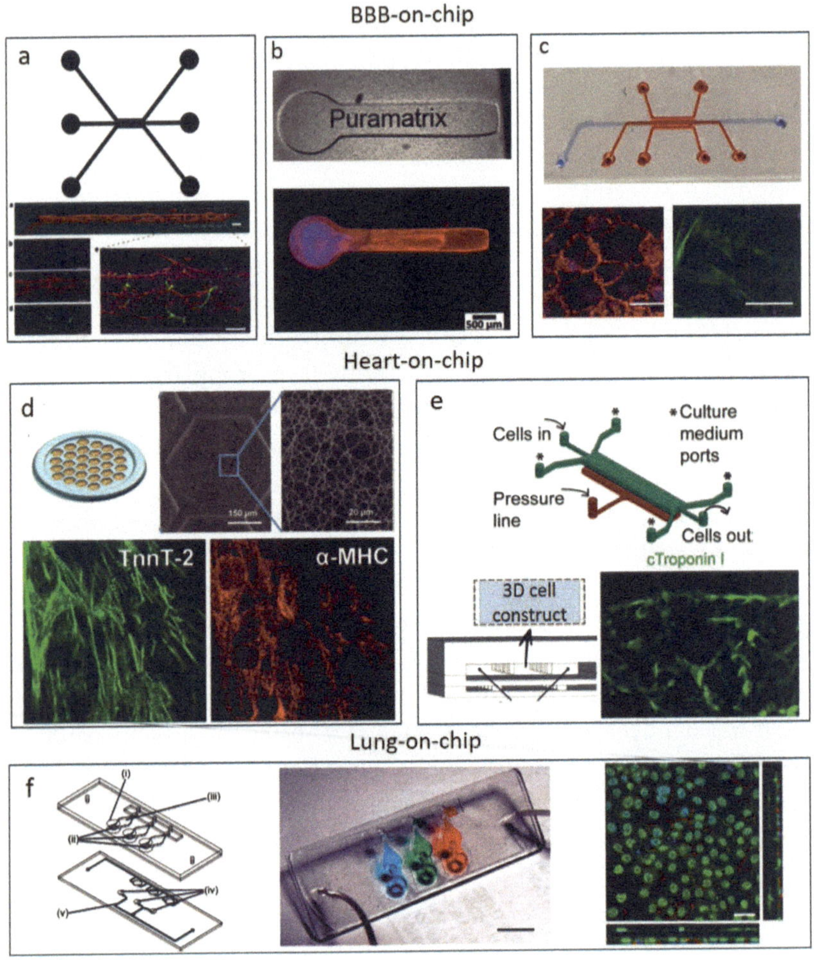

Fig. 7 Main human tissues and organs developed within microfabricated biochips: blood-brain barrier BBB-on-chip (A) 207, (B) 208, (C) 209, cardiomyocytes-on-chip (D) 215, heart-on-chip (E) 216, lung-on-chip (F) 222. *(Panel A: Reprinted with permission © Royal Society of Chemistry; Panel B: Reprinted with permission © Royal Society of Chemistry; Panel C: Reprinted with permission © Springer Nature; Panel D: Reprinted with permission © Royal Society of Chemistry; PanelE: Reprinted with permission © Royal Society of Chemistry; Panel F: Reprinted with permission © Royal Society of Chemistry.)*

comparison to PDMS. Nevertheless, the collagen–elastin solution produced thick layers, 5–10μm, as compared to alveolar ECM. To overcome this drawback, Radiom *et al.* introduced an array of suspended monolayers of gelatin nanofibers (thickness less than 1μm) for replicating the alveolar air–tissue interface.[225]

4.4 Blood microvessels on chip

The efficient vascularization of tissues and organs on chip is a major scientific and technical challenge since a constant blood supply is necessary for tissues and organs to survive in vivo.[226] The vasculature plays a crucial role in many diseases. For example, leaky tumor vasculature is responsible for a hostile tumor microenvironment (TME) in vivo and irrigates solid tumors, resulting in cancer progression and drug resistance.[227] Considering the emerging evidence from patients with COVID-19, the endothelial cells and vascular leakage are essential contributors to the initiation and propagation of severe pulmonary edema.[228] This highlights the main microfluidics-based models developed using two distinct approaches: via endothelial-lined patterned channels or via self-assembled and spontaneous networks.[229-231] Several works reported on the generation of endothelial lined microchannels[156,232,233] and microvascular networks formation in engineered microsystems by combining microfluidics, biomaterials, and tissue engineering.[59,234] The main contribution has been given by Noo Li Jeon's group (Seoul National University), developing microfluidic-based platforms to form perfusable networks of 3D microvessels as well as tumor vasculatures.[235] Other microfluidic platforms, made of PDMS[236] or collagen gel,[140] have been introduced to form 3D in vitro microvessel models, as vascularized tissue arrays and multistep angiogenesis, respectively. Finally, Nashimoto *et al.* presented a vascularized tumor-on-a-chip platform with intraluminal flow, preserving the perfusability of the vascular network during 3D cell culture.[198]

4.5 Liver and gut on chip

The liver and gut play crucial roles in both homeostasis and diseases. Their modeling on-chip has been discussed in several reviews.[237,238] Advanced microfluidic liver systems represent models useful to be employed in drug discovery and development, human in vivo pharmacokinetic and pharmacodynamic studies. Microfluidic platforms based on membrane integration[239] and concave microwell-based PDMS structures[240] have been presented as a liver model, respectively. The research group of Jianhua Pei (Shanghai Jiao Tong University) differentiated functional liver organoids from human iPSCs in a 3D perfusable microfluidic chip, combining the intrinsic stem cell self-organization principle with microengineering technology.[231] The more sophisticated mechanically active gut on chip model has been developed by the Ingber group (Wyss Institute), enabling human intestinal epithelium, capillary endothelium, immune cells, and even commensal microbial cells to grow.[241,242]

4.6 Automated microfluidic 3D cell culture

The automation of experiments in biology laboratories is gaining attention to develop advanced healthcare solutions. Automated microfluidic stem cell culture and differentiation can improve the quality of experiments enabling long-term cell differentiation on engineered microsystems, with reduced manual labor. First, a platform for automatic hiPSC differentiation has been reported, based on robotic manipulation of culture wares.[243] Recently, Yong Chen's group (ENS Paris) proposed and validated a low-cost, versatile, and compact platform to automate 3D cell culture and hiPSC differentiation.[48] This automatic laboratory platform for stem cells differentiation is a useful alternative to manual protocols and robotic culture platforms, widening its application to a variety of microengineered substrates.

5. Microfluidic 3D cell culture for drug development

Traditional preclinical models (2D cell culture and animals) need to be improved for a high efficient development of new biologically active molecules to treat human diseases. Indeed, novel preclinical models such as organ and tissue on chip can efficiently predict the effects of drug candidate compounds in the body.[244] Compared with traditional preclinical drug screening methods, a microfluidic-based system has superior advantages in terms of sample consumption, reaction time, visualization and operational costs.[26] 3D microfluidic cell culture models represent a less expensive and reliable platform for investigation as a potential high throughput screening (HTS).[58] These innovative platforms can be essential to make crucial "go/no-go" decisions before proceeding toward further preclinical and human studies, which help to safeguard financial resources.[19]

Microfluidic chips are designed to model a large number of human disorders, mainly tumors, central Nervous System (CNS) pathologies, heart and immune diseases. Many microfluidic tumor models have been applied for anti-cancer drug development and discovery.[252] Irreversibly and reversibly sealed microfluidic platforms have been presented for growth of 3D tumor spheroids and their analysis after drug treatment.[183,188] As 3D tumor spheroids analysis has gained in popularity, there is an increasing request for tools providing big amounts of data on single cells within the spheroids.

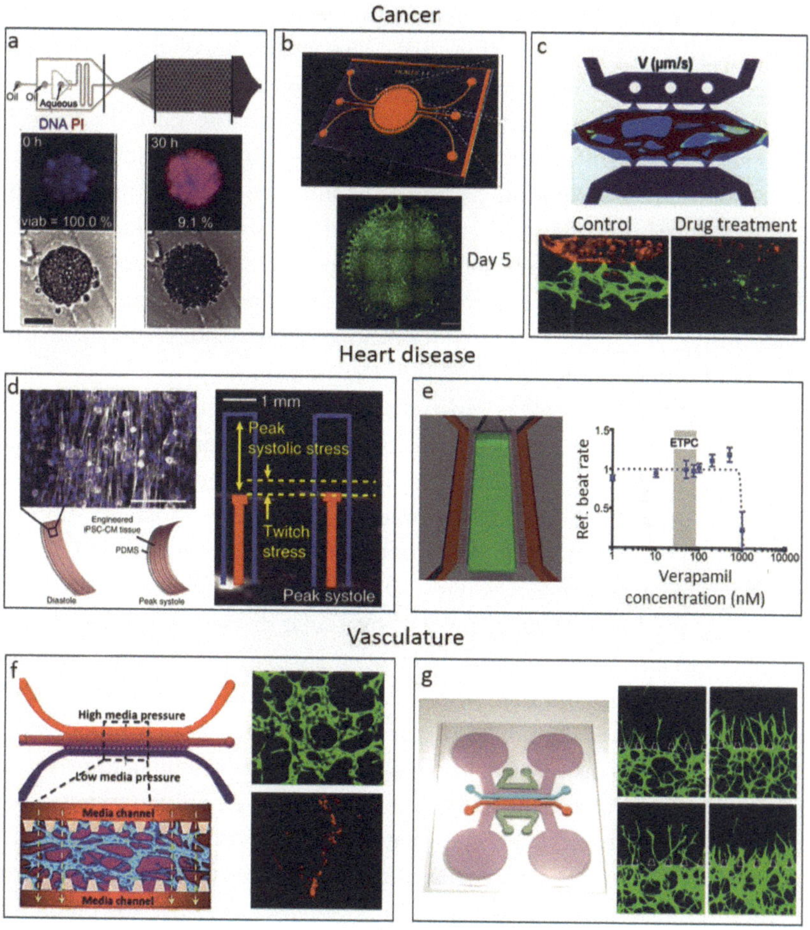

Fig. 8 Example of platforms for drug effect analysis on chip. Anticancer drugs tested on chip (A) 245, (B) 246, (C) 247. PDMS-based interfaces for cardiomyopathy studies and pathogenicity of titin gene variants (D) 248, (E) 249. Vasculature on chip for nanoparticles, drugs uptake and influence on interstitial flow (F) 250, (G) 251. *(Panel A: Reprinted with permission © Elsevier; Panel B: Reprinted with permission © American Association for the Advancement of Science; Panel C: Reprinted with permission © Royal Society of Chemistry; Panel D: Reprinted with permission © Springer Nature; Panel E: Reprinted with permission © Springer Nature; Panel F: Reprinted with permission © Royal Society of Chemistry; Panel G: Reprinted with permission © Royal Society of Chemistry.)*

Tomasi and colleagues presented a microfluidic chip for HTS by parallelizing the manipulation of individual spheroids in Fig. 8A within anchored droplets.[245] In this platform, several concentrations of a drug model were tested on the hepatic-like spheroids. A "tumor-on-a-chip" model to test the effects of drug-loaded nanoparticles in a gradient fashion has been also developed[246] (Fig. 8B). The validation of the platform was carried out by the investigation of the viability with immunostaining to confirm the dose–response action of cells exposed to the CMCht/PAMAM nanoparticle gradient. To faithfully mimic the tumor on chip, Nashimoto *et al.* proposed a microfluidic platform enabling the study of tumor progression with intraluminal flow in an engineered vascular network.[196] A biochip mimicking mass transport in the capillaries of tumor microenvironment (Fig. 8C) and the delivery of multiple nutrients and/or drugs to the tumor has been also proposed.[247] To overcome one of the major challenge in mimicking cancer on-chip, including the host immunological setting, Nguyen *et al.* integrated four cell populations on a single chip: cancer, immune, endothelial and fibroblasts, reconstituting *ex vivo* a human tumor ecosystem (HER2 breast cancer).[197] This work showed that tumor-on-chip can be able to recapitulate the biological complexity of cancer associated fibroblast (CAF) with the direct visualization of the microenvironment to understand the crucial interplay of CAFs with cancer cells and immune cells.

Due to the high mortality for cardiovascular diseases, heart functionality on chip and circulatory systems have been also widely explored by microfluidic researchers. However, it is still a challenge to reproduce the human heart in vitro for the complex structural architecture of its chambers and the electrical functionality. Recent progresses on heart-on-chip and its application for drug development have been covered in extensive reviews.[253,254] The majority of heart-on-a-chip platforms for drug investigation include biological, electrical, mechanical, and topographical cues to obtain the tissue maturation. It is therefore of utmost importance to improve the predictivity of a chamber-specific therapy targeting adult human cells. For example, a heart-on-a-chip platform has been proposed to monitor the contractile behavior of 3D cardiac tissue constructs, including frequency, contraction force, and synchronization,[255] where an image processing system and a piezoelectric-based device were combined. In addition, a "heart-on-a-chip" which is based on an automated fabrication method to process thin film cantilevers of soft elastomers was also introduced for drug effect analysis.[256] Applying iPSCs protocols and mimicking the complexity of the human heart into microfluidic platforms, in future it could be possible to generate cell-based chips showing atrial and ventricular tissue, sinoatrial node, atrioventricular node, and Purkinje fibers. iPSCs cells with PDMS-based interfaces were developed to investigate the pathophysiology underlying the cardiomyopathy of Barth syndrome (BTHS) that represents a disease caused by a specific gene encoding tafazzin (TAZ) mutation (Fig. 8D).[248] Finally, a PDMS pillars-based device was applied to engineer cardiac microtissues from human-induced pluripotent stem (iPS) cells[249] to evaluate the pathogenicity of titin gene variants in Fig. 8E. This work claimed that certain missense mutations, like truncating variants in the Titin gene TTNtvs, diminish contractile performance and are pathogenic.

The blood vasculature that is constituted by vessels of variable shapes and functions is responsible for blood cells, nutrients, and oxygen distribution maintaining physiological tissue homeostasis. Some diseases are caused by a damage of the vasculature and its change in permeability.[257] The tumor-vasculature-on-a-chip was developed to investigate the nanoparticle extravasation and tumor accumulation.[258,259] For example, the solid tumors are characterized by a leaky tumor vasculature and an altered drainage of lymph vessels, promoting accumulation of nanoparticles in the zone near the tumor. A functional human 3D microvascular network was used to analyze the toxicity after exposure to ambient fine particulate matter (FPM)[250] (Fig. 8F). The research group of Noo Li Jeon (Seoul National University) investigated the flow-mediated endothelial dynamics and phenotype changes as a crucial phenomenon involved in the remodeling of vascular networks[251] (Fig. 8G).

Currently available OOC and TME systems have been developed to study the beneficial effects and toxicity of drugs. Still, the in vivo pharmacokinetics and pharmacodynamics of the potential drug candidates have not been extensively considered. Thus, a microfluidic platform able to connect 3D cell culture with the gut-liver axis can represent an advanced in vitro ADME-Tox culture model, for analysis of drug metabolism.[64] The development of liver and gut-on-chip and their application in drug metabolism studies was reviewed in several works.[237,238] For example, a human Gut-on-a-Chip (Gut Chip) platform composed by human intestinal epithelial cells and other cell types was successful applied to recapitulate the damage to tissues caused by exposure to ionizing radiation and assess the efficacy of relative drugs in vitro.[260]

6. Concluding remarks

In this chapter, recent advances in microfluidic 3D cell culture, extensively applied worldwide for preclinical drug discovery and development, have been presented. The in vitro microphysiological systems constitute the foundation upon which was built the organ-on-a-chip technology.

The main geometries and materials used to fabricate laboratory and industrial-scale microfluidic devices have been introduced. Traditional preclinical models, such as animal testing, still present several issues mainly genetic differences and ethical problems. The need to develop human tissue on chip with proven functionality based on the integration of hydrogels to perfectly mimic the in vivo microenvironment has been evidenced. However, if for the evaluation of the safety and efficacy of drugs, it may be still necessary to rely on animal models in preclinical trials, organ-on-chip can be an ideal method for screening before going to the animal tests, in order to reduce them. Finally, to definitively consecrate human organ-on-a-chip technology in the drug development process, collaborative projects are needed to fix cell quality standards guaranteeing reproducibility and robustness of these models. The scientific community operating in this field should work with regulatory scientists, animal researcher and federal agencies (i.e., FDA, EMA, etc.) to compare organ-on-a-chip results with animal models.

References

1. Amir-Aslani A, Mangematin V. The future of drug discovery and development: shifting emphasis towards personalized medicine. *Technol Forecast Soc Change*. 2010;77:203–217.
2. DiMasi JA, Grabowski HG, Hansen RW. Innovation in the pharmaceutical industry: new estimates of R&D costs. *J Health Econ*. 2016;47:20–33.
3. Kaitin KI. Deconstructing the drug development process: the new face of innovation. *Clin Pharmacol Ther*. 2010;87:356–361.
4. Kaitin KI. Published correction on deconstructing the drug development process: the new face of innovation. *Clin Pharmacol Ther*. 2011;89:148.
5. FDA. The drug development process. https://www.fda.gov/patients/learn-about-drug-and-device-approvals/drug-development-process.
6. FDA. http://www.fda.gov/NewsEvents/Newsroom/PressAnnouncements/2006/ucm108576.htm.
7. Zhang BY, Radisic M. Organ-on-a-chip devices advance to market. *Lab Chip*. 2017;17:2395–2420.
8. Hay M, Thomas DW, Craighead JL, Economides C, Rosenthal J. Clinical development success rates for investigational drugs. *Nat Biotechnol*. 2014;32:40–51.
9. Schuhmacher A, Gassman O, Hinder M. Changing R&D models in research-based pharmaceutical companies. *J Transl Med*. 2016;14:105.
10. Arrowsmith J, Miller P. Phase II and phase III attrition rates 2011–2012. *Nat Rev Drug Discov*. 2013;12:569.
11. Berggren R, Moller M, Moss R, Poda P, Smietana K. Outlook for the next 5 years in drug innovation. *Nat Rev Drug Discov*. 2012;11:435–436.
12. Hutchinson L, Kirk R. High drug attrition rates–where are we going wrong? *Nat Rev Clin Oncol*. 2011;8:189–190.
13. Coppeta JR, Mescher MJ, Isenberg BC, et al. A portable and reconfigurable multi-organ platform for drug development with onboard microfluidic flow control. *Lab Chip*. 2016;17:134–144.

14. Langhans SA. Three-dimensional in vitro cell culture models in drug discovery and drug repositioning. *Front Pharmacol.* 2018;9:6.
15. Kapalczynska M, Kolenda T, Przybyla W, et al. 2D and 3D cell cultures–a comparison of different types of cancer cell cultures. *Arch Med Sci.* 2018;14:910–919.
16. Begley C, Ellis L. Raise standards for preclinical cancer research. *Nature.* 2012; 483:531–533.
17. van Norman GA. Limitations of animal studies for predicting toxicity in clinical trials: Part 2: potential alternatives to the use of animals in preclinical trials. *JACC Basic Transl Sci.* 2020;5(4):387–397.
18. Duval K, Grover H, Han LH, et al. Modeling physiological events in 2D vs. 3D cell culture. *Physiology.* 2017;32:266–277.
19. Castiaux AD, Spence DM, Scott Martin R. Review of 3D cell culture with analysis in microfluidic systems. *Anal Methods.* 2019;11(33):4220–4232.
20. Doke SK, Dhawale SC. Alternatives to animal testing: a review. *Saudi Pharm J.* 2015;23:223–229.
21. Tyers M, Mann M. From genomics to proteomics. *Nature.* 2003;422:193–197.
22. RWM S, Burch RL. *The principles of Humane Experimental Technique.* John Hopkins University; 1959. https://caat.jhsph.edu/principles/the-principles-of-humane-experimental-technique.
23. Kojima H, Seidle T, Spielmann H. *Alternatives to Animal Testing: Proceedings of Asian Congress 2016.* Springer Open; 2019.
24. Huh D, Hamilton GA, Ingber DE. From 3D cell culture to organs-on-chips. *Trends Cell Biol.* 2011;21(12):745–754.
25. Liu H, Roy K. Biomimetic three-dimensional cultures significantly increase hematopoietic differentiation efficacy of embryonic stem cells. *Tissue Eng.* 2005;11(1–2): 319–330.
26. Sun J, Warden AR, Ding X. Recent advances in microfluidics for drug screening. *Biomicrofluidics.* 2019;13(6):061503.
27. Esch E, Bahinski A, Huh D. Organs-on-chips at the frontiers of drug discovery. *Nat Rev Drug Discov.* 2015;14:248–260.
28. Jensen C, Teng Y. Is it time to start transitioning from 2D to 3D cell culture? *Front Mol Biosci.* 2020;7:33.
29. Zuppinger C. 3D cardiac cell culture: a critical review of current technologies and applications. *Front Cardiovasc Med.* 2019;6:87.
30. Edmondson R, Broglie JJ, Adcock AF, Yang L. Three-dimensional cell culture systems and their applications in drug discovery and cell-based biosensors. Assay Drug Dev Technol. 12; 1(4):207–218.
31. Maimouni I, Cejas CM, Cossy J, Tabeling P, Russo M. Microfluidics mediated production of foams for biomedical applications. *Micromachines.* 2020;11:83.
32. Cui X, Hartanto Y, Zhang H. Advances in multicellular spheroids formation. *J R Soc Interface.* 2017;14(127):20160877.
33. Ramadhan W, Kagawa G, Moriyama K, et al. Construction of higher-order cellular microstructures by a self-wrapping co-culture strategy using a redox-responsive hydrogel. *Sci Rep.* 2020;10:6710.
34. Ho WJ, Pham EA, Kim JW, et al. Incorporation of multicellular spheroids into 3-D polymeric scaffolds provides an improved tumor model for screening anticancer drugs. *Cancer Sci.* 2010;101(12):2637–2643.
35. He Y, Huang B, Rofaani E, et al. Fabrication of micro-cages and caged tumor spheroids for microfluidic chip-based assays. *Microelectron Eng.* 2020;225:111256.
36. Ravi M, Paramesh V, Kaviya SR, Anuradha E, Solomon FD. 3D cell culture systems: advantages and applications. *J Cell Physiol.* 2015;230:16–26.
37. Dhaliwal A. 3D cell culture: a review. *Mater Methods.* 2012;2:162.

38. Dutta RC, Dutta AK. Cell-interactive 3D-scaffold; advances and applications. *Biotechnol Adv*. 2009;27(4):334–339.
39. Bissell MJ, Radisky DC, Rizki A, Weaver VM, Petersen OW. The organizing principle: microenvironmental influences in the normal and malignant breast. *Differentation*. 2002;70:537–547.
40. Hakkinen KM, Harunaga JS, Doyle AD, Yamada KM. Direct comparisons of the morphology, migration, cell adhesions, and actin cytoskeleton of fibroblasts in four different three-dimensional extracellular matrices. *Tissue Eng A*. 2011;17:713–724.
41. Kumar S, Geiger H. HSC niche biology and HSC expansion Ex vivo. *Trends Mol Med*. 2017;23(9):799–819.
42. Whitesides GM. The origins and the future of microfluidics. *Nature*. 2006;442 (7101):368–373.
43. Russo M, Bevilacqua P, Netti PA, Torino E. A microfluidic platform to design crosslinked hyaluronic acid nanoparticles (cHANPs) for enhanced MRI. *Sci Rep*. 2016;6:37906.
44. Wu MH, Huang SB, Lee GB. Microfluidic cell culture systems for drug research. *Lab Chip*. 2010;10(8):939–956.
45. Li KJ, Valadez AV, Zuo P, Nie Z. Microfluidic 3D cell culture: potential application for tissue- based bioassays. *Bioanalysis*. 2012;4:1509–1525.
46. Yong AW, Li K, Merajver SD, Fu J. Tracking the tumor invasion front using long-term fluidic tumoroid culture. *Sci Rep*. 2017;7:10784.
47. Montanez-Sauri SI, Beebe DJ, Sung KE. Microscale screening systems for 3D cellular microenvironments: platforms, advances, and challenges. *Cell Mol Life Sci*. 2015; 72(2):237–249.
48. Pitingolo G, He Y, Huang B, Wang L, Shi J, Chen Y. An automatic cell culture platform for differentiation of human induced pluripotent stem cells. *Microelectron Eng*. 2020;231:111371.
49. Vickerman V, Blundo J, Chung S, Kamm RD. Design, fabrication and implementation of a novel multi-parameter control microfluidic platform for three-dimensional cell culture and real-time imaging. *Lab Chip*. 2008;8(9):1468–1477.
50. Rothbauer M, Zirath H, Ertl P. Recent advances in microfluidic technologies for cell-to-cell interaction studies. *Lab Chip*. 2018;18(2):249–270.
51. Ling Y, Rubin J, Deng Y, et al. A cell-laden microfluidic hydrogel. *Lab Chip*. 2007; 7(6):756–762.
52. Ballermann BJ, Dardik A, Eng E, Liu A. Shear stress and the endothelium. *Kidney Int Suppl*. 1998;67:S100–S108.
53. Buchanan CF, Verbridge SS, Vlachos PP, Rylander MN. Flow shear stress regulates endothelial barrier function and expression of angiogenic factors in a 3D microfluidic tumor vascular model. *Cell Adh Migr*. 2014;8(5):517–524.
54. Yu F, Selva Kumar ND, Choudhury D, Foo LC, Ng SH. Microfluidic platforms for modeling biological barriers in the circulatory system. *Drug Discov Today*. 2018; 23(4):815–829.
55. Cheng SY, Heilman S, Wasserman M, Archer S, Shuler ML, Wu M. A hydrogel-based microfluidic device for the studies of directed cell migration. *Lab Chip*. 2007;7 (6):763–769.
56. Chen YA, King AD, Shih HC, et al. Generation of oxygen gradients in microfluidic devices for cell culture using spatially confined chemical reactions. *Lab Chip*. 2011; 11(21):3626–3633.
57. Young EWK, Beebe DJ. Fundamentals of microfluidic cell culture in controlled microenvironments. *Chem Soc Rev*. 2010;39(3):1036–1048.
58. Gupta N, Liu JR, Patel B, Solomon DE, Vaidya B, Gupta V. Microfluidics-based 3D cell culture models: utility in novel drug discovery and delivery research. *Bioeng Transl Med*. 2016;1(1):63–81.

59. Zhang X, Li L, Luo C. Gel integration for microfluidic applications. *Lab Chip*. 2016;16 (10):1757–1776.
60. Rosser JM, Olmos Calvo I, Schlager M, Purtscher M, Jenner F, Ertl P. Recent advances of biologically inspired 3D microfluidic hydrogel cell culture systems. *J Cell Biol Cell Metab*. 2015;2:5.
61. Terell JA, Jones CG, Kabandana GKM, Chen C. From cells-on-a-chip to organs-on-a-chip: scaffolding materials for 3D cell culture in microfluidics. *J Mater Chem B*. 2020;8:6667–6685.
62. Ingber DE. Is it time for reviewer 3 to request human organ chip experiments instead of animal validation studies? *Adv Sci*. 2020;7:2002030. https://doi.org/10.1002/advs.202002030.
63. Zhang C, Zhao Z, Abdul Rahim NA, Noort D, Yu H. Towards a human-on-chip: culturing multiple cell types on a chip with compartmentalized microenvironments. *Lab Chip*. 2009;9(22):3185–3192.
64. Ishida S. Organs-on-a-chip: current applications and consideration points for in vitro ADME-Tox studies. *Drug Metab Pharmacokinet*. 2018;33(1):49–54.
65. Ronaldson-Bouchard K, Vunjak-Novakovic G. Organs-on-a-chip: a fast track for engineered human tissues in drug development. *Cell Stem Cell*. 2018;22(3):310–324.
66. Zhang J, Wei X, Zeng R, Xu F, Li X. Stem cell culture and differentiation in microfluidic devices toward organ-on-a-chip. *Future Sci OA*. 2017;3(2):FSO187. https://doi.org/10.4155/fsoa-2016–0091.
67. Chen YC, Lou X, Chang Z, Ingram P, Yoon E. High throughout cancer cell sphere formation for characterizing the efficacy of photo dynamic therapy in 3D cell culture. *Sci Rep*. 2015;5:12175.
68. Guild J, Haque A, Gheibi P, et al. Embryonic stem cells cultured in microfluidic chambers take control of their fate by producing endogenous signals including LIF. *Stem Cells*. 2016;34:1501–1512. https://doi.org/10.1002/stem.2324.
69. Huang SB, Wu MH, Wang SS, Lee GB. Microfluidic cell culture chip with multiplexed medium delivery and efficient cell/scaffold loading mechanisms for high-throughput perfusion 3-dimensional cell culture-based assays. *Biomed Microdevices*. 2011;13(3):415–430. https://doi.org/10.1007/s10544-011-9510-1.
70. Kolnik M, Tsimring LS, Hasty J. Vacuum-assisted cell loading enables shear-free mammalian microfluidic culture. *Lab Chip*. 2012;12:4732–4737. https://doi.org/10.1039/c2lc40569e.
71. Dimov IK, Kijanka G, Park Y, Ducrée J, Kang T, Lee LP. Integrated microfluidic array plate (iMAP) for cellular and molecular analysis. *Lab Chip*. 2011;11:2701–2710. https://doi.org/10.1039/c1lc20105k.
72. Wei YC, Chen F, Zhang T, et al. Vascular smooth muscle cell culture in microfluidic devices. *Biomicrofluidics*. 2014;8:046504. https://doi.org/10.1063/1.4893914.
73. Kellogg RA, Gomez-Sjoberg R, Leyrat AA, Tay S. High-throughput microfluidic single-cell analysis pipeline for studies of signaling dynamics. *Nat Protoc*. 2014; 9:1713–1726.
74. Baker BM, Chen CS. Deconstructing the third dimension—how 3D culture microenvironments alter cellular cues. *J Cell Sci*. 2012;125:3015–3024.
75. Cuchiara MP, Allen AC, Chen TM, Miller JS, West JL. Multilayer microfluidic pegda hydrogels. *Biomaterials*. 2010;31:5491–5497.
76. Sung KE, Su G, Pehlke C. Control of 3-dimensional collagen matrix polymerization for reproducible human mammary fibroblast cell culture in microfluidic devices. *Biomaterials*. 2009;30:4833–4841.
77. Toh YC, Lim TC, Tai D, Xiao G, van Noort D, Yu H. A microfluidic 3D hepatocyte chip for drug toxicity testing. *Lab Chip*. 2009;9:2026–2035.
78. Gottwald E, Giselbrecht S, Augspurger C. chip-based platform for the in vitro generation of tissues in three-dimensional organization. *Lab Chip*. 2007;7:777–785.

79. Jang K, Sato K, Igawa K, Chung UI, Kitamori T. Development of an osteoblast-based 3D continuous-perfusion microfluidic system for drug screening. *Anal Bioanal Chem.* 2008;390:825–832.
80. Borenstein JT, Tandon V, Tao SL, Charest JL. *Microfluidic Cell Culture Systems.* Elsevier; 2018.
81. Vyawahare S, Griffiths AD, Merten CA. Miniaturization and parallelization of biological and chemical assays in microfluidic devices. *Chem Biol.* 2010;17:1052–1065.
82. Haycock JW. *3D Cell Culture.* Humana Press; 2011.
83. Hung PJ, Lee P, Sabounchi P, Lin R, Lee PL. Continuous perfusion microfluidic cell culture array for high-throughput cell-based assays. *Biotechnol Bioeng.* 2005;89(1):1–8.
84. Coluccio ML, Perozziello G, Malara N, et al. Microfluidic platforms for cell cultures and investigations. *Microelectron Eng.* 2019;208:14–28.
85. Meddens MBM, Lui S, Finnegan PS, Edwards TL, James CD, Lidke KA. Single objective light-sheet microscopy for high-speed whole-cell 3D super-resolution. *Biomed Opt Exp.* 2016;7:2219–2236.
86. Tehranirokh M, Kouzani AZ, Francis PS, Kanwar JR. Microfluidic devices for cell cultivation and proliferation. *Biomicrofluidics.* 2013;7(5):51502. https://doi.org/10.1063/1.4826935.
87. Chan CY, Goral VN, DeRosa ME, Huang TJ, Yuen PK. A polystyrene-based microfluidic device with three-dimensional interconnected microporous walls for perfusion cell culture. *Biomicrofluidics.* 2014;8:046505.
88. Becker H, Schulz I, Mosig A, Jahn T, Gartner C. Microfluidic devices for cell culture and handling in organ-on-a-chip applications. *Proc SPIE.* 2014;8976:89760N.
89. Li R, Lv X, Hasan M, et al. A rapidly fabricated microfluidic chip for cell culture. *J Chromatogr Sci.* 2015;54:523–530.
90. Shen C, Li Y, Wang Y, Meng Q. Non-swelling hydrogel-based microfluidic chips. *Lab Chip.* 2019;19:3962–3973.
91. Paguirigan A, Beebe DJ. Gelatin based microfluidic devices for cell culture. *Lab Chip.* 2006;6:407–413.
92. Derda R, Laromaine A, Mammoto A, et al. Paper-supported 3D cell culture for tissue-based bioassays. *PNAS.* 2009;106:18457–18462.
93. Chiang HJ, Yeh SL, Peng CC, Liao WH, Tung YC. Polydimethylsiloxane-polycarbonate microfluidic devices for cell migration studies under perpendicular chemical and oxygen gradients. *J Vis Exp.* 2017;120:55292.
94. Macown RJ, Veraitch FS, Szita N. Robust, microfabricated culture devices with improved control over the soluble microenvironment for the culture of embryonic stem cells. *Biotechnol J.* 2014;9:805–813.
95. Nge PN, Rogers CI, Woolley AT. Advances in microfluidic materials, functions, integration, and applications. *Chem Rev.* 2013;113.
96. Titmarsh DM, Ovchinnikov DA, Wolvetang EJ, Cooper-White JJ. Full factorial screening of human embryonic stem cell maintenance with multiplexed microbioreactor arrays. *Biotechnol J.* 2013;8:822–834.
97. Halldorsson S, Lucumi E, Gómez-Sjöberg R, Fleming RM. *Advantages and challenges of microfluidic cell culture in polydimethylsiloxane devices.* 2015;63:218–231. https://doi.org/10.1016/j.bios.2014.07.029.
98. Torino S, Corrado B, Iodice M, Coppola G. PDMS-based microfluidic devices for cell culture. *Inventions.* 2018;3(3):65.
99. Park J, Li Y, Berthiaume F, Toner M, Yarmush ML, Tilles AW. Radial flow hepatocyte bioreactor using stacked microfabricated grooved substrates. *Biotechnol Bioeng.* 2008;99:455–467.
100. Chen Y, Zhang L, Chen G. Fabrication, modification, and application of poly(methyl methacrylate) microfluidic chips. *Electrophoresis.* 2008;29:1801–1814.

101. Jena RK, Yue CY. Cyclic olefin copolymer based microfluidic devices for biochip applications: ultraviolet surface grafting using 2-methacryloyloxyethyl phosphorylcholine. *Biomicrofluidics*. 2012;6(1):12822–1282212. https://doi.org/10.1063/1.3682098.
102. Gencturk E, Mutlu S, Ulgen KO. Advances in microfluidic devices made from thermoplastics used in cell biology and analyses. *Biomicrofluidics*. 2017;11:051502.
103. Super A, Jaccard N, Cardoso Marques MP, et al. Real-time monitoring of specific oxygen uptake rates of embryonic stem cells in a microfluidic cell culture device. *Biotechnol J*. 2016;11:1179–1189.
104. Chang W, Lin S, Wang C, et al. PMMA/PDMS valves and pumps for disposable microfluidics. *Lab Chip*. 2009;9:3088–3094.
105. Shiroma LS, Piazzetta MHO, Duarte-Junior GF, et al. Self-regenerating and hybrid irreversible/reversible PDMS microfluidic devices. *Sci Rep*. 2016;6:26032.
106. Ziolkowska K, Jedrych E, Kwapiszewski R, Lopacinska J, Skolimowski M, Chudy M. PDMS/glass microfluidic cell culture system for cytotoxicity tests and cells passage. *Sens Actuators B Chem*. 2010;145:533–542.
107. Lin JL, Wang SS, Wu MH, Oh-Yang CC. Development of an integrated microfluidic perfusion cell culture system for real-time microscopic observation of biological cells. *Sensors*. 2011;11:8395–8411.
108. Webster A, Dyer CE, Haswell SJ, Greenman JA. A microfluidic device for tissue biopsy culture and interrogation. *Anal Methods*. 2010;2:1005–1007.
109. Rissin DM, Kan CW, Campbell TG, et al. Single-molecule enzyme-linked immunosorbent assay detects serum proteins at subfemtomolar concentrations. *Nat Biotechnol*. 2010;28:595–599.
110. Yadavali S, Lee D, Issadore D. Robust microfabrication of highly parallelized three-dimensional microfluidics on silicon. *Sci Rep*. 2019;9:12213.
111. Pitingolo G, Riaud A, Nastruzzi C, Taly V. Tunable and reversible gelatin based bonding for microfluidic cell culture. *Adv Eng Mater*. 2019;21:1900145.
112. Tao FF, Xiao X, Lei KF, Lee IC. Paper-based cell culture microfluidic system. *BioChip J*. 2015;9:97–105.
113. Rousset N, Monet F, T G. Simulation-assisted design of microfluidic sample traps for optimal trapping and culture of non-adherent single cells, tissues, and spheroids. *Sci Rep*. 2017;7:1–12.
114. Lazar IM, Deng J, Stremler MA, Ahuja S. Microfluidic reactors for advancing the ms analysis of fast biological responses. *Microsyst Nanoeng*. 2019;5:1–16.
115. Kane BJ, Zinner MJ, Yarmush ML, Toner M. Liver-specific functional studies in a microfluidic array of primary mammalian hepatocytes. *Anal Chem*. 2006;78:4291–4298.
116. Lin CH, Hsiao YH, Chang HC, et al. A microfluidic dual-well device for high-throughput single-cell capture and culture. *Lab Chip*. 2015;15:2928–2938.
117. Di Carlo D, Aghdam N, Lee PL. Single-cell enzyme concentrations, kinetics, and inhibition analysis using high density hydrodynamic cell isolation arrays. *Anal Chem*. 2006;78:4925–4930.
118. Ong SM, Zhang C, Toh YC. A gel-free 3D microfluidic cell culture system. *Biomaterials*. 2008;29:3237–3244.
119. Sonnen KF, Merten CA. Microfluidics as an emerging precision tool in developmental biology. *Dev Cell*. 2019;48:293–311.
120. Lin L, Chiu Y-S, Thiery JP, Lim CT. Microfluidic cell trap array for controlled positioning of single cells in adhesive micropatterns. *Lab Chip*. 2013;13:714–721.
121. Gomez-Sjoberg R, Leyrat AA, Pirone DM. Versatile, fully automated, microfluidic cell culture system. *Anal Chem*. 2007;79(22):8557–8563.

122. Lecault V, Vaninsberghe M, Sekulovic S. High throughput analysis of single hematopoietic stem cell purification in microfluidic cell culture arrays. *Nat Methods*. 2011;8:581–586.
123. Kwon T, Prentice H, Oliveira JD, et al. Microfluidic Cell Retention device for perfusion of mammalian suspension culture. *Sci Rep*. 2017;7:6703. https://doi.org/10.1038/s41598-017-06949-8.
124. Kim HS, Devarenne TP, Han A. A high-throughput microfluidic single-cell screening platform capable of selective cell extraction. *Lab Chip*. 2015;15:2467.
125. Huang YL, Tung CK, Zheng A, Kim BJ, Wu M. Interstitial flows promote amoeboid over mesenchymal motility of breast cancer cells revealed by a three dimensional microfluidic model. *Integr Biol*. 2015;7:1402–1411.
126. Kim L, Toh YC, Voldman J, Yu H. A practical guide to microfluidic perfusion culture of adherent mammalian cells. *Lab Chip*. 2007;7:681–694.
127. Figallo E, Cannizzaro C, Gerecht S. Micro-bioreactor array for controlling cellular microenvironments. *Lab Chip*. 2007;7:710–719.
128. Guckenberger DJ, de Groot TE, Wan AMD, Beebe DJ, Young EWK. Micromilling: a method for ultra-rapid prototyping of plastic microfluidic devices. *Lab Chip*. 2015;15:2364.
129. Mohanty S, Larsen LB, Trifol J, et al. Fabrication of scalable and structured tissue engineering scaffolds using water dissolvable sacrificial 3D printed moulds. *Mater Sci Eng C*. 2015;55:569–578.
130. Gale BK, Jafek AR, Lambert CJ, et al. A review of current methods in microfluidic device fabrication and future commercialization prospects. *Inventions*. 2018;3:60.
131. Brinskmeier E, Preuss W. Micro-machining. *Philos Trans R Soc A*. 2012;370: 3973–3992.
132. Perng BY, Wu CW, Shen YK, Lin Y. Microfluidic chip fabrication using hot embossing and thermal bonding of cop. *Polym Adv Technol*. 2010;21:457–466.
133. Jeon JS, Chung S, Kamm RD, Charest JL. Hot embossing for fabrication of a microfluidic 3D cell culture platform. *Biomed Microdevices*. 2011;13:325–333.
134. Karasek P, Grym J, Planeta J, Foret F. Etching of glass microchips with supercritical water. *Lab Chip*. 2014;15:311–318.
135. Awaja F, Wong TT, Arhatari B. Lab-on-a-chip device made by autohesion-bonded polymers. *Biomed Microdevices*. 2018;20:7.
136. Andrieux S, Drenckhan W, Stubenrauch C. Generation of solid foams with controlled polydispersity using microfluidics. *Langmuir*. 2018;34:1581–1590.
137. Andrieux S, Quell A, Stubenrauch C, Drenckhan W. Liquid foam templating–a route to tailor-mad epolymer foams. *Adv Colloids Interface Sci*. 2018;256:276–290.
138. Andrieux S, Medina L, Herbst M, Berglund LA, Stubenrauch C. Monodisperse highly ordered chitosan/cellulose nanocomposite foams. *Compos A Appl Sci Manuf*. 2019; 125:105516.
139. Steigert J, Haeberle S, Brenner T, et al. Rapid prototyping of microfluidic chips in COC. *J Micromech Microeng*. 2007;17:333–341.
140. Lee Y, Choi JW, Yu J, et al. Microfluidics within a well: an injection-molded plastic array 3D culture platform. *Lab Chip*. 2018;18:2433–2440.
141. Smolka M, Haase A, Palfinger U, et al. Roll-to-roll pilot line for large-scale manufacturing of microfluidic devices. In: Mahajan K, Lye G, Eibl-Schindler R, eds; 2017. *ECI Symposium Series*;.
142. Krebs A, Knoll T, Nussbaum D, Velten T. Polymer-based fabrication techniques for enclosed microchannels in biomedical applications. *Microsyst Technol*. 2011;18: 1871–1877.
143. Toh AGG, Wang Z, Wang Z. Modular membrane valves for universal integration within thermoplastic devices. *Microfluid Nanofluid*. 2016;20:85.

144. Bhattacharya S, Datta A, Berg JM, Gangopadhay S. Studies on surface wettability of poly(dimethyl)siloxane (PDMS) and glass under oxygen-plasma treatment and correlation with bond strength. *J Microelectromech Syst.* 2005;14:590–597.
145. Lee NY, Chung BH. Novel poly(dimethylsiloxane) bonding strategy via room temperature "chemical gluing". *Langmuir.* 2009;25:3861–3866.
146. Tsao CW, DeVoe DL. Bonding of thermoplastic polymer microfluidics. *Microfluid Nanofluid.* 2009;6:1–16.
147. Harrison DJ, Fan Z, Seiler K, Manz A, Widmer HM. Rapid separation of fluorescein derivatives using a micromachined capillary electrophoresis system. *Anal Chim Acta.* 1993;15:361–366.
148. Jia ZJ, Fang Q, Fand ZL. Bonding of glass microfluidic chips at room temperatures. *Anal Chem.* 2004;76:5597–5602.
149. Silvestrini S, Ferraro D, Toth T, et al. Tailoring the wetting properties of thiolene microfluidic materials. *Lab Chip.* 2012;12:4041–4043.
150. Tang L, Lee NY. A facile route for irreversible bonding of plastic-PDMS hybrid microdevices at room temperature. *Lab Chip.* 2010;10:1274–1280.
151. Gajasinghe RWRL, Senveli SU, Rawal S, et al. Experimental study of PDMS bonding to various substrates for monolithic microfluidic applications. *J Micromech Microeng.* 2014;24:075010.
152. Sunkara V, Park DK, Hwang H, Chantiwas R, Soper SA, Cho YK. Simple room temperature bonding of thermoplastics and poly(dimethylsiloxane). *Lab Chip.* 2020; 11:962–965.
153. Wu J, Lee NY. One-step surface modification for irreversible bonding of various plastics with a poly(dimethylsiloxane) elastomer at room temperature. *Lab Chip.* 2014;14:1564–1571.
154. Chen PC, Duong LH. Novel solvent bonding method for thermoplastic microfluidic chips. *Sens Actuators B Chem.* 2016;(556–562).
155. Agostini M, Greco G, Cecchini M. Polydimethylsiloxane (PDMS) irreversible bonding to untreated plastics and metals for microfluidics applications. *APL Mater.* 2019; 7:081108.
156. Pitingolo G, Vecchione R, Falanga AP, Guarnieri D, Netti PA. Fabrication of a modular hybrid chip to mimic endothelial-lined microvessels in flow conditions. *J Micromech Microeng.* 2017;27(3):035014.
157. Grygoryev K, Herzog G, Jackson N, et al. Reversible integration of microfluidic devices with microelectrode arrays for neurobiological applications. *BioNanoScience.* 2014;4:263–275.
158. Xie X, Maharjan S, Liu S, Zhang YS, Livermore C. A modular, reconfigurable microfabricated assembly platform for microfluidic transport and multitype cell culture and drug testing. *Micromachines.* 2020;11:2.
159. Owens CE, John Hart A. High-precision modular microfluidics by micromilling of interlocking injection-molded blocks. *Lab Chip.* 2018;18:890–901.
160. Vera RH, O'Callaghan P, Fatsis-Kavalopoulos N, Kreuger J. Modular microfluidic systems cast from 3D-printed molds for imaging leukocyte adherence to differentially treated endothelial cultures. *Sci Rep.* 2019;9:11321.
161. Geckil H, Xu F, Zhang X, Moon SJ, Demirci U. Engineering hydrogels as extracellular matrix mimics. *Nanomedicine (Lond).* 2010;5:469–484.
162. Frantz C, Stewart KM, Weaver VM. The extracellular matrix at a glance. *J Cell Sci.* 2010;123:4195–4200.
163. Yue B. Biology of the extracellular matrix: an overview. *J Glaucoma.* 2014;23: S20–S23.
164. Kyburz KA, Anseth KS. Synthetic mimics of the extracellular matrix: how simple is complex enough? *Ann Biomed Eng.* 2015;43:489–500.

165. Hoshiba T, Yamaoka T. Extracellular matrix scaffolds for tissue engineering and biological research. In: Royal Society of Chemistry; 2019:1–14. Decellularized Extracellular Matrix: Characterization, Fabrication and Applications;.
166. Han S, Yan JJ, Shin Y, et al. A versatile assay for monitoring in vivo-like transendothelial migration of neutrophils. *Lab Chip*. 2012;12:3861–3865.
167. Astashkina AI, Mann BK, Prestwich GD, Grainger DW. Comparing predictive drug nephrotoxicity biomarkers in kidney 3-D primary organoid culture and immortalized cell lines. *Biomaterials*. 2012;33:4712–4721.
168. Hansen NUB, Genovese F, Leeming DJ, Karsdal MA. The importance of extracellular matrix for cell function and in vivo likeness. *Exp Mol Pathol*. 2015;98:286–294.
169. Nicolas J, Magli S, Rabbachin L, Sampaolesi S, Nicotra F, Russo L. 3D extracellular matrix mimics: fundamental concepts and role of materials chemistry to influence stem cell fate. *Biomacromolecules*. 2020;21:1968–1994.
170. Huang GY, Zhou LH, Zhang QZ, et al. Microfluidic hydrogels for tissue engineering. *Biofabrication*. 2011;3:012001.
171. Khademhosseini A, Langer R. Microengineered hydrogels for tissue engineering. *Biomaterials*. 2007;28:5087–5092.
172. Valdez J, Cook CD, Chopko Ahrens C, et al. On-demand dissolution of modular, synthetic extracellular matrixreveals local epithelial-stromal communication networks. *Biomaterials*. 2017;130:90–103.
173. Hwang CM, Khademhosseini A, Park Y, Sun K, Lee SH. Microfluidic chip-based fabrication of PLGA microfiber scaffolds for tissue engineering. *Langmuir*. 2008;24: 6845–6851.
174. Lu YC, Song W, An D, et al. Designing compartmentalized hydrogel microparticles for cell encapsulation and scalable 3D cell culture. *J Mater Chem B*. 2015;3:353–360.
175. Shin Y, Han S, Jeon JS, et al. Microfluidic assay for simultaneous culture of multiple cell types on surfaces or within hydrogels. *Nat Protoc*. 2012;7:1247–1259.
176. Yeh J, Ling Y, Karp JM, et al. Micromolding of shape-controlled, harvestable cell-laden hydrogels. *Biomaterials*. 2006;27:5391–5398.
177. Chen FM, Liu X. Advancing biomaterials of human origin for tissue engineering. *Prog Polym Sci*. 2016;53:86–168.
178. Place ES, Evans ND, Stevens MM. Complexity in biomaterials for tissue engineering. *Nat Mater*. 2009;8457–470.
179. van Duinen V, Trietsch SJ, Joore J, Vulto P, Hankemeier T. Microfluidic 3D cell culture: from tools to tissue models. *Curr Opin Biotechnol*. 2015;35:118–126.
180. Sutherland RM, Rodger Inch W, McCredie JA. A multi-component radiation survival curve using an in vitro tumor model. *Int J Radiat Biol Relat Stud Phys Chem Med*. 1970;18(5):491–495.
181. Weiswald LB, Bellet D, Dangles-Marie V. Spherical cancer models in tumor biology. *Neoplasia*. 2015;17(1):1–15.
182. Mehta G, Hsiao AY, Ingram M, Luker GD, Takayama S. Opportunities and challenges for use of tumor spheroids as models to test drug delivery and efficacy. *J Control Release*. 2018;164(2):192–204.
183. Moshksayan K, Kashaninejad N, Warkiani ME, et al. Spheroids-on-a-chip: recent advances and design considerations in microfluidic platforms for spheroid formation and culture. *Sens Actuators B Chem*. 2018;263:151–176.
184. Fukuda J, Kohji N. Orderly arrangement of hepatocyte spheroids on a microfabricated chip. Tissue Eng. 11; 7–8(1254–1262).
185. Hsiao AY, Torisawa YS, Tung YC, et al. Microfluidic system for formation of Pc-3 prostate cancer co-culture spheroids. *Biomaterials*. 2009;30(16):3020–3027.
186. Kwapiszewska K, Michalczuk A, Rybka M, Kwapiszewski R, Brzozka Z. A microfluidic-based platform for tumour spheroid culture, monitoring and drug screening. *Lab Chip*. 2014;14(12):2096–2104.

187. Chen Y, Gao D, Liu H, Lin S, Jiang Y. Drug cytotoxicity and signaling pathway analysis with three-dimensional tumor spheroids in a microwell-based microfluidic chip for drug screening. *Anal Chim Acta*. 2015;898:85–92.
188. Pitingolo G, Nizard P, Riaud A, Taly V. Beyond the on/off chip trade-off: a reversibly sealed microfluidic platform for 3D tumor microtissue analysis. *Sens Actuators B Chem*. 2018;274:393–401.
189. Sart S, Tomasi RFX, Amselem G, Baroud CN. Multiscale cytometry and regulation of 3D cell cultures on a chip. *Nat Commun*. 2017;8:469.
190. Katt ME, Placone AL, Wong AD, Xu ZS, Searson PC. In vitro tumor models: advantages, disadvantages, variables, and selecting the right platform. *Front Bioeng Biotechnol*. 2016;4:12.
191. Young EWK. Cells, tissues, and organs on chips: challenges and opportunities for the cancer tumor microenvironment. *Integr Biol*. 2013;5(9):1096–1109.
192. Wan L, Neumann CA, LeDuc PR. Tumor-on-a-chip for integrating a 3D tumor microenvironment: chemical and mechanical factors. *Lab Chip*. 2020;20(5):873–888.
193. Sontheimer-Phelps A, Hassell BA, Ingber DE. Modelling cancer in microfluidic human organs-on-chips. *Nat Rev Cancer*. 2019;19(2):65–81.
194. Zervantonakis IK, Hughes-Alford SK, Charest JL, Condeelis JS, Gertler FB, Kamm RD. Three-dimensional microfluidic model for tumor cell intravasation and endothelial barrier function. *PNAS*. 2012;109(34):13515–13520.
195. Hassell BA. Human organ chip models recapitulate orthotopic lung cancer growth, therapeutic responses, and tumor dormancy in vitro. *Cell Rep*. 2017;21(2):508–516.
196. Nashimoto Y, Okada R, Hanada S, et al. Vascularized cancer on a chip: the effect of perfusion on growth and drug delivery of tumor spheroid. *Biomaterials*. 2020;229:119547.
197. Nguyen M, De Ninno A, Mencattini A, et al. Dissecting effects of anti-cancer drugs and cancer-associated fibroblasts by on-chip reconstitution of immunocompetent tumor microenvironments. *Cell Rep*. 2018;25(13):3884–3893.
198. Nashimoto Y, Hayashi T, Kunita I, et al. Integrating perfusable vascular networks with a three-dimensional tissue in a microfluidic device. *Integr Biol*. 2017;9(6):506–518.
199. Thomson JA, Itskovitz-Eldor J, Shapiro SS, et al. Embryonic stem cell lines derived from human blastocysts. *Science*. 1998;282(5391):1145–1147.
200. Takahashi K, Tanabe K, Ohnuki M, et al. Induction of pluripotent stem cells from adult human fibroblasts by defined factors. *Cell*. 2007;131(5):861–872.
201. Pouton CW, Haynes JM. Embryonic stem cells as a source of models for drug discovery. *Nat Rev Drug Discov*. 2007;6(8):605–616.
202. Rowe RG, Daley GQ. Induced pluripotent stem cells in disease modelling and drug discovery. *Nat Rev Genet*. 2019;20(7):377–388.
203. Jahromi MAM, Abdoli A, Rahmanian M, et al. Microfluidic brain-on-a-chip: perspectives for mimicking neural system disorders. *Mol Neurobiol*. 2019;56(12):8489–8512.
204. Bang S. Brain-on-a-chip: a history of development and future perspective. *Biomicrofluidics*. 2019;13(5):051301.
205. Vatine GD, Barrile R, Workman MJ, et al. Human iPSC-derived blood-brain barrier chips enable disease modeling and personalized medicine applications. *Cell Stem Cell*. 2019;24(6):995–1005.
206. S M, Song YH, McCrary MW, Schmidt CE. Advances in Ex vivo models and lab-on-a-chip devices for neural tissue engineering. *Biomaterials*. 2019;198:146–166.
207. Moreno EL, Hachi S, Hemmer K, et al. Differentiation of neuroepithelial stem cells into functional dopaminergic neurons in 3D microfluidic cell culture. *Lab Chip*. 2015;15(11):2419–2428.
208. Huval RM, Miller OH, Lowry Curly J, Fan Y, Hall BJ, Moore MJ. Microengineered peripheral nerve-on-a-chip for preclinical physiological testing. *Lab Chip*. 2015;15(10):2221–2232.

209. Ahn SI. Microengineered human blood-brain barrier platform for understanding nanoparticle transport mechanisms. *Nat Commun.* 2020;11(1):1–12.
210. Devalla HD, Passier R. Cardiac differentiation of pluripotent stem cells and implications for modeling the heart in health and disease. *Sci Transl Med.* 2018;10:435.
211. Jastrzebska E, Tomecka E, Jesion I. Heart-on-a-chip based on stem cell biology. *Biosens Bioelectron.* 2016;75:67–81.
212. Giobbe GG, Michielin F, Luni C, et al. Functional differentiation of human pluripotent stem cells on a chip. *Nat Methods.* 2015;12(7):637–640.
213. Wan CR, Chung S, Kamm RD. Differentiation of embryonic stem cells into cardiomyocytes in a compliant microfluidic system. *Ann Biomed Eng.* 2011;39(6): 1840–1847.
214. Sokolowska P, Zukowski K, Lasocka I, Szulc-Dabrowska L, Jastrzebska E. Human mesenchymal stem cell (HMSC) differentiation towards cardiac cells using a new microbioanalytical method. *Analyst.* 2020;145:3017–3028.
215. Tang Y, Liu L, Li J, et al. Induction and differentiation of human induced pluripotent stem cells into functional cardiomyocytes on a compartmented monolayer of gelatin nanofibers. *Nanoscale.* 2016;8(30):14530–14540.
216. Marsano A, Conficconi C, Lemme M, et al. Beating heart on a chip: a novel microfluidic platform to generate functional 3D cardiac microtissues. *Lab Chip.* 2016;16 (3):599–610.
217. Si L, Bai H, Rodas M, et al. Human organs-on-chips as tools for repurposing approved drugs as potential influenza and covid19 therapeutics in viral pandemics. *bioRxiv preprint.* 2020. https://doi.org/10.1101/2020.04.13.039917.
218. Clevers H. Covid-19: organoids go viral. *Nat Rev Mol Cell Biol.* 2020;21:355–356.
219. Ainslie GR. Microphysiological lung models to evaluate the safety of new pharmaceutical modalities: a biopharmaceutical perspective. *Lab Chip.* 2019;19(9):3152–3161.
220. Huh D, Matthews BD, Mammoto A, Montoya-Zavala M, Hsin HY, Ingber DE. Reconstituting organ-level lung functions on a chip. *Science.* 2010;328(5986): 1662–1668.
221. Stucki JD, Hobi N, Galimov A, et al. Medium throughput breathing human primary cell alveolus-on-chip model. *Sci Rep.* 2018;8:14359.
222. Stucki AO, Stucki JD, Hall SRR, et al. A lung-on-a-chip array with an integrated bio-inspired respiration mechanism. *Lab Chip.* 2015;15:1302–1310.
223. Jain A, Barrile R, van der Meer AD, et al. Primary human lung alveolus on a chip model of intravascular thrombosis for assessment of therapeutics. *Clin Pharmacol Ther.* 2018;103(2):332–340.
224. Zamprogno P, Wuthrich S, Achenbach S, et al. Second-generation lung-on-a-chip array with a stretchable biological membrane. *bioRxiv preprint.* 2019. https://doi.org/10.1101/608919.
225. Radiom M, He Y, Peng-Wang J, Baeza-Squiban A, Berret JF, Chen Y. Alveolar mimics with periodic strain and its effect on the cell layer formation. *Biotechnol Bioeng.* 2020;117(9):2827–2841.
226. Wang WY, Lin D, Jarman EH, Polacheck WJ, Baker BM. Functional angiogenesis requires microenvironmental cues balancing endothelial cell migration and proliferation. *Lab Chip.* 2020;20(6):1153–1166.
227. Siemann DW. The unique characteristics of tumor vasculature and preclinical evidence for its selective disruption by tumor-vascular disrupting agents. *Cancer Treat Rev.* 2011;37(1):63–74.
228. Teuwen LA, Geldhof V, Pasut A, Carmeliet P. COVID-19: the vasculature unleashed. *Nat Rev Immunol.* 2020;20:389–391.
229. Bogorad MI, DeStefano J, Karlsson J, Wong AD, Gerecht S, Searson PC. Review: in vitro microvessel models. *Lab Chip.* 2015;15(22):4242–4255.

230. Haase K, Kamm RD. Advances in on-chip vascularization. *Regen Med.* 2017;12 (3):285–302.
231. Wang X, Sun Q, Pei J. Microfluidic-based 3D engineered microvascular networks and their applications in vascularized microtumor models. *Micromachines.* 2018; 9(10):493.
232. Fiddes LK, Raz N, Srigunapalan S, et al. A circular cross-section PDMS microfluidics system for replication of cardiovascular flow conditions. *Biomaterials.* 2010;31 (13):3459–3464.
233. Esch MB, Post DJ, Shuler ML, Stokol T. Characterization of in vitro endothelial linings grown within microfluidic channels. *Tissue Eng A.* 2011;17(23–24):2965–2971.
234. Miller JS, Stevens KR, Yang MT, et al. Rapid casting of patterned vascular networks for perfusable engineered three-dimensional tissues. *Nat Mater.* 2012;11(9): 768–774.
235. Kim S, Lee H, Chung M, Jeon NL. Engineering of functional, perfusable 3D microvascular networks on a chip. *Lab Chip.* 2013;13(8):1489–1500.
236. Hsu YH, Moya ML, Hughes CCW, George SC, Lee AP. A microfluidic platform for generating large-scale nearly identical human microphysiological vascularized tissue arrays. *Lab Chip.* 2013;13(15):2990–2998.
237. Beckwitt CH, Clark AM, Wheeler S, et al. Liver 'organ on a chip'. *Exp Cell Res.* 2018;363(1):15–25.
238. Bein A, Shin W, Jalili-Firoozinezhad S, et al. Microfluidic organ-on-a-chip models of human intestine. *Cell Mol Gastroenterol Hepatol.* 2018;5(4):659–668.
239. Du Y, Li N, Luo C, et al. Mimicking liver sinusoidal structures and functions using a 3D-configured microfluidic chip. *Lab Chip.* 2017;17:782–794.
240. Ma LD, Wang YT, Wang JR, et al. Design and fabrication of a liver-on-a-chip platform for convenient, highly efficient, and safe in situ perfusion culture of 3D hepatic spheroids. *Lab Chip.* 2018;18(17):2547–2562.
241. Kim HJ, Ingber DE. Gut-on-a-chip microenvironment induces human intestinal cells to undergo villus differentiation. *Integr Biol.* 2013;5(9):1130–1140.
242. Kim HJ, Collins JJ, Ingber DE. Contributions of microbiome and mechanical deformation to intestinal bacterial overgrowth and inflammation in a human gut-on-a-chip. *PNAS.* 2016;113(1):E7–E15.
243. Kane KIW, Moreno EL, Hachi S, et al. Automated microfluidic cell culture of stem cell derived dopaminergic neurons. *Sci Rep.* 2019;9:1796.
244. Skardal A, Shupe T, Atala A. Organoid on a chip and body on a chip systems for drug screening and disease modeling. *Drug Discov Today.* 2016;21(9):1399–1411.
245. Tomasi RFX, Sart S, Champetier T, Baroud CN. Individual control and quantification of 3D spheroids in a high-density microfluidic droplet array. *Cell Rep.* 2020;31 (8):107670.
246. Carvalho MR, Barata D, Teixeira LM, et al. Colorectal tumor-on-a-chip system: a 3D tool for precision onco-nanomedicine. *Sci Adv.* 2019;5(5):eaaw1317.
247. Shirure V, Bi Y, Curtis MB, et al. Tumor-on-a-chip platform to investigate progression and drug sensitivity in cell lines and patient-derived organoids. *Lab Chip.* 2018;18(23). 3687-2702.
248. Wang G, McCain ML, Yang L, et al. Modeling the mitochondrial cardiomyopathy of Barth syndrome with induced pluripotent stem cell and heart-on-chip technologies. *Nat Med.* 2014;20(6):616–623.
249. Mathur A, Loskill P, Shao K, et al. Human iPSC-based cardiac microphysiological system for drug screening applications. *Sci Rep.* 2015;5:8883.
250. Li Y, Pi QM, Wang PC, et al. Functional human 3D microvascular networks on a chip to study the procoagulant effects of ambient fine particulate matter. *RSC Advances.* 2017;7(88):56108–56116.

251. Kim S, Chung M, Ahn J, Lee S, Jeon NL. Interstitial flow regulates the angiogenic response and phenotype of endothelial cells in a 3D culture model. *Lab Chip*. 2016;16(21):4189–4199.
252. Shang M, Soon RH, Lim CT, Khoo BL, Han J. Microfluidic modelling of the tumor microenvironment for anti-cancer drug development. *Lab Chip*. 2019;19(3):369–386.
253. Ribas J, Sadeghi H, Manbachi A, et al. Cardiovascular organ-on-a-chip platforms for drug discovery and development. *Appl Vitro Toxicol*. 2016;2(2):82–96.
254. Zhao Y, Rafatian N, Wang EY, et al. Towards chamber specific heart-on-a-chip for drug testing applications. *Adv Drug Deliv Rev*. 2020;165-166:60–76. https://doi.org/10.1016/j.addr.2019.12.002.
255. Sakamiya M, Fang Y, Mo X, Shen J, Zhang T. A heart-on-a-chip platform for online monitoring of contractile behavior via digital image processing and piezoelectric sensing technique. *Med Eng Phys*. 2020;75:36–44.
256. Agarwal A, Goss JA, Cho A, McCain ML, Parker KK. Microfluidic heart on a chip for higher throughput pharmacological studies. *Lab Chip*. 2013;13(18):3599–3608.
257. Claesson-Welsh L. Vascular permeability–the essentials. *Ups J Med Sci*. 2015;120(3):135–143.
258. Wang HF, Ran R, Liu Y, et al. Tumor-vasculature-on-a-chip for investigating nanoparticle extravasation and tumor accumulation. *ACS Nano*. 2018;12(11):11600–11609.
259. Vu MN, Rajasekhar P, Poole DP, et al. Rapid assessment of nanoparticle extravasation in a microfluidic tumor model. *ACS Appl Nano Mater*. 2019;2(4):1844–1856.
260. Jalili-Firoozinezhad S, Prantil-Baun R, Jiang A, et al. Modeling radiation injury-induced cell death and countermeasure drug responses in a human gut-on-a-chip. *Cell Death Dis*. 2018;9(2):1–14.

Further reading

261. Baudoin R, Griscom L, Prot JM, Legallais C, Leclerc E. Behavior of HepG2/C3A cell cultures in a microfluidic bioreactor. *Biochem Eng J*. 2011;53(2):172–181.
262. Qi Z, Xu L, Xu Y, et al. Disposable silicon-glass microfluidic devices: precise, robust and cheap. *Lab Chip*. 2018;18:3872–3880.
263. Weibel DB, DiLuzio WR, Whitesides GM. Microfabrication meets microbiology. *Nat Rev Microbiol*. 2007;5(3):209–218.
264. Peng L, Wu H, Shu Y, Yi P, Deng Y, Lai X. Roll-to-roll hot embossing system with shape preserving mechanism for the large-area fabrication of microstructures. *Rev Sci Instrum*. 2016;87:105120.

CHAPTER SEVEN

Cultivating human tissues and organs over lab-on-a-chip models: Recent progress and applications

Stuti Bhagat[a,b] **and Sanjay Singh**[a,b,*]

[a]Nanomaterials and Toxicology Laboratory, Biological and Life Sciences, School of Arts and Sciences, Ahmedabad University, Ahmedabad, Gujarat, India
[b]DBT-National Institute of Animal Biotechnology (DBT-NIAB), Hyderabad, Telangana, India
*Corresponding author: e-mail addresses: sanjay.singh@ahduni.edu.in; sanjay@niab.org.in

Contents

Progress in Molecular Biology and Translational Science, Volume 187
ISSN 1877-1173
https://doi.org/10.1016/bs.pmbts.2021.07.023
Copyright © 2022 Elsevier Inc.
All rights reserved.

Abstract

In vivo models are indispensable for preclinical studies for various human disease modeling and drug screening, however, face several obstacles such as animal model species differences and ethical clearance. Additionally, it is difficult to accurately predict the organ interaction, drug efficacy, and toxicity using conventional *in vitro* two-dimensional (2D) cell culture models. The microfluidic-based systems provide excellent opportunity to recapitulate the human organ/tissue functions under *in vitro* conditions. The organ/tissue-on-chip models are one of best emerging technologies that offer functional organs/tissues on a microfluidic chip. This technology has potential to noninvasively study the organ physiology, tissue development, and diseases etymology. This chapter comprises the benifits of 2D and three-dimensional (3D) *in vitro* cultures as well as highlights the importance of microfluidic-based lab-on-a-chip technique. The development of different organs/tissues-on-chip models and their biomedical application in various diseases such as cardiovascular diseases, neurodegenerative diseases, respiratory-based diseases, cancers, liver and kidney diseases, etc., have also been discussed.

1. Introduction

The most relevant method to study the physiology of human organs or to understand the pharmacokinetics and pharmacodynamics of a novel drug formulation is through the *in vivo* experimental models. Being a low-throughput testing, *in vitro* systems involving cell culture models are preferred. For decades, the conventional 2D cell culture techniques are being used to investigate the functions of tissues, new drug discovery, identification of biomarkers for diseases, etc. However, 2D experimental systems have failed to accurately estimate the physiological condition of the living organ/tissue/system. Cell culture matrix used in 3D cell cultures provides biologically relevant functions and give rise an opportunity to study a biomimic organ/tissue-on-chip. Also, 3D culture models are more physiologically relevant and predictive than 2D cultures. The 3D cell culture technique helps to mimic *in vivo* microenvironment to understand the cell-cell interaction, different molecular functions, drug resistance, and drug penetration.[1] Microfluidic system is one of the best technologies that are being tested to mimic the architectures of biological system. Microfluidic is scientific technology, which precisely process the microscale fluids (10^{-9} to 10^{-18} L) using microchannels of diameter ranging from hundreds to micron size. Also, microchannels have large surface area and high mass transfer, which favors its usage in microfluidic chips. The major goal of organ/tissue-on-chip is to recapitulate the microenvironment of organ and regulate the key parameters such as shear force, oxygen, CO_2, temperature, osmolality, cell patterning, tissue-boundaries, and organ-tissue interactions, etc. It also provides continuous nutrient fluid flow and separate organ compartment to study the interface of the tissue.[2]

The microfluidic-based lab-on-chip models are suitable for the development of *in vitro* engineered tissue mimics. The tissue-on-chip devices are designed to mimic the functions of the organs of the body and react in physiological manner when exposed to hormones, drugs, and cell signaling molecules.[3] Over the past few decades, microfluidics-based tissue-on-chip research has seen tremendous growth leading to develop organs-on-chip models. Organ-on-chip models comprise multiple organ specific cell types, which are arranged to resemble the organ-specific morphology and function. Most organ-on-chip models use porous membranes as cell substrate to define cell layers in which cells can arrange and even allow crosstalk between various cell types such as endothelial and epithelial cells.[4] There have been several reports on developing organ-on-chip-based study models for various human organs including liver, kidney, lung, heart, intestine, brain, etc.[5] Such models also allow to integrate the circulation of immune cells and monitor their interaction with other tissues/organs, therefore, used as excellent disease models. Bottom-up approach (stem cell-based) is majorly used to build a disease-on-chip model to study pathophysiology, differentiation, cell-cell interaction, and complexity of disease, etc.[6] Various disease-based models have been developed to investigate the mechanism, and screening of drugs or vaccines. For example, cancer-on-chip-based system of different organs/tissues has enabled us to understand the role of different cell types in cancer progression, and monitoring of treatment efficacy.[7]

2. Traditional cell culture techniques

Organ/tissue specific cell culture models has been the preferred method to understand the mechanism, formation, function, pathology of tissue, and screening of treatment agents. Based on the growth conditions, the cell lines could be broadly of two types: (1) adherent cells—these cells are attached to the bottom surface of a plastic or glass dish coated with sterile membranes. HeLa (Henrietta Lacks, cervical adinocarcinoma) cells, NIH/3T3 (mouse embryo) cells, and WRL-68 (normal human liver) cells (Fig. 1A) are some of the examples of adherent cell lines. (2) non-adherent cells—these cells prefer growing in suspension form in the culture medium.

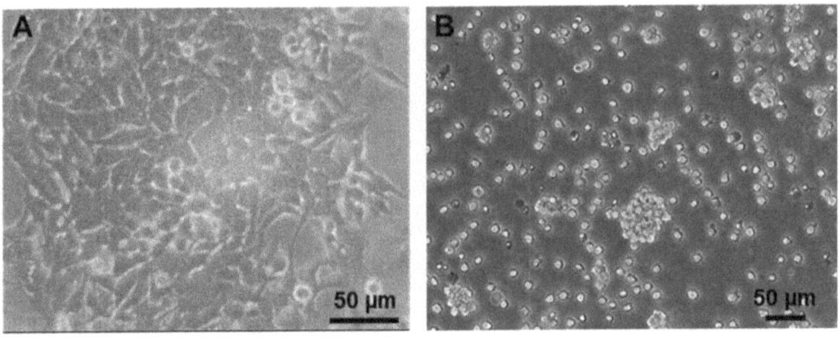

Fig. 1 Microscopic image showing the typical morphology of *WRL-68 [adherent (A)] and Jurkat [non-adherent cell line (B)].*

Jurkat (pseudodiploid human) cells and CHO (Chinese hamster ovary) (Fig. 1B) cells are the examples of suspension cell lines. 2D cell cultures are most commonly used techniques but recently 3D cell culture has also gaining attention because of possessing several advantages over 2D culture.

2.1 2D cell culture

Since 1900, 2D cell culture techniques are being used for various applications. The *in vitro* 2D cells grow on flat dishes/flasks, which typically made of plastic coating of synthetic polymers such as polystyrene, poly-D-lysine, collagen, etc. The cells for 2D cultures are directly obtained from the animal or plant tissues (primary culture).[8] For example, HeLa cell line was established as immortal cell line from a cervical adenocarcinoma tumor of young woman named Henrietta Lacks. This cell line was also used for developing vaccines for Polio and acquired immunodeficiency syndrome (AIDS).[9] HeLa cells also showed inter and intra-species contamination, which further helped to overgrow other cell lines.[10]

2.1.1 Advantages

This cell culture technique is simple and low cost. It is easy to maintain the cells in cell culture medium. The 2D cell line can be immortalized thus are able to regrow for multiple times. The results from this technique is reproducible and easy to interpret due to simplicity of culture protocol.[8] This technique also helps to study the structure and morphology, and division of cells. The 2D cell culture technique also used for the determination of the mechanism of the physiology and biochemistry of cells that further helps to understand the organ/tissue functions. For example, primary cells resemble to the parental cells from organ/tissue; thus, it is easy to study cells' physiology, molecular biology, and biochemistry.

2.1.2 Disadvantages

2D cell culture does not mimic the actual structure and functions of biological tissues or tumors. Generally, these cells are isolated from the organs/tissues/tumors, which causes altered morphology as well as physiology. Therefore, these cells lack the typical phenotype and functions including cell-cell interaction, and extracellular matrix (ECM)-cell interactions (cell differentiation, cell proliferation, cell response to a stimuli, drug metabolism, etc.), gene and protein expression.[1] Frequently, the isolated cells lose their polarity leading to altered responses of different phenomena such as apoptosis, proliferation, etc. Additionally, monolayer cells have unlimited access to different ingredients such as oxygen, metabolites, and nutrients, etc., however, do not provide microenvironment to the cells.[1]

2.2 3D cell culture

The first 3D culture was carried out in soft agar solution, in 1970s. Since then, various new techniques have been found and well researched. According to the method of preparation, it can be divided into (1) Suspension culture on non-adherent plate, (2) Culture in gel-like substance or in concentrated medium, and, (3) Cultures on scaffolds. Table 1 contains comparison of these techniques with associated advantages and disadvantages.

2.2.1 Importance of organ/tissue-on-chip technique

Organ/tissue-on-chip technique offers various advantages over conventional 2D and 3D cell culture methods. (1) It provides fluid flow, relevant mechanical cues, tissue-organ interface and separate access to vascular and parenchymal compartments,[5] (2) It also offers constant contact to apical lumen of the vascular and epithelial channels and recreates the *in vivo* microenvironment,[15] (3) Supports long term co-culture of gut microbes,[15] (4) The continues fluid flow allows to recruit the immune cells in the model,[15] (5) Allows to investigate the same location at the tissue interface every day,[16] (6) Multi-electrode arrays analyzing cell electrical potentials, and electrodes for measuring transepithelial electrical resistance (used for evaluation of the differentiation of epithelial monolayer cells in conventional static cultures) can be integrated in microfluidic chips.[5]

Table 1 List of different 3D cell culture techniques and their advantages and disadvantages.

3D cell culture techniques	Description	Advantages	Disadvantages	Applications	References
Suspension culture on non-adherent plate	▪ Single or co-cultured cells can be seeded on the non-adherent plates with appropriate medium ▪ The polystyrene plates incubated with non-ionic surfactant such as Pluronic F-68, Pluronic F127 to make surface non-adherent ▪ Alternate day media change is required ▪ The spheroids formation can be seen in 5–6 days	▪ Simple, and easy culture ▪ Easy to extract spheroids from plates	▪ This technique is not compatible with all types of cell lines ▪ Also, cell aggregation is very frequent in this technique	▪ Preparation of single or co-cultured cancer spheroids ▪ Tumor spheroid models used to study the pharmacokinetics of drug ▪ To study cell-cell interaction	1,8,11,12
Culture in gel-like substance	▪ Agarose used as to coat the top and bottom layer of the plate ▪ Cells are flooded in Matrigel with appropriate media and allowed to settle ▪ Matrigel is consist of collagen, perlecan, laminin, and nidogen ▪ Media change is required in every alternate day ▪ Matrigel is removed after 1–2 weeks to obtain spheroids	▪ Cells in Matrigel have 3D interaction thus, it behaves like tissue ▪ Used to study metastasis of cancer cells ▪ Matrigel can be easily recovered for analysis	▪ Difficult to obtain spheroids for some cell lines ▪ Plate preparation is tedious, two-layer agarose gel required long time for culture ▪ For certain experiment (immunohistochemistry) it is difficult to extract spheroids embedded in gel ▪ The endogenous bioactive material in Matrigel, influence structure formation	▪ To obtain single and co-culture spheroids ▪ For the development of cell-based biosensors ▪ In drug discovery application ▪ To monitor cell-cell interaction, and signaling pathways	1,13

Culture on scaffold	▪ Seed the cells in scaffold with an appropriate medium ▪ The cells can migrate and attach to scaffold (made of collagen, silk, alginate,) to grow and divide ▪ After few days dissolve the scaffold using dissociation solution to obtain spheroids ▪ Further, neutralizing solution will be used to neutralize the dissociation solution ▪ 3D structure can be prepared within a week	▪ System is compatible with commercially available kit used for immunofluorescence staining, DNA/RNA/Protein extraction ▪ Multiple cell lines can be used to develop organ/tissue-like structure ▪ Diseases modeling can be possible with this method	▪ Materials used in scaffold/chip may interfere with the cell adhesion, growth and behavior ▪ Cell observation and extraction for some experiments is restricted	▪ Majorly used in stem cells and its differentiation research ▪ For the development of multiorgan-on-chip models ▪ In drug discovery application ▪ To monitor cell-cell interaction, and signaling pathways	1,14

3. Emerging organ/tissue-on-chip techniques

In microfluidic-based organ/tissue-on-chip, living cells are cultured in continuously perfused and micrometer sized chambers to recapitulate the physiological conditions of organs and tissues. The simplest system contains single type of cell, which is cultured in perfused microfluidic chamber to exhibit tissue-like functions. In a more complex system, two or more microchannels, multiple cell lines, and porous membranes are used to recreate the interface between two or more tissues. The microfluidic systems incorporate physical force to maintain fluid shear stress, mechanical compression, and cyclic strain. Various techniques are available for fabrication of microfluidic devices such as reactive ion etching, plasma etching, wet etching, soft lithography, photolithography, laser ablation, hot embossing, injection molding, and 3D printing, etc.[17] Among these, soft lithography is the most common technique used for the fabrication of lab-on-chip device. This technique encompasses replica molding using elastomeric materials [poly-dimethysiloxane (PDMS), a thermostat polyester]. The PDMS has excellent optical transparency and thus suitable for different cellular/biological applications. The microchannels of the device are fabricated using patterned silicon master/template consisting of photo resistance material such as SU-8. In this method, silicon wafer is coated with SU-8 and exposed to UV light through mask with desired design of channels. Further, PDMS is poured on silicon wafer containing photomask (at 80 °C for $\sim$2 h) to create PDMS replica. The PDMS replica can be sealed with different materials such as PDMS and glass, etc., using oxygen plasma treatment[18] (Fig. 2). Using this method, multiple layers of thin membrane patterned PDMS can be stacked together to create a multilayered 3D microfluidic channel.[19] Another most recent technique is 3D printing of functional elements of microfluidic chip. There are four different approaches used for the fabrication of microfluidic devices using 3D printing.[20] (1) *Direct printing approach*—it is also called one-step fabrication approach in which microfluidic device is printed by incorporating valves, inlets, outlets, microchannels, etc. (2) *Mold-based approach*—it is indirect fabrication method in which microfluidic device is fabricated by 3D printing approach. Subsequently, replica molding technique (PDMS stamp or glass) is used for the bonding of the device. (3) *Hybrid method*—PDMS and other thermoplastic polymers are used together to overcome the limitations of low optical transparency and gas impermeability. (4) *Modular approach*—it is also known

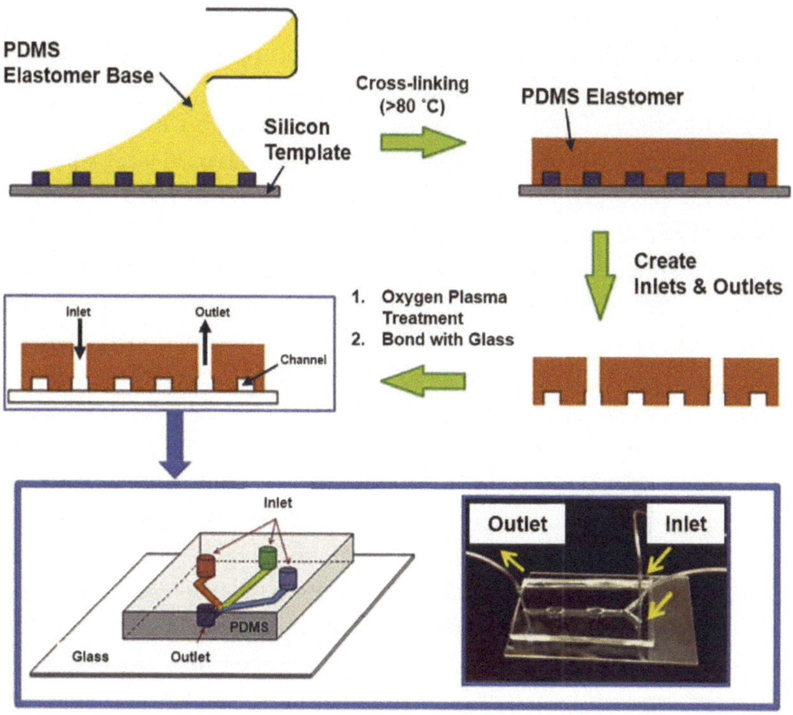

Fig. 2 Fabrication of microfluidic device using soft lithography method. The bottom image illustrates the example of microfluidic device. *Reprinted with permission from Cha C, Piraino F, Khademhosseini A. Chapter 9—Microfabrication technology in tissue engineering. In: Blitterswijk CAV, De Boer J, eds.* Tissue Engineering. *2nd ed. Academic Press; 2014:283–310 Copyright © Elsevier.*

as Lego®-like structure. The different parts of microfluidic device can be assembled like Lego® model to fabricate the device.[20] Several microfluidic chips are fabricated to recapitulate the functions of different organs/tissues such as kidney, liver, intestine, brain, lung, etc.[19]

3.1 Liver-on-chip

Liver is the major compact organ in human body with high metabolic activity performing over 500 roles. The liver tissues are highly regenerative, however, prone to viral infections and chronic diseases including fibrosis, cirrhosis, chronic hepatitis, and hepatocellular carcinoma, etc. Therefore, in order to understand the physiology and pathology of this organ, several strategies are developed to reconstruct human liver under *in vitro* conditions. Ortega-Prieto et al.[21] used collagen-coated polystyrene scaffolds for the development of hepatic microtissues by using mouse fibroblast cell line (NIH/3T3-J2) and primary human hepatocytes (PHH) in 2D and 3D cell culture system. In conventional 2D cell culture system, the co-culture of PHH and NIH/3T3-J2 cells showed dedifferentiation over 10–13 days. For 3D culture, LiverChip (CNBio Innovations) was used with 1 μL/s flow rate for the recirculation of culture media and oxygen. After 2 weeks, 3D culture of hepatic microtissues was produced with ~10-fold higher levels of albumin (Hepatocyte-specific marker). It was also found that albumin and lactate dehydrogenase levels remain stable for over 40 days. Further, the developed 3D hepatic microtissues were infected with patient-derived hepatitis B virus (HBV). This 3D system was able to recreate all the stages of HBV life cycle. Another microfluidic liver-on-chip was developed by combining four major types of hepatic cells to replicate the hepatic functions, microenvironment and liver sinusoids.[22] Additionally, liver-on-chip microfluidic-based models are feasible to create liver models, which are closer to the actual physiological environment *in vivo*.[23] Liver plays major role in drug metabolism; therefore, it is important to predict the accurate toxicity and drug metabolism during drug discovery process. In this context, various liver-on-chip models were developed to test the drug toxicity.[24,25] Also, liver-on-chip-based models can be used for first pass drug metabolism and pharmacokinetic research.[26] For example, Choe et al.[27] developed a microfluidics-based gut-liver-on-chip model to study the first pass drug metabolism of flavonoids and apigenin. This microfluidic chip contained two different chambers: (1) gut chamber containing gut epithelial cells (Caco-2) and (2) liver chamber containing liver cancer cells (HepG2). In this chip, drug sequentially absorbed in the gut chamber and then metabolic reaction occurred in the liver chamber. This model was close to the reported *in vivo* metabolic profiles of gut.

3.2 Kidney-on-chip

Kidney is the most complex and sophisticated organ responsible for metabolism and excretion *in vivo*. Kidney-on-chip based *in vitro* or *ex vivo* models are emerging to provide the new insights and study the mechanism and progression of various kidney diseases. Several of these strategies are utilized for the screening of drugs for kidney-related diseases. Zhou et al.[28] developed a model ofhypertensive glomerulopathy where mouse podocytes and immortalized glomerular endothelial cells were lined in close proximity to create layers of these cells. The changes in fluid flow rate affected the arrangement of kidney cytoskeleton which further resulted in glomerular leakage and cellular damage. The simplest kidney-on-chip model was first developed in 2013, where Madin-Darby canine kidney epithelial cells and human kidney-2 cells were adhered at bottom of the microchannel and loaded with physiological shear stress.[29,30] In this model, authors reported the elevated expression of Na^+/K^+ ATPase and formation of cilia on the cells. These finding suggested that kidney-on-chip can be developed by maintaining shear stress load. Further, Sateesh et al.[31] designed kidney-on-chip model to study the fluid shear stress that is responsible for reabsorption of required solutes, blood homeostasis and rearrangement of cytoskeletons. They proposed that 0.1 Pa of pressure needed at a cell site, however, increase in temperature (350 K) could lead to decrease in pressure requirement. Additionally, Yin et al.[32] developed three layered microfluidic-based kidney-on-chip model for drug screening. This model contained two parts, (1) drug concentration gradient generator, and (2) a flow and temperature-controller. In this model, peritubular capillary endothelial cells and renal proximal tubular epithelial cells were used to assess the drug toxicity in nephrocytes. Fig. 3 shows the hardware diagram from top (Fig. 3A) and side view (Fig. 3B) of the developed kidney model. This microfluidic-based culture platform provided an automated process to control the temperature and flow rate of culture media.

3.3 Heart-on-chip

In vitro models of cardiovascular cells only create monolayer tissue under static condition. The monolayer tissues have flat orientation and no flow condition, which does not mimic the *in vivo* physiological conditions. Under *in vivo* condition, cardiac tissues are under constant physical and mechanical stimulation where shear stress and blood flow play major role in defining the cardiac cell structure, alignment, and phenotype. Heart-on-chip models can be used for regeneration of heart tissues, drug screening, and characterization of different cardiac diseases.[33] Additionally, heart-on-chip devices built with electrodes, serve two important purposes: (1) Stimulate the cardiac cells for the tissue development and regeneration; (2) Used as a readout tool to measure electric response of cells for a particular simulation.[33] Kamei et al.[34] have developed an integrated microfluidic heart/cancer-on-chip device by using healthy human cardiac and cancer cells. Structurally, the device contained three sets of artificial blood circulation loop, pneumatic valve, and peristaltic micro-pump for the controlled and precise fluid flow. The developed

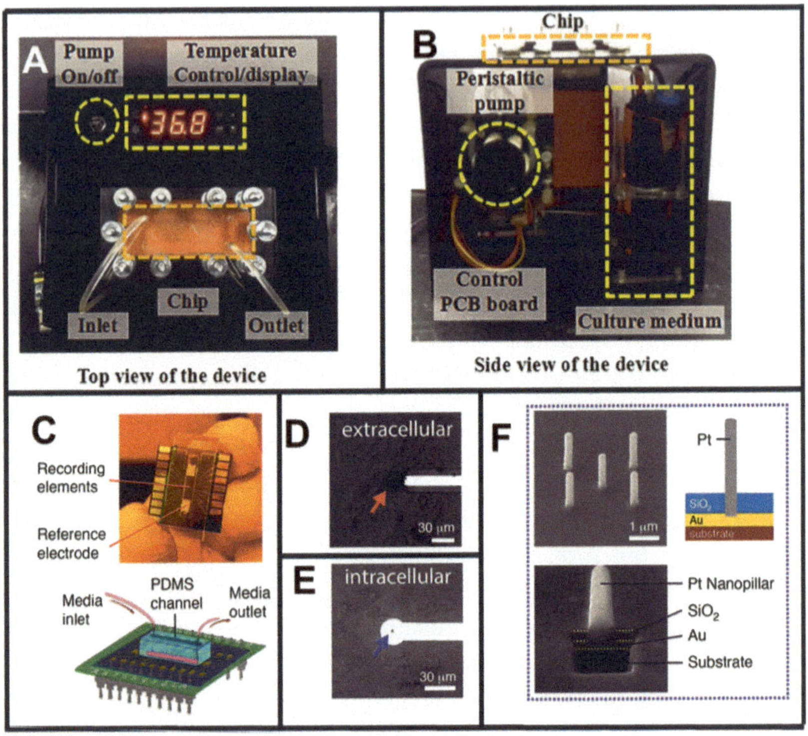

Fig. 3 Hardware illustration of the top view (A) and side view (B) of the kidney-on-chip platform to control fluid flow and temperature. (C) Hardware illustration of the fully assembled heart-on-chip containing reference electrode, and PDMS channel for fluid flow. Extracellular (D) and intercellular (E) recording element coated with Pt black (red arrow) and Au pad with five vertical Pt nanopillars (blue arrow), respectively. (F) SEM image of vertical Pt nanopillars. Inset shows a cross section of single nanopillar. Bottom SEM image shows a cross section of a single nanopillar after etching in SiO_2, Au pad and substrate. *Panels (A and B): Reprinted with permission from Yin L, Du G, Zhang B, et al. Efficient drug screening and nephrotoxicity assessment on co-culture microfluidic kidney chip.* Sci Rep. *2020;10(1):6568. doi:10.1038/s41598-020-63096-3 Copyright © Springer Nature. Panel (C–F): Reprinted with permission from Liu H, Bolonduro OA, Hu N, et al. Heart-on-a-chip model with integrated extra- and intracellular bioelectronics for monitoring cardiac electrophysiology under acute hypoxia.* Nano Lett. *2020;20(4):2585–2593. doi:10.1021/acs.nanolett.0c00076 Copyright © American Chemical Society.*

heart/cancer-on-chip device was used for the screening of doxorubicin drug that showed toxic effects on normal cardiac cells. Recently, Liu et al.[35] developed a heart-on-chip nanodevice, which could mimic the hypoxial condition during heart attack (blocked arteries in the heart). This device was made up of multiple arrays of electronic sensors placed outside and inside of the cardiac cells to detect the rise and fall of voltage through individual cell membranes

(Fig. 3C–E). Also, voltage waves were moved across the cell layers, which recreate the beat in harmony in the chip, just like heart. The intracellular bioelectronics was made up with Pt-nanopillers, which had a base of Au and SiO_2 (Fig. 3F).

3.4 Gut-on-chip

Gut accomplishes major body function, specially transportation and absorption of nutrients and drugs and also plays essential role in immune and endocrine system. Animal models have been primarily used for the study of gut; however, these models are unable to represent human physiology. Gut-on-chip device offers an innovative *in vitro* platform to study human gut pathology, physiology and pharmacology.[5] Gut-on-chip models provide unique opportunity to understand the host-bacterial interactions under *ex vivo* experimental conditions.[36] For example, Shah et al.[37] fabricated a microfluidic chip device using *in vitro* model to study the interaction of human gastrointestinal system with microbiome. In this model, human intestinal epithelial cells were co-cultured with a commensal bacterial species *Lactobacillus rhamnosus* under anaerobic conditions. This model facilitated the investigation of host-microbe molecular interactions, which was further used to answer the questions related to the human health and diseases linked with gastrointestine-microbe interaction. Dysregulation of microbes in gut can affect the mucosal immune response, which causes diseases such as inflammatory bowel disease, chronic inflammation, and infection, etc.[38] To understand the mechanism of interaction between mucosal immune response and gut, Mosig and co-workers developed a microfluidic chip-based model, which resembled organotypic microanatomical structure of human intestine with innate immune cells (Mucosal microphage and dendritic cells).[39] Fig. 4A shows the model of the intestine-on-chip containing porous membrane, which is suspended in a biochip. Microchannels are combined into the biochip and separately perfuse both endothelial and epithelial layers at 50 μL/min. Within 7 days, the intestinal cells self-organized in 3D cell layer with columnar epithelium and increased microvilli formation (Fig. 4B). The endothelial (Fig. 4C and D) and epithelial cells (Fig. 4E–M) of this chip was easily differentiated with specific biomarkers.

3.5 Neuron-on-chip

The nerve cells (neuron) are the basic unit of human central and peripheral nervous system. These cells do not undergo cell division; therefore, shocking injuries and degenerative diseases of central nervous system arise due to

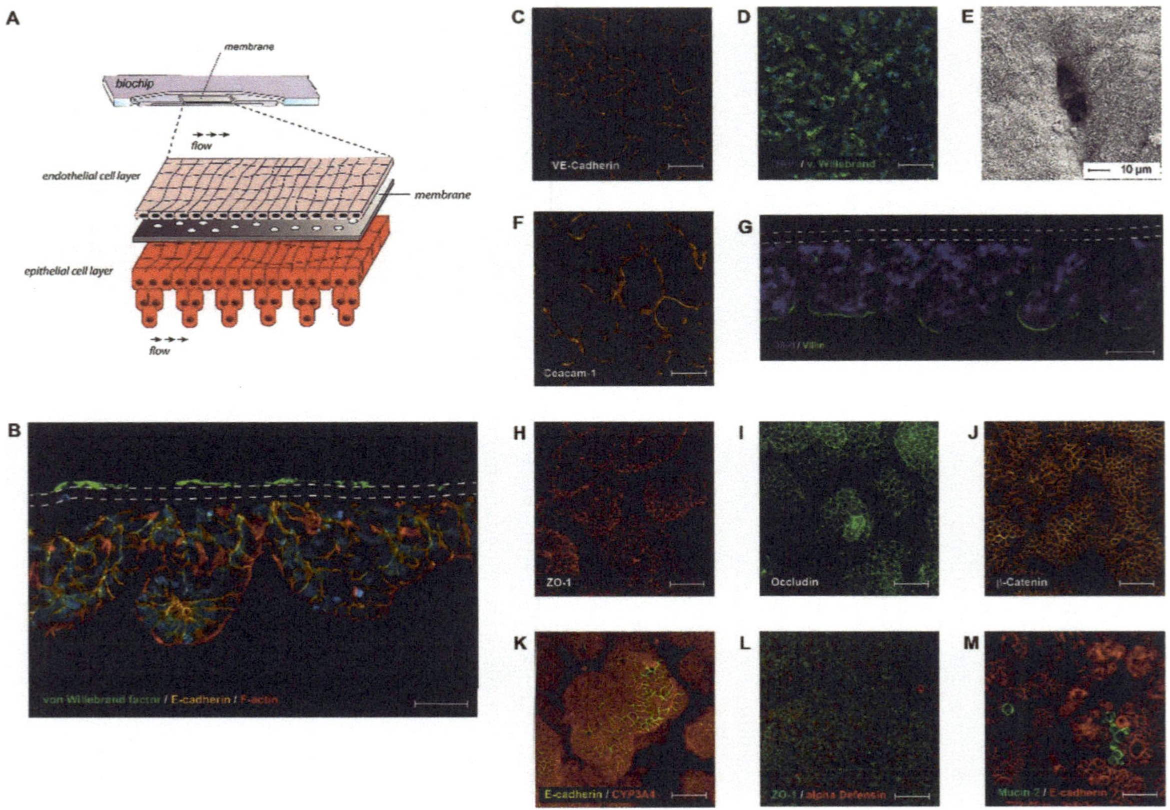

Fig. 4 See figure legend on opposite page.

their damage. In *in vitro* models, neuronal cells are generated by neural stem/progenitor cells (NSPCs), which differentiate into desired neural cell types. However, traditional 2D cell culture methods are unable to create central nervous system mimicking the required microenvironment. Microfluidic-based system has created new opportunities to develop the complex neuronal tissues and their immediate microenvironment to effectively study different neurodegenerative diseases. Kilic et al.[40] developed a brain-on-chip model to recapitulate the *in vivo* conditions of neural differentiation of human pluripotent cells (NTera 2). The chip was made up of multi-layer silicone elastomer and all layers were coated with PDMS membrane. Within 4 weeks, these cells differentiated into neuronal cluster with astrocytes mimicking as brain parenchyma. This microfluidic-based model facilitated easy study of various neurotoxicity and drug discovery studies. Kamudzandu et al.[41] have also developed a microfluidic-based *in vitro* complex brain circuit that was involved in neurodegenerative diseases. The microchannels were populated by specifically cultured basal ganglia neuronal subtype cells to represent the parts of inter-communicating five-population neural circuit. It was observed that the electrical activity of the circuitry was comparatively similar to *in vivo* basal circuitry. The functional connectivity of the neuronal cells in this circuit was immature. Therefore, more mature neuronal circuit is needed to be developed for long term functional measurement.

Fig. 4 Design of intestine-on-chip model. (A) Intestine-on-chip model was composed of multi-layered scaffold containing endothelial and epithelial cells. (B) Cross-section of the 3D intestinal model: endothelial and epithelial cells express von Willebrand factor (green) and E-cadherin (orange), respectively. The cell layers are separated by a porous membrane (dashed line). F-actin filaments and Nuclei are stained with phalloidin (red) and DAPI (blue), respectively. Scale bar 100 μm. (C and D) Endothelial cells form a confluent monolayer and express (C) VE-cadherin (orange) and (D) von Willebrand factor (green). (E–M) Epithelial cell layer: (E) SEM image of the crypt-like structures. Expression of (F) CEACAM-1 (orange); (G) villin (green) (DAP blue, dashed lines marks membrane); (H) ZO-1 (red); (I) occluding (green); (J) β-catenin (orange); (K) E-cadherin (orange); CYP3A4 (red); (L) α-defensin (red); ZO-1 (green); (M) mucin-2 (green); E-cadherin (red); (C and D, F–J) Scale bar 50 μm. Nuclei were stained with DAPI (blue). *Reprinted with permission from Maurer M, Gresnigt MS, Last A, et al. A three-dimensional immunocompetent intestine-on-chip model as in vitro platform for functional and microbial interaction studies.* Biomaterials. *2019;220:119396. doi:10.1016/j.biomaterials.2019.119396 Copyright © Elsevier.*

3.6 Lung-on-chip

Since the respiratory tract is the major entry port of the body, several disease-causing microorganisms also take entry through this route and cause infections. The gas exchange mechanism in lungs is regulated by alveoli (offer thin mucosal layer and vast absorptive surface area), which has also been used for the drug administration purposes. *In vitro* systems are unable to reproduce the complexity of alveoli, however, microfluidics-based systems have shown ability to create artificial extracorporeal lung model with accurate fluid flow and continuous gas exchange. Xu et al.[42] developed a microfluidic-based lung-on-chip device to assess the pulmonary risk of fine particulate matter (PM) (<2.5 μm) exposure. This model replicated the features of alveolar-blood barrier, which helped to estimate the exogenous toxicity. The lung-on-chip model was developed using co-culture of human alveolar epithelial cells and human pulmonary vascular endothelial cells. The microfluidic system was made up of three parallel channels (Fig. 5A–C). The Matrigel was infused in middle channel to impart a niche to the endothelial and epithelial cells. After 3 days, the channel layers resembled to the functions of alveolar-capillary barrier (Fig. 5D and E). The particulate matter was applied to the lung epithelial cells containing channels to model the pulmonary exposure of the matter. It was observed that 200 μg/mL concentration of PM (>2.5 μm) caused significant increase in reactive oxygen species (ROS) generation, inflammation and apoptosis in lung endothelial and epithelial cells. Recent studies have shown that more complex lung-on-chip models could be developed that include microvasculature-network or interstitial fibroblast. Mejías et al.[43] developed unique lung-on-chip model which resembled human lung interstitium-airway interface using 3D vascular network. Here, the normal and diseased fibroblasts were cultured within fibrin collagen hydrogel under an airlifted airway epithelium in three-compartment device resembling the cystic fibrosis.

3.7 Body-on-chip

The body-on-chip is also referred as multiorgan-on-chip or human-on-chip. Since, single organ-on-chip has been not very successful in representing total functional changes, complexity, and integrity of various organs of the body[44] multiorgan construction on-a-chip has been currently attracting the researcher's attention worldwide. The multiorgan-chip contains cells of different organs and tissues that are connected by blood vessel

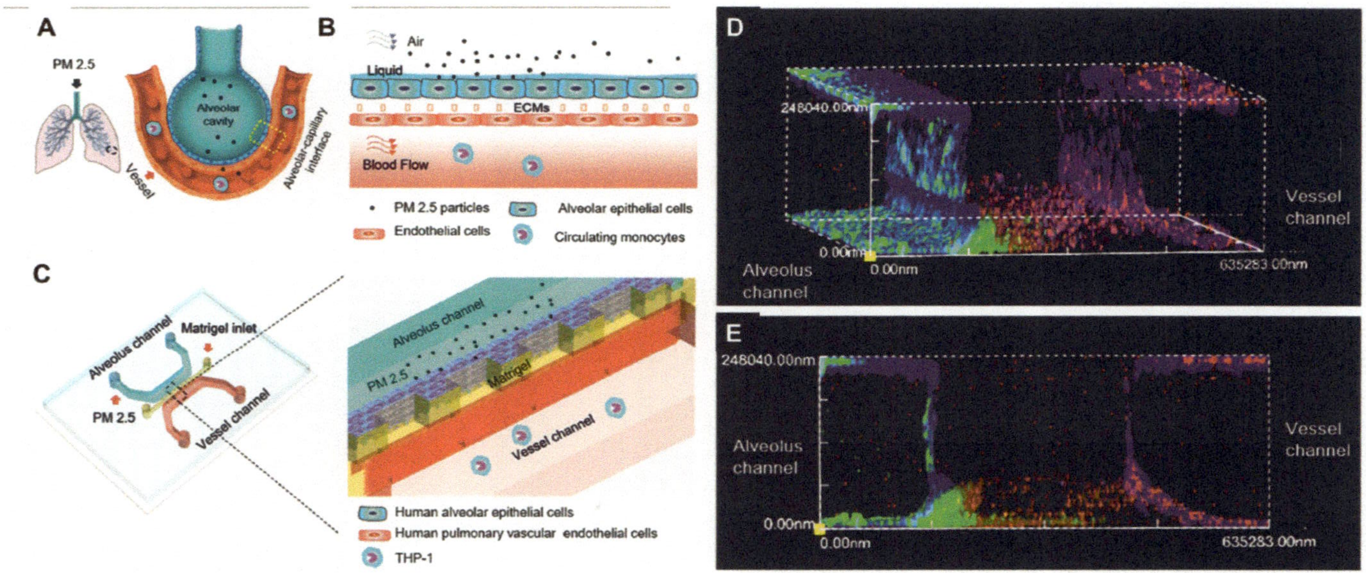

Fig. 5 Schematic shows the PM2.5 exposure on the *in vivo* (A) and lung-on-a-chip model. (B) >PM2.5 deposition on the alveolar-capillary barrier *in vivo*. (C) Lung-on-a-chip model exposed with PM2.5. Confocal image of over-all (D) and side view (E) of lung-on-chip model with alveolar-capillary barrier, where, E-cadherin (HPAEpiCs) and VE-cadherin (HUVECs) were stained with green and red color, respectively. Nuclei was stained with DAPI (blue). *Reprinted with permission from Xu C, Zhang M, Chen W, Jiang L, Chen C, Qin J. Assessment of air pollutant PM2.5 pulmonary exposure using a 3D lung-on-chip model.* ACS Biomater Sci Eng. *2020;6(5):3081–3090. doi:10.1021/acsbiomaterials.0c00221 Copyright © American Chemical Society.*

channels to achieve the integration and interaction between the established organ systems.[45] The channels can be separated into static, semi-static and flexible patterns.[46] In static method, organs are connected by single device, whereas in semi-static system multiple organs are joined by fluidic network, and in flexible system single organ is connected with flexible microchannels. Researchers have made few breakthroughs by developing 2 organs,[26] 3 organs,[47] 4 organs,[48] and up to 10 organs-on-a-chip.[49] Oleaga et al.[50] have developed two organ-on-chip (heart and liver) device which contained cardiomyocytes and primary hepatocytes. The cells were cultured on the glass coverslip (hepatocytes) and cantilever (cardiomyocytes). Fluid exchange was established through two (R1 and R2) reservoirs. The drug (Cyclophosphamide and Terfenadine) was passed through R1 reservoir. The amplifier with printed circuit board and elastomeric connector was used for the detection of beating signals of the heart. The Laser deflection-based apparatus was used to detect cardiac contractile. This heart-liver system can be used in preclinical studies to determine the drug toxicity on the heart upon the transformation through hepatic metabolism. Gregor et al.[51] developed the 3D liver-intestine-on-chip model to study the first pass metabolism of orally administered drug. This model was suitable to study the gastrointestinal tract, effects of nutraceutical substances, and uptake pattern of compounds. This kind of microfluidic device can be projected to decrease the number of drug candidates in pre-clinical screening processes.

4. Organ-on-chip and associated biomedical applications

Organ-on-chip models are being used to investigate the pathogenesis of various human diseases, and related drug screening. In this context, several methods have been developed that are based on microfluidic system to create organ-on-chip models. These models are expected to be used for the investigation of novel drugs and vaccines for efficient management of various human diseases. Certain popular and well-studied disease models are being discussed in the following section.

4.1 Models for cancer study

Lack of preclinical models slows down the novel developments in cancer study and thus, approval of new anti-cancer therapies. Most of the drugs show promising results under *in vitro* experiments on cancer models, however, fail under clinical trials. More complex *in vitro* models such as trans-well culture, organoids and spheroids have shown better mimicking of cancerous tissues than conventional 2D cell culture models. However, the former models do not have organ level structure, established tissue-tissue interface, continuous fluid flow, and mechanical cues (that are experienced by the cells in a living organ), etc. It is well known that chemotherapy results in various side effects that sometimes lead to the failure of the treatment. Thus, patient-specific treatment strategy is necessary to achieve the optimal results. In this context, microfluidic-based model systems have shown better results. It offers to study the cell behavior within the relevant tissue/organ microenvironment under *in vitro* condition. Additionally, organ-on-chip based models can be used to study the local cellular, chemical, molecular and biophysical parameters of the cancerous tissues/organs.[52] It also enables to process the samples [that are not available in large quantity (e.g., patient's biopsies)] to develop the cancer model for study.[53] Hassell et al.[54] developed a lung cancer model on chip using an alveolar-capillary interface microsystem (Fig. 6A and B). The developed system consisted of two closely related channels separated by flexible and thin PDMS membrane (coated with ECM). Non-small cell lung cancer cells were cultured on the PDMS membrane containing lung microvascular cells and human small airway epithelial cells, which formed a hollow vascular lumen-like structure. The physiological breathing motion was created by cyclin suction. This lung cancer-on-chip model was reflecting the microenvironment of specific lung cancer,as observed in patients. It was also found that in the presence of mechanical stimulation, lung cancer cells decreased the sensitivity toward tyrosine kinase inhibitor by reducing EGFR expression and phosphorylation. Thus, in this study, it was concluded that mechanical cue can influence the lung cancer growth, sensitivity and tumor dormancy. Cancer cells are known to hijack the immune checkpoints to evade the immune surveillance. The tumor microenvironment influences the tumor-immune interaction and responses to immunotherapies. Jenkins et al.[55] developed a cancer tumor chip using murine/patient derived organotypic tumor spheroids to check the response of the immune checkpoint blockade. This tumor chip revealed that TBK1 (Tank binding kinase 1) inhibition enhances the PD-1 blockade. Microfluidic-based system also helps in real-time tracking of cell movement and behavior, which further assists to determine the metastasis

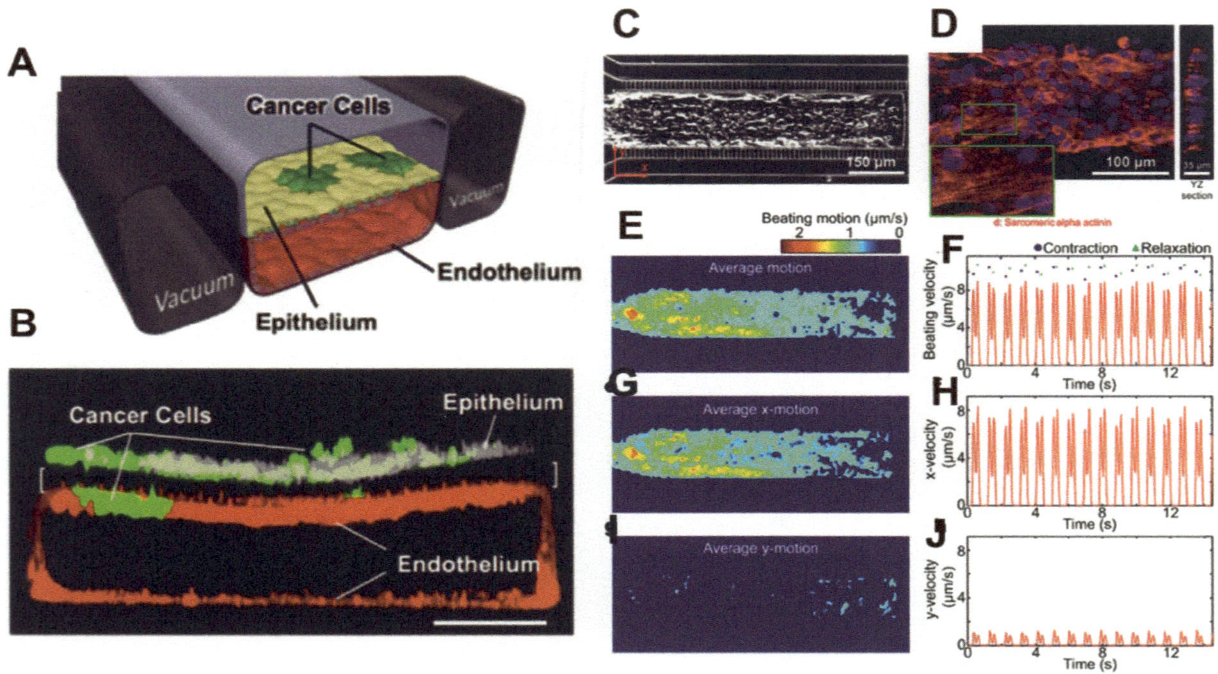

Fig. 6 (A) Schematic showing a cross-section of alveolus-on-chip model in which human lung epithelial cells and a low density of non-small cell lung cancer cells were cultured on the ECM coated PDMS membrane. (B) Confocal fluorescence micrograph of a cross-section alveolus chip showing green fluorescence protein-labeled lung cancer cell with primary lung alveolar epithelial cells (white). (C) Optical micrograph of a cardiac tissue in the heart-on-chip model. (D) Confocal fluorescence micrograph showing multiple layer thick cell structure. Inset shows a magnified micrograph the sarcomeric alpha actinin and DAPI staining. (E and F) Heat map of the time-averaged beating motion and corresponding average beating kinetics, (G and H) along the channel axis, and (I and J) perpendicular to the channel axis. *Panel (A–B): Reprinted with permission from Hassell BA, Goyal G, Lee E, et al. Human organ chip models recapitulate orthotopic lung cancer growth, therapeutic responses, and tumor dormancy in vitro.* Cell Rep. *2017;21(2):508–516. doi:10.1016/j.celrep.2017.09.043 Copyright © Elsevier. Panels (C–J): Reprinted with permission from Mathur A, Loskill P, Shao K, et al. Human iPSC-based cardiac microphysiological system for drug screening applications.* Sci Rep. *2015;5:8883. doi:10.1038/srep08883 Copyright © Springer Nature.*

behavior of cancer cells. For example, Wang et al.[56] developed metastasis-on-chip model to determine the progression of kidney cancer metastasis in the liver. In this chip, kidney cancer cells were co-culture with hepatocytes in decellularized liver matrix. These kinds of chips are valuable tool for assessing treatment regimens and developing new anticancer agents.

4.2 Models to study cardiovascular diseases

Cardiovascular diseases are most common causes of death worldwide. In 2019, ~116.4 million population is estimated with heart associated diseases globally.[57] Several preclinical studies are closed due to the severe side effects on the heart associated vasculatures. Since, the preclinical models are poor predictors of human responses, several stem-cell, tissue-engineering and organ-on-chip-based *in vitro* models are generated to systematically study the cardiac diseases. Other cardiac diseases such as cardiomyopathy and rhythm disorders are also envisaged to be studied by pluripotent stem cells-based chip models. For example, Mathur et al.[58] have developed human induced pluripotent stem cells-based microfluidic heart-on-chip system to screen drugs for rhythm disorder. It was observed that the developed chip could produce 3D cardiac tissue within 1 day of loading and started to beat [55–80 bpm (beats per minute)] without any external stimuli. Interestingly, after 7 days, the beat rate was aligned in uniaxial manner with multiple cellular layers (Fig. 6C–J). Computational motion tracking system confirmed that beating was aligned along the axis of the microphysiological system. The cardiac microphysiological system can help to study the rhythm disorder also effectively predict accurate IC_{50} and EC_{50} values of the tested drugs. Several chronic heart diseases such as hypertension and aortic valve stenosis have also been studied by using mechanotransduction method where external mechanical forces are translated into the internal electrical and chemical cellular response. McCain et al.[59] created a muscular thin film platform to develop the failing myocardium-on-chip. The model was quantified for failed myocardium using gene expression, calcium handling, contractile function, and myocyte architecture. The α- and β-myosin heavy chain expression was decreased, which further changed the shapes of myocytes and alignment of sarcomere. In stretched tissues, the peak of systolic stress was significantly reduced that resembled to the failed myocytes condition. Myocardial tissue mimicking myocardium-on-chip models have helped to study valvular tissue and related diseases. Lee et al.[60] developed a microfluidic-chip to study the mechanobiological responses of valvular epithelial cells using pneumatically actuated micropumps. In the microfluidic cardiac profile generator, two sets of microvalves (different size) mimicking the atrium and the ventricle were developed. Along with microvalves, other two axillary microvalves were used to generate fluidic shear stress pattern. Using this chip, it was found that positive to negative shear stress played an important role in determining porcine aortic valvular endothelial cells trans-differentiation and adaptation.

4.3 Models for neurodegenerative diseases

Neurodegenerative diseases occur due to the loss of neural cell functions or neuron death. For example, Parkinson's disease occurs when neuron losses its function in the part of the brain called substantia nigra, and Huntington's disease occurs when basal ganglia are affected.[61] Various factors such as oxidative stress, protein aggregation and misfolding in peripheral and central nervous system (PNS and CNS, respectively) are responsible to cause damage to the neuronal cells. Such issues are also associated with the alteration in microenvironment and interaction between neurons, astrocytes, endothelial cells, and pericytes. The mechanism of neurodegenerative disorders such as Parkinson's diseases, Alzheimer's diseases, and amyotrophic lateral sclerosis is not fully understood so far. These issues have also impressed to develop microfluidic-based organ models to address the neurodegenerative disorders. Microfluidic-based models recapitulate certain aspects of spatial complexity of CNS and PNS through co-culture system, endothelial lineage, and muscle cells. Blood-brain barrier mimicking system has been developed using the cultivation of stem cells on organ-on-chip system. Vatine et al.[62] have developed a human blood-brain barrier-chip using healthy induced pluripotent stem cell-derived brain microvascular endothelial-like cells, neurons, and astrocytes with collagen-IV-fibronectin (Fig. 7A). The developed system produced a monolayer that expressed the endothelial marker PECAM-1; the tight junction markers ZO-1 (zona occludens-1), OCCLUDIN (OCLN), and CLAUDIN-5 (CLDN-5); and GLUT-1 (glucose transporter) (Fig. 7B). The endothelial cells formed continuous ZO-1, which suggested the formation of blood vessel-like structures (Fig. 7C). The RNA sequencing and principle component analysis revealed that

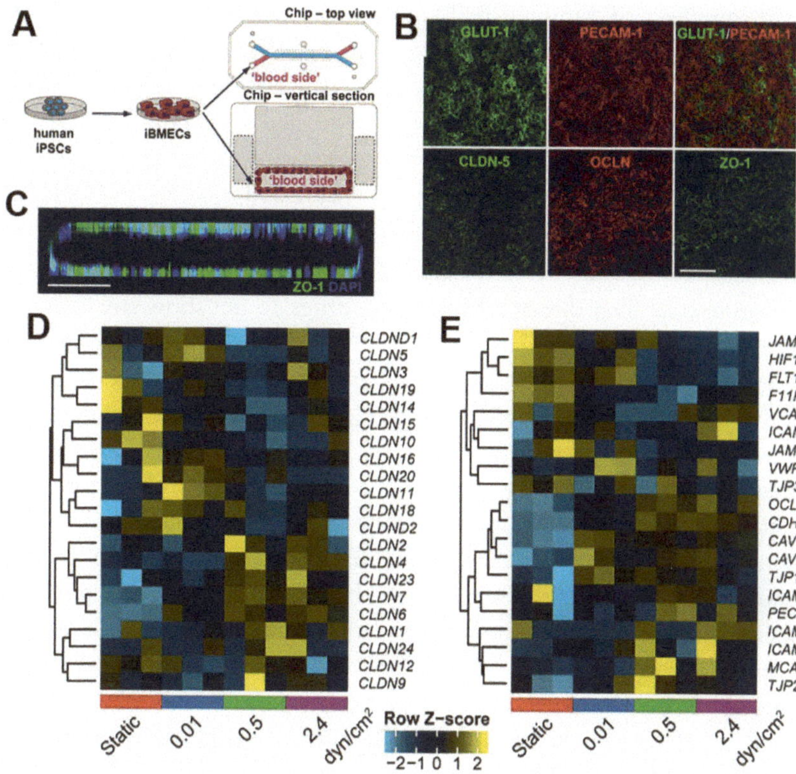

Fig. 7 (A) Schematic showing microfluidic neuron-on-chip model in which brain microvascular endothelial-like cells (iBEMCs) seeded on the bottom of the chip. (B) Neuron-on-chip model expressed endothelial marker: GLUT-1 (green), CLDN-5 (green), ZO-1 (green) PECAM-1 (red), and OCLN (red). (C) The continuous formation of ZO-1 (green) at the periphery of the blood side showed blood vessel like formation (DAPI-blue). (D) Relative expression of CLDN genes in response to shear stress. (E) Relative expression of junctional-related genes in response to shear stress. *Reprinted with permission from Vatine GD, Barrile R, Workman MJ, et al. Human iPSC-derived blood-brain barrier chips enable disease modeling and personalized medicine applications.* Cell Stem Cell. *2019;24(6):995–1005.e6. doi:10.1016/j.stem.2019.05.011 Copyright © Elsevier.*

expression of the tight junctions CLDN, OCLN, PECAM1, Von Willebrand factor) are flow dependent (Fig. 7D and E). This chip could predict inter-personal variability in blood-brain barrier performance, which can help in Parkinson's disease modeling and screening ofdrugs. Park et al.[63] created neuro-spheroids to recapitulate the *in vivo* brain microenvironment and tested the effect of Amyloid-β (Major contributor in Alzheimer diseases). The Amyloid-β (treatment was given in static condition) exhibited significant damage in neural network and reduced the viability of neurospheroids.

4.4 Models to study liver diseases

Various *in vivo* models are established to study liver functions and screening of drugs to treat liver diseases. Most of the drugs do not show any liver injuries or damage when tested in *in vivo* models.[64] These models do not exhibit the exact human liver function,[65] thus, making it inconvenient to use the highly dynamic processes of the liver. The innovative technology of liver-on-chip can overcome these problems by recapitulating the liver function. For example, Gori et al.[66] have developed a liver-on-chip model to study the pathogenesis of non-alcoholic fatty liver diseases. Here, the liver cancer cells were cultured under fatty acid free medium (without palmitic and oleic acid) for 48 h. The developed model mimicked endothelial-parenchymal interface of a liver sinusoid, allowed the diffusion of nutrients, and removed waste products like hepatic microvasculature. Non-significant increase in ROS generation was observed, similar to benign chronic steatosis, when treated with 2:1 ratio of oleic acid: palmitic acid. The oleic acid showed protective action against palmitic acid's toxicity. The high level of palmitic acids is known to cause steatosis and thus promote β-oxidation of free fatty acids leading to the oxidative stress. Hepatic inflammation is a leading cause of the liver tissue damage and also associated with viral hepatitis and hepatocellular carcinoma. A vascularized liver-on-chip was designed to study the hepatic acinus.[67] This liver-on-chip model was treated with lipopolysaccharide to alter the immune defense system.[68] Under *in vitro* conditions, increasing number of Kupffer cells were found in hepatic inflammation condition.[69] In this model, hepatocytes and Kupffer cells ratio was increased from 10:1 to 10:4 that led to enhance the inflammatory cytokines expression as confirmed by the hepatic inflammation. Viral infection in case of hepatitis B and hepatitis C have shown several types of chronic liver diseases. Ortega-Prieto et al.[21] developed a liver-on-chip model to study hepatitis B infection by replicating the life cycle of HBV mimicking the pattern observed in patients. Additionally, Kang et al.[70] have developed a liver-on-chip model using co-culture of primary rat adrenal medullary endothelial cells and hepatocytes that were infected by recombinant adenovirus with the HBV genome. The incorporation of the endothelial cells with hepatocytes improved and sustained the expression of liver specific differentiation markers such as albumin, transferrin hepatocyte nuclear factor-4, and β-actin for 21 days than hepatocytes alone.

4.5 Models for respiratory diseases

Respiratory diseases occur due to smoking tobacco, air pollution, and respiratory tract infections. Chronic obstructive pulmonary disease (COPD), asthma, cystic fibrosis, pneumonia, lung cancer, etc., are categorised as respiratory diseases. Microfluidic-based models to study these diseases have been very well developed and used for various applications.[71] In order to study the complex process of asthma and COPD, Benam et al.[72] have developed a lung small airway-on-chip consisting of differentiated muco-ciliary, and bronchiolar epithelium with an underlying lung microvascular endothelium. The Goblet cells of this model induced the secretion of inflammatory cytokines when treated with interleukin-13. Also, the cilia beating frequency was found to be decreased and comparable with asthma patient's mucosa airways. According to the epidemiological studies, idiopathic pulmonary fibrosis is related with gastroesophageal reflux diseases.[73] Felder et al.[74] created epithelial wound framework on a lung-on-chip to evaluate the effects of gastric contents such as pepsin, and hydrochloric acid, etc. In this model, alveolar micro-injuries were created to recapitulate the damaged features of the lungs. It was observed that gastric content damaged the alveolar epithelium barrier integrity, which led to the pulmonary fibrosis. In another study, breathing lung-on-chip model showed the significant impact of physiological cyclic mechanical stretch on human hepatic growth factors.[75] These findings offered better understanding of the complex system of pulmonary fibrosis. Lung-on-chip model also facilitated to discover safer and more efficient drugs for respiratory diseases. For example, Huh et al.[76] have investigated angiopoietin-1 on lung-on-chip model to prevent interleukin-2 induced vascular leakage. This device was used to recapitulate the drug toxicity-induced pulmonary edema (observed in cancer patients). It was observed that angiopoietin-1 and new transient receptor potential vanilloid 4 (TRPV4) ion channel inhibitor (GSK2193874) prevented the formation of paracellular gaps in spite of rhythmic mechanical strain. Jain et al.[77] have developed a human primary alveolar epithelium interfacing with endothelium cells with the use of lung alveolus-on-chip model. This model recapitulated the *in vivo* platelet-endothelial dynamics and found that the lipopolysaccharide endotoxin activates the alveolar epithelium and promotes intravascular thrombosis. The system was further tested to study the response of Parmodulin-2 (Protease activated receptor-1 inhibitor), which showed cytoprotective and antithrombotic activities.

4.6 Models to study digestive diseases

The digestive system is an extensive and intricate part of the body; therefore, it is imperative to study the various types of digestive diseases including food intolerance, chronic constipation, gastroesophageal reflux diseases, and inflammatory bowel diseases, etc. Only limited reports are there to develop models to study digestive diseases. The gut microbiome of animal models and humans are different (~85% human microbiome is absent in mouse model). In this context, developing a human intestine mimic on microfluidic chip will have tremendous impact on the study of digestive system-based diseases. Kim et al.[78] have created an intestine-on-chip model to study the relation between intestinal bacterial overgrowth and associated inflammation. The developed model was able to recapitulate the features of ileus and inflammatory bowel disease. Pro-inflammatory cytokines were found to be produced after the interaction of immune cells of model with lipopolysaccharide (endotoxin), which induced villus injury and compromised the functions of intestinal barrier. For digestive disease modeling and drug discovery, a commercially available, OrganPlate® (developed by MIMETAS, an organ-on-chip manufacturing company) has been used by researchers. OrganPlate® have major three compartment which can be used for cell seeding. In a study by Lo Sasso et al.,[79] intestine inflammation was studied using OrganPlate®. Here, Caco-2 and mucus-secreting HT29-MTX cells were used in the top compartment of the chip. The middle and bottom compartment contained collagen-I ECM, and co-culture of immune cells (THP-1 and MUTZ-3), respectively. Eventually, the top compartment of chip showed tubular structure against the collagen-I ECM and also secreted mucus. The exposure of tumor necrosis factor-α led to induce cytokine production and other inflammatory conditions (due to the loss of epithelial barrier function). Overall, this model showed bowel disease pathophysiology. Using this model, anti-inflammatory drug, TPCA-1 was studied to observe the prevention of cytokine induced-inflammation. Table 2 summarizes the development of various organ-on-a-chip models and their utility in various biomedical applications.

Table 2 List of developed organ-on-chip models and their applications in drug screening.

Organ-on-chip models	Cell types	Model diseases or conditions	Drugs used	Important observations	Reference
Liver-on-chip	Primary hepatocytes	Toxicity evaluation	Acetaminophen	The drug was toxic at the concentration of 10 and 30 mM. Production of lactate was found to be decreased in a concentration-dependent manner after the treatment of Acetaminophen	80
	HepG2/C3A	Drug induced liver injury	Troglitazone	A 28 μM concentration of drug induced toxicity in model in 21 days	81
Kidney-on-chip	Caco-2	Drug absorption	SN-38 (7-ethyl-10-hydroxycamptothecin)	This model can be used to determine the permeability co-efficient. Strong correlation was found between logP value and experimentally determined permeability co-efficient.	82
	Kidney tissue isolated from rat	Chronic kidney diseases	Aristolochic acid I (AA-I)	AA-I increased the cytotoxicity in kidney's proximal epithelial cells, and formed aristo-lactam adducts.	83
Intestine-on-chip	Caco-2	Gut radiation injury	Dimethyl oxaloylglycin (DMOG)	DMOG showed radio-protective effect by elevating the levels of HIF-1α and HIF-2α (cell survival promotor).	84
	Pluripotent stem cells	To check response to external stimuli	Interferon-γ (IFN-γ)	A 1 hr exposure of interferon-γ, phosphorylated the STAT1 gene, whereas 3 days exposure, upregulate the interferon-γ related genes leading to the increase in intestinal permeability.	85

Continued

Table 2 List of developed organ-on-chip models and their applications in drug screening.—cont'd

Organ-on-chip models	Cell types	Model diseases or conditions	Drugs used	Important observations	Reference
Neuron-on-chip	Pluripotent stem cells	Brain tumor	Bortezomib, Paclitaxel	Both drugs showed poor penetration across blood-brain barrier into brain tumor.	86
	Pluripotent stem cells derived neurons	Alzheimer's disease	β-Secretase inhibitor	β-Secretase inhibitor was found to be responsible to rescue the reduction of endocytosis.	87
Lung-on-chip	Non-small cell lung cancer cells	Lung cancer	Tyrosine kinase inhibitor (TKI)	TKI therapy responses to physical cues related to breathing motion and changed signaling through epidermal growth factor receptor and MET protein kinase.	54

5. Organ-on-chip and nanomedicine

The emerging field of nanomedicine has achieved significant advancement to realize the applications of nano formulation in biomedical field such as drug delivery,[88] gene delivery,[89] photothermal therapy,[90] biosensors,[91] antioxidant treatment,[91] and contrast agent,[92] etc. Nanomedicines exhibit unique advantages compared to the systematic administration of free molecules or drugs, such as prolonged blood circulation time, reduced side effects, enhanced permeable and retention effect leading to better therapeutic index, and targeted treatment, etc. For example, Albanese et al.[93] created tumor spheroids-on-chip using Breast cancer cell line (MDA-MB-435) to study the uptake of fluorescent gold nanoparticles (AuNPs). The microfluidic device provided precise flow condition for the transportation of media and AuNPs. The size, surface coating molecules, and the flow condition of AuNPs was found to alter the uptake of nanoparticles in tumor tissues. PEGylated and transferrin conjugated AuNPs showed accumulation only in the periphery of the tumor, however, failed to penetrate deep inside of the tumor spheroids.[94] In another study, lung-on-chip model was developed for the real-time monitoring of pH-responsive human serum albumin coated zinc oxide quantum dots (HSA-ZnO QDs).[95] This model was made up of lung cancer and fibroblast cells with collagen ECM to closely recapitulate the cancerous alveolus epithelium. The 60 nm sized HSA-ZnO QDs were successfully internalized as studied by the trans-epithelial electrical resistance and optical pH sensors to monitor their real-time uptake.

6. Limitations

Recently, there have been tremendous surge in the strategies of developing successful models of human tissues and organs-on-a-chip to study the mechanism of associated diseases as well as growth and developments during therapeutic interventions. However, developing new models of organ-on-chip is a challenging technique with significant batch-to-batch variation. Due to the complex architecture of the human tissues/organs, it becomes difficult to reproduce the construct in miniaturized form. Different organ-on-chip devices require different growth medium, which is one of the major limitations. The multiorgan-on-chip devices are reported by several research groups; however, only limited studies have shown successful angiogenesis models. Additionally, it would be difficult to exactly mimic all the functions of the targeted human tissues/organs in 3D models.

Although expected but no success so far that these techniques would completely replace the experiments on animal models in preclinical phase of the study. The development of 3D models of tissues and organs is time consuming and associated with high cost, and low success rate.[96] There are several major obstacles in using stem cells to develop organ-on-chip model due to their variable efficiency and lack of robust differentiation protocol. Also, it is hard to recapitulate the morphology of cardiac fibroblast using stem cells. Additionally, there are few reports on interaction between drugs and PDMS membrane, which can give the false results.[97,98] Also, the sustainability of 3D model is short-term, may be only a few weeks, thus, long term effects of drug cannot be studied.[97] It is also reported that 3D models sometimes exhibit different phenotype than *in vivo* condition.[99]

7. Conclusion and future aspect

Microfluidics-based tissue/organ culture devices have potential to be used for healthcare applications. The success of this technology can radically transform the basic research and provide acceleration to the drug development program. Almost every important human organ has been successfully developed on a chip model with tissues showing almost similar functions of natural organs. The microfluidic systems are miniaturized, fast, accurate, and suitable for developing quick disease models, drug screening, and diagnosis.

Various research groups have developed new types of scaffolds and culturing technique to improve vascularization, and produced several other disease models. In future, the use of human stem cells cultured in the organ shaped models could lead to the successful generation of intended human organs in the laboratory. Additionally, induced pluripotent stem cells may also serve the purpose of reprogramming of human organs and thus hold promise in the field of regenerative medicines. Although the field has shown some initial promises to significantly improve the human health, however, the real potential of this technology has not been realized so far. In summary, microfluidic-based technique for organ/tissue-on-chip is powerful technique which will advance the field of biomedicine, leading to screen the novel therapeutics strategies and overall improve the human health.

Acknowledgments

The financial assistance received from the Gujarat Institute for Chemical Technology (GICT), Gandhinagar for the establishment of Centre for Nanotechnology Research and Applications (CENTRA), Science and Engineering Research Board (SERB), India (grant number ILS/SERB/2015-2016/01), and Indian Council of Medical Research (ICMR) (grant number 35/9/2019-Nano/BMS) to S. Singh is gratefully acknowledged. S. Bhagat thanks ICMR for providing senior research fellowship.

Conflict of interest

Authors declare no conflict of interest.

References

1. Kapałczyńska M, Kolenda T, Przybyła W, et al. 2D and 3D cell cultures—a comparison of different types of cancer cell cultures. *Arch Med Sci*. 2018;14(4):910. https://doi.org/10.5114/aoms.2016.63743.
2. Perestrelo AR, Águas AC, Rainer A, Forte G. Microfluidic organ/body-on-a-chip devices at the convergence of biology and microengineering. *Sensors*. 2015;15(12):31142–31170. https://doi.org/10.3390/s151229848.
3. Low LA, Tagle DA. Tissue chips—innovative tools for drug development and disease modeling. *Lab Chip*. 2017;17(18):3026–3036. https://doi.org/10.1039/c7lc00462a.
4. Mosig AS. Organ-on-chip models: new opportunities for biomedical research. *Future Sci OA*. 2017;3(2):FSO130. https://doi.org/10.4155/fsoa-2016-0038.
5. Sosa-Hernandez JE, Villalba-Rodriguez AM, Romero-Castillo KD, et al. Organs-on-a-chip module: a review from the development and applications perspective. *Micromachines (Basel)*. 2018;9(10):536. https://doi.org/10.3390/mi9100536.
6. Yin X, Mead BE, Safaee H, Langer R, Karp JM, Levy O. Engineering stem cell organoids. *Cell Stem Cell*. 2016;18(1):25–38. https://doi.org/10.1016/j.stem.2015.12.005.
7. Hachey SJ, Hughes CCW. Applications of tumor chip technology. *Lab Chip*. 2018;18(19):2893–2912. https://doi.org/10.1039/c8lc00330k.
8. Duval K, Grover H, Han L-H, et al. Modeling physiological events in 2D vs. 3D cell culture. *Phys Ther*. 2017;32(4):266–277. https://doi.org/10.1152/physiol.00036.2016.
9. Crotty S, Andino R. Poliovirus vaccine strains as mucosal vaccine vectors and their potential use to develop an AIDS vaccine. *Adv Drug Deliv Rev*. 2004;56(6):835–852. https://doi.org/10.1016/j.addr.2003.10.042.
10. Lucey BP, Nelson-Rees WA, Hutchins GM. Henrietta lacks, HeLa cells, and cell culture contamination. *Arch Med Sci*. 2009;133(9):1463–1467. https://doi.org/10.1043/1543-2165-133.9.1463.
11. Teimouri A, Yeung P, Agu R. 2D vs. 3D cell culture models for in vitro topical (dermatological) medication testing. In: *Cell Culture*. IntechOpen; 2018.
12. Ryu NE, Lee SH, Park H. Spheroid culture system methods and applications for mesenchymal stem cells. *Cells*. 2019;8(12):1620. https://doi.org/10.3390/cells8121620.
13. Edmondson R, Broglie JJ, Adcock AF, Yang L. Three-dimensional cell culture systems and their applications in drug discovery and cell-based biosensors. *Assay Drug Dev Technol*. 2014;12(4):207–218. https://doi.org/10.1089/adt.2014.573.
14. Campuzano S, Pelling AE. Scaffolds for 3D cell culture and cellular agriculture applications derived from non-animal sources. Mini review. *Front Sustain Food Syst*. 2019;3(38):1–9. https://doi.org/10.3389/fsufs.2019.00038.
15. Ingber DE. Developmentally inspired human "organs on chips". *Development*. 2018;145(16). https://doi.org/10.1242/dev.156125.
16. Sakolish CM, Esch MB, Hickman JJ, Shuler ML, Mahler GJ. Modeling barrier tissues in vitro: methods, achievements, and challenges. *EBioMedicine*. 2016;5:30–39. https://doi.org/10.1016/j.ebiom.2016.02.023.
17. Fiorini GS, Chiu DT. Disposable microfluidic devices: fabrication, function, and application. *BioTechniques*. 2005;38(3):429–446. https://doi.org/10.2144/05383RV02.

18. Cha C, Piraino F, Khademhosseini A. Chapter 9—Microfabrication technology in tissue engineering. In: De Boer J, ed. *Engineering*. 2nd ed. Academic Press; 2014:283–310.
19. Bhatia SN, Ingber DE. Microfluidic organs-on-chips. *Nat Biotechnol*. 2014;32 (8):760–772. https://doi.org/10.1038/nbt.2989.
20. Mehta V, Rath SN. 3D printed microfluidic devices: a review focused on four fundamental manufacturing approaches and implications on the field of healthcare. *Bio-Des Manuf*. 2021;4(2):311–343. https://doi.org/10.1007/s42242-020-00112-5.
21. Ortega-Prieto AM, Skelton JK, Wai SN, et al. 3D microfluidic liver cultures as a physiological preclinical tool for hepatitis B virus infection. *Nat Commun*. 2018;9 (1):682. https://doi.org/10.1038/s41467-018-02969-8.
22. Du Y, Li N, Yang H, et al. Mimicking liver sinusoidal structures and functions using a 3D-configured microfluidic chip. *Lab Chip*. 2017;17(5):782–794. https://doi.org/10.1039/C6LC01374K.
23. Deng J, Wei W, Chen Z, et al. Engineered liver-on-a-chip platform to mimic liver functions and its biomedical applications: a review. *Micromachines (Basel)*. 2019;10 (10):676. https://doi.org/10.3390/mi10100676.
24. Yu F, Deng R, Hao Tong W, et al. A perfusion incubator liver chip for 3D cell culture with application on chronic hepatotoxicity testing. *Sci Rep*. 2017;7(1):14528. https://doi.org/10.1038/s41598-017-13848-5.
25. Lin N, Zhou X, Geng X, et al. Repeated dose multi-drug testing using a microfluidic chip-based coculture of human liver and kidney proximal tubules equivalents. *Sci Rep*. 2020;10(1):8879. https://doi.org/10.1038/s41598-020-65817-0.
26. van Midwoud PM, Merema MT, Verpoorte E, Groothuis GMM. A microfluidic approach for in vitro assessment of interorgan interactions in drug metabolism using intestinal and liver slices. *Lab Chip*. 2010;10(20):2778–2786. https://doi.org/10.1039/C0LC00043D.
27. Choe A, Ha SK, Choi I, Choi N, Sung JH. Microfluidic gut-liver chip for reproducing the first pass metabolism. *Biomed Microdevices*. 2017;19(1):4. https://doi.org/10.1007/s10544-016-0143-2.
28. Zhou M, Zhang X, Wen X, et al. Development of a functional glomerulus at the organ level on a chip to mimic hypertensive nephropathy. *Sci Rep*. 2016;6(1):31771. https://doi.org/10.1038/srep31771.
29. Huang H-C, Chang Y-J, Chen W-C, Harn HI-C, Tang M-J, Wu C-C. Enhancement of renal epithelial cell functions through microfluidic-based coculture with adipose-derived stem cells. *Tissue Eng A*. 2013;19(17–18):2024–2034. https://doi.org/10.1089/ten.tea.2012.0605.
30. Zhou M, Ma H, Lin H, Qin J. Induction of epithelial-to-mesenchymal transition in proximal tubular epithelial cells on microfluidic devices. *Biomaterials*. 2014;35(5):1390–1401. https://doi.org/10.1016/j.biomaterials.2013.10.070.
31. Sateesh J, Guha K, Dutta A, Sengupta P, Srinivasa RK. Regenerating re-absorption function of proximal convoluted tubule using microfluidics for kidney-on-chip applications. *SN Appl Sci*. 2019;2(1):39. https://doi.org/10.1007/s42452-019-1840-2.
32. Yin L, Du G, Zhang B, et al. Efficient drug screening and nephrotoxicity assessment on co-culture microfluidic kidney chip. *Sci Rep*. 2020;10(1):6568. https://doi.org/10.1038/s41598-020-63096-3.
33. Zhang YS, Aleman J, Arneri A, et al. From cardiac tissue engineering to heart-on-a-chip: beating challenges. *Biomed Mater*. 2015;10(3):034006. https://doi.org/10.1088/1748-6041/10/3/034006.
34. K-i K, Kato Y, Hirai Y, et al. Integrated heart/cancer on a chip to reproduce the side effects of anti-cancer drugs in vitro. *RSC Adv*. 2017;7(58):36777–36786. https://doi.org/10.1039/C7RA07716E.

35. Liu H, Bolonduro OA, Hu N, et al. Heart-on-a-chip model with integrated extra- and intracellular bioelectronics for monitoring cardiac electrophysiology under acute hypoxia. *Nano Lett.* 2020;20(4):2585–2593. https://doi.org/10.1021/acs.nanolett.0c00076.
36. Belkaid Y, Hand TW. Role of the microbiota in immunity and inflammation. *Cell.* 2014;157(1):121–141. https://doi.org/10.1016/j.cell.2014.03.011.
37. Shah P, Fritz JV, Glaab E, et al. A microfluidics-based in vitro model of the gastrointestinal human-microbe interface. *Nat Commun.* 2016;7:11535. https://doi.org/10.1038/ncomms11535.
38. Abt MC, Artis D. The intestinal microbiota in health and disease: the influence of microbial products on immune cell homeostasis. *Curr Opin Gastroenterol.* 2009;25(6):496–502. https://doi.org/10.1097/MOG.0b013e328331b6b4.
39. Maurer M, Gresnigt MS, Last A, et al. A three-dimensional immunocompetent intestine-on-chip model as in vitro platform for functional and microbial interaction studies. *Biomaterials.* 2019;220:119396. https://doi.org/10.1016/j.biomaterials.2019.119396.
40. Kilic O, Pamies D, Lavell E, et al. Brain-on-a-chip model enables analysis of human neuronal differentiation and chemotaxis. *Lab Chip.* 2016;16(21):4152–4162. https://doi.org/10.1039/C6LC00946H.
41. Kamudzandu M, Köse-Dunn M, Evans M, Fricker R, Roach P. A micro-fabricated in vitro complex neuronal circuit platform. *Biomed Phys Eng Express.* 2019;5(4):045016.
42. Xu C, Zhang M, Chen W, Jiang L, Chen C, Qin J. Assessment of air pollutant PM2.5 pulmonary exposure using a 3D lung-on-chip model. *ACS Biomater Sci Eng.* 2020;6(5):3081–3090. https://doi.org/10.1021/acsbiomaterials.0c00221.
43. Mejías JC, Nelson MR, Liseth O, Roy K. A 96-well format microvascularized human lung-on-a-chip platform for microphysiological modeling of fibrotic diseases. *Lab Chip.* 2020;20:3601–3611. https://doi.org/10.1039/D0LC00644K.
44. Marx U, Walles H, Hoffmann S, et al. 'Human-on-a-chip' developments: a translational cutting-edge alternative to systemic safety assessment and efficiency evaluation of substances in laboratory animals and man? *Altern Lab Anim.* 2012;40(5):235–257. https://doi.org/10.1177/026119291204000504.
45. Palaninathan V, Kumar V, Maekawa T, et al. Multi-organ on a chip for personalized precision medicine. *MRS Commun.* 2018;8(3):652–667. https://doi.org/10.1557/mrc.2018.148.
46. Rogal J, Probst C, Loskill P. Integration concepts for multi-organ chips: how to maintain flexibility?! *Future Sci OA.* 2017;3(2):FSO180. https://doi.org/10.4155/fsoa-2016-0092.
47. Skardal A, Murphy SV, Devarasetty M, et al. Multi-tissue interactions in an integrated three-tissue organ-on-a-chip platform. *Sci Rep.* 2017;7(1):8837. https://doi.org/10.1038/s41598-017-08879-x.
48. Maschmeyer I, Hasenberg T, Jaenicke A, et al. Chip-based human liver-intestine and liver-skin co-cultures—a first step toward systemic repeated dose substance testing in vitro. *Eur J Pharm Biopharm.* 2015;95(Pt A):77–87. https://doi.org/10.1016/j.ejpb.2015.03.002.
49. Edington CD, Chen WLK, Geishecker E, et al. Interconnected microphysiological systems for quantitative biology and pharmacology studies. *Sci Rep.* 2018;8(1):4530. https://doi.org/10.1038/s41598-018-22749-0.
50. Oleaga C, Riu A, Rothemund S, et al. Investigation of the effect of hepatic metabolism on off-target cardiotoxicity in a multi-organ human-on-a-chip system. *Biomaterials.* 2018;182:176–190. https://doi.org/10.1016/j.biomaterials.2018.07.062.
51. De Gregorio V, Telesco M, Corrado B, et al. Intestine-liver axis on-chip reveals the intestinal protective role on hepatic damage by emulating ethanol first-pass metabolism. Original research. *Front Bioeng Biotechnol.* 2020;8:163. https://doi.org/10.3389/fbioe.2020.00163.

52. Sontheimer-Phelps A, Hassell BA, Ingber DE. Modelling cancer in microfluidic human organs-on-chips. *Nat Rev Cancer.* 2019;19(2):65–81. https://doi.org/10.1038/s41568-018-0104-6.
53. Mathur L, Ballinger M, Utharala R, Merten CA. Microfluidics as an enabling technology for personalized cancer therapy. *Small.* 2020;16(9):1904321. https://doi.org/10.1002/smll.201904321.
54. Hassell BA, Goyal G, Lee E, et al. Human organ chip models recapitulate orthotopic lung cancer growth, therapeutic responses, and tumor dormancy in vitro. *Cell Rep.* 2017;21(2):508–516. https://doi.org/10.1016/j.celrep.2017.09.043.
55. Jenkins RW, Aref AR, Lizotte PH, et al. Ex vivo profiling of PD-1 blockade using organotypic tumor spheroids. *Cancer Discov.* 2018;8(2):196–215. https://doi.org/10.1158/2159-8290.CD-17-0833.
56. Wang Y, Wu D, Wu G, et al. Metastasis-on-a-chip mimicking the progression of kidney cancer in the liver for predicting treatment efficacy. *Theranostics.* 2020;10(1):300–311. https://doi.org/10.7150/thno.38736.
57. American Heart Association. *Heart Disease and Stroke Statistics—2019 Update. Proffesional Heart Daily*; 2019. Updated 31 January. Accessed 20 October, 2020 https://professional.heart.org/en/science-news/heart-disease-and-stroke-statistics-2019-update#:~:text=The%20Heart%20Disease%20and%20Stroke,of%20CVD%20every%2038%20seconds.
58. Mathur A, Loskill P, Shao K, et al. Human iPSC-based cardiac microphysiological system for drug screening applications. *Sci Rep.* 2015;5:8883. https://doi.org/10.1038/srep08883.
59. McCain ML, Sheehy SP, Grosberg A, Goss JA, Parker KK. Recapitulating maladaptive, multiscale remodeling of failing myocardium on a chip. *Proc Natl Acad Sci U S A.* 2013;110(24):9770–9775. https://doi.org/10.1073/pnas.1304913110.
60. Lee J, Estlack Z, Somaweera H, Wang X, Lacerda CMR, Kim J. A microfluidic cardiac flow profile generator for studying the effect of shear stress on valvular endothelial cells. *Lab Chip.* 2018;18(19):2946–2954. https://doi.org/10.1039/C8LC00545A.
61. Berman T, Bayati A. What are neurodegenerative diseases and how do they affect the brain. *Front Young Minds.* 2018;6:70. https://doi.org/10.3389/frym.2018.00070.
62. Vatine GD, Barrile R, Workman MJ, et al. Human iPSC-derived blood-brain barrier chips enable disease modeling and personalized medicine applications. *Cell Stem Cell.* 2019;24(6):995–1005.e6. https://doi.org/10.1016/j.stem.2019.05.011.
63. Park J, Lee BK, Jeong GS, Hyun JK, Lee CJ, Lee S-H. Three-dimensional brain-on-a-chip with an interstitial level of flow and its application as an in vitro model of Alzheimer's disease. *Lab Chip.* 2015;15(1):141–150. https://doi.org/10.1039/C4LC00962B.
64. Olson H, Betton G, Robinson D, et al. Concordance of the toxicity of pharmaceuticals in humans and in animals. *Regul Toxicol Pharmacol.* 2000;32(1):56–67. https://doi.org/10.1006/rtph.2000.1399.
65. van Midwoud PM, Verpoorte E, Groothuis GM. Microfluidic devices for in vitro studies on liver drug metabolism and toxicity. *Integr Biol (Camb).* 2011;3(5):509–521. https://doi.org/10.1039/c0ib00119h.
66. Gori M, Simonelli MC, Giannitelli SM, Businaro L, Trombetta M, Rainer A. Investigating nonalcoholic fatty liver disease in a liver-on-a-chip microfluidic device. *PLoS One.* 2016;11(7):e0159729. https://doi.org/10.1371/journal.pone.0159729.
67. Chen WLK, Edington C, Suter E, et al. Integrated gut/liver microphysiological systems elucidates inflammatory inter-tissue crosstalk. *Biotechnol Bioeng.* 2017;114(11):2648–2659. https://doi.org/10.1002/bit.26370.
68. Cosgrove BD, King BM, Hasan MA, et al. Synergistic drug-cytokine induction of hepatocellular death as an in vitro approach for the study of inflammation-associated idiosyncratic drug hepatotoxicity. *Toxicol Appl Pharmacol.* 2009;237(3):317–330. https://doi.org/10.1016/j.taap.2009.04.002.

69. McIntyre N. *Textbook of Hepatology: From Basic Science to Clinical Practice.* John Wiley & Sons; 2008.
70. Kang YB, Rawat S, Duchemin N, Bouchard M, Noh M. Human liver sinusoid on a chip for hepatitis B virus replication study. *Micromachines.* 2017;8(1):27. https://doi.org/10.3390/mi8010027.
71. Konar D, Devarasetty M, Yildiz DV, Atala A, Murphy SV. Lung-on-a-chip technologies for disease modeling and drug development. *Biomed Eng Comput Biol.* 2016;7 (Suppl 1):17–27. https://doi.org/10.4137/BECB.S34252.
72. Benam KH, Villenave R, Lucchesi C, et al. Small airway-on-a-chip enables analysis of human lung inflammation and drug responses in vitro. *Nat Methods.* 2016;13 (2):151–157. https://doi.org/10.1038/nmeth.3697.
73. Allaix ME, Fisichella PM, Noth I, Mendez BM, Patti MG. The pulmonary side of reflux disease: from heartburn to lung fibrosis. *J Gastrointest Surg.* 2013;17(8):1526–1535. https://doi.org/10.1007/s11605-013-2208-3.
74. Felder M, Stucki AO, Stucki JD, Geiser T, Guenat OT. The potential of microfluidic lung epithelial wounding: towards in vivo-like alveolar microinjuries. *Integr Biol.* 2014;6 (12):1132–1140. https://doi.org/10.1039/C4IB00149D.
75. Felder M, Trueeb B, Stucki AO, et al. Impaired wound healing of alveolar lung epithelial cells in a breathing lung-on-a-chip. *Front Bioeng Biotechnol.* 2019;7:3. https://doi.org/10.3389/fbioe.2019.00003.
76. Huh D, Leslie DC, Matthews BD, et al. A human disease model of drug toxicity-induced pulmonary edema in a lung-on-a-chip microdevice. *Sci Transl Med.* 2012;4 (159):159ra147. https://doi.org/10.1126/scitranslmed.3004249.
77. Jain A, Barrile R, van der Meer AD, et al. Primary human lung alveolus-on-a-chip model of intravascular thrombosis for assessment of therapeutics. *Clin Pharmacol Ther.* 2018;103(2):332–340. https://doi.org/10.1002/cpt.742.
78. Kim HJ, Li H, Collins JJ, Ingber DE. Contributions of microbiome and mechanical deformation to intestinal bacterial overgrowth and inflammation in a human gut-on-a-chip. *Proc Natl Acad Sci U S A.* 2016;113(1):E7–15. https://doi.org/10.1073/pnas.1522193112.
79. Lo Sasso G, Gijzen L, Marescotti D, et al. P006 3D Multicellular intestine-on-a-chip model for disease modelling and drug discovery. *J Crohn's Colitis.* 2020;14(Suppl_1): S132–S133. https://doi.org/10.1093/ecco-jcc/jjz203.135.
80. Freyer N, Greuel S, Knospel F, et al. Microscale 3D liver bioreactor for in vitro hepatotoxicity testing under perfusion conditions. *Bioengineering (Basel).* 2018;5(1):24. https://doi.org/10.3390/bioengineering5010024.
81. Vernetti LA, Senutovitch N, Boltz R, et al. A human liver microphysiology platform for investigating physiology, drug safety, and disease models. *Exp Biol Med (Maywood).* 2016;241(1):101–114. https://doi.org/10.1177/1535370215592121.
82. Pocock K, Delon L, Bala V, et al. Intestine-on-a-chip microfluidic model for efficient in vitro screening of oral chemotherapeutic uptake. *ACS Biomater Sci Eng.* 2017;3 (6):951–959. https://doi.org/10.1021/acsbiomaterials.7b00023.
83. Chang SY, Weber EJ, Sidorenko VS, et al. Human liver-kidney model elucidates the mechanisms of aristolochic acid nephrotoxicity. *JCI Insight.* 2017;2(22). https://doi.org/10.1172/jci.insight.95978, e95978.
84. Jalili-Firoozinezhad S, Prantil-Baun R, Jiang A, et al. Modeling radiation injury-induced cell death and countermeasure drug responses in a human gut-on-a-chip. *Cell Death Dis.* 2018;9(2):223. https://doi.org/10.1038/s41419-018-0304-8.
85. Workman MJ, Gleeson JP, Troisi EJ, et al. Enhanced utilization of induced pluripotent stem cell-derived human intestinal organoids using microengineered chips. *Cell Mol Gastroenterol Hepatol.* 2018;5(4):669–677 e2. https://doi.org/10.1016/j.jcmgh.2017.12.008.

86. Qi D, Wu S, Lin H, et al. Establishment of a human iPSC- and nanofiber-based microphysiological blood-brain barrier system. *ACS Appl Mater Interfaces*. 2018;10 (26):21825–21835. https://doi.org/10.1021/acsami.8b03962.
87. Woodruff G, Reyna SM, Dunlap M, et al. Defective transcytosis of APP and lipoproteins in human iPSC-derived neurons with familial Alzheimer's disease mutations. *Cell Rep*. 2016;17(3):759–773. https://doi.org/10.1016/j.celrep.2016.09.034.
88. Patra JK, Das G, Fraceto LF, et al. Nano based drug delivery systems: recent developments and future prospects. *J Nanobiotechnol*. 2018;16(1):71. https://doi.org/10.1186/s12951-018-0392-8.
89. Suk JS, Xu Q, Kim N, Hanes J, Ensign LM. PEGylation as a strategy for improving nanoparticle-based drug and gene delivery. *Adv Drug Deliv Rev*. 2016;99(Pt A): 28–51. https://doi.org/10.1016/j.addr.2015.09.012.
90. Liu Y, Crawford BM, Vo-Dinh T. Gold nanoparticles-mediated photothermal therapy and immunotherapy. *Immunotherapy*. 2018;10(13):1175–1188. https://doi.org/10.2217/imt-2018-0029.
91. Singh S. Nanomaterials exhibiting enzyme-like properties (nanozymes): current advances and future perspectives. Mini review. *Front Chem*. 2019;7:46. https://doi.org/10.3389/fchem.2019.00046.
92. Paunesku T, Gutiontov S, Brown K, Woloschak GE. Radiosensitization and nanoparticles. *Cancer Treat Res*. 2015;166:151–171. https://doi.org/10.1007/978-3-319-16555-4_7.
93. Albanese A, Lam AK, Sykes EA, Rocheleau JV, Chan WC. Tumour-on-a-chip provides an optical window into nanoparticle tissue transport. *Nat Commun*. 2013;4: 2718. https://doi.org/10.1038/ncomms3718.
94. Bhise NS, Ribas J, Manoharan V, et al. Organ-on-a-chip platforms for studying drug delivery systems. *J Control Release*. 2014;190:82–93. https://doi.org/10.1016/j.jconrel.2014.05.004.
95. Meghani N, Kim KH, Kim SH, Lee SH, Choi KH. Evaluation and live monitoring of pH-responsive HSA-ZnO nanoparticles using a lung-on-a-chip model. *Arch Pharm Res*. 2020;43(5):503–513. https://doi.org/10.1007/s12272-020-01236-z.
96. Mittal R, Woo FW, Castro CS, et al. Organ-on-chip models: implications in drug discovery and clinical applications. *J Cell Physiol*. 2019;234(6):8352–8380. https://doi.org/10.1002/jcp.27729.
97. Huh D, Kim HJ, Fraser JP, et al. Microfabrication of human organs-on-chips. *Nat Protoc*. 2013;8(11):2135–2157. https://doi.org/10.1038/nprot.2013.137.
98. Huh D, Hamilton GA, Ingber DE. From 3D cell culture to organs-on-chips. *Trends Cell Biol*. 2011;21(12):745–754. https://doi.org/10.1016/j.tcb.2011.09.005.
99. van der Meer AD, van den Berg A. Organs-on-chips: breaking the in vitro impasse. *Integr Biol (Camb)*. 2012;4(5):461–470. https://doi.org/10.1039/c2ib00176d.

CHAPTER EIGHT

Microfluidics based point-of-care for disease diagnostics

Gargi Bhattacharjee[a], Rupesh Maurya[a], Khalid J. Alzahrani[b], Nisarg Gohil[a], Navya L. Lam[c], and Vijai Singh[a,*]

[a]Department of Biosciences, School of Science, Indrashil University, Rajpur, Mehsana, Gujarat, India
[b]Department of Clinical Laboratories Sciences, College of Applied Medical Sciences, Taif University, Taif, Saudi Arabia
[c]The J. David Gladstone Institutes, San Francisco, CA, United States
*Corresponding author: e-mail addresses: vijaisingh15@gmail.com; vijai.singh@indrashiluniversity.edu.in

Contents

Abstract

Microfluidics platform is widely used for several basic biological to advanced biotechnological applications. It reduces the expenditure of reagent consumption by readily reducing the volume of the reaction system. It is being used for early diagnosis of diseases, detection of pathogens, cancer markers, high-throughput screening and many such applications. Currently, microfluidics and lab-on-chip is integrated together with sample preparation, extraction, analysis and detection of biomarkers for disease diagnosis. This technology offers low-cost, rapid, sensitive and paper-based lateral flow mode of detection which is user-friendly and scalable. In this chapter, we highlight recent developments in microfluidics platform for disease diagnosis.

1. Introduction

A healthy life-style, availability of enriched food quality, potent drugs, clean energy and environment are utmost essential to fit and survive in the alarming environmental condition. On the other hand, with surge in the world population, a rapid increase in the incidences of communicable and non-communicable (cancer, cardiovascular, diabetes etc.) diseases have been observed worldwide which has led to increase in morbidity and

Progress in Molecular Biology and Translational Science, Volume 187
ISSN 1877-1173
https://doi.org/10.1016/bs.pmbts.2021.07.024
Copyright © 2022 Elsevier Inc.
All rights reserved.

mortality. This has been a major challenge and a matter of great concern amongst scientists working in this area to tackle and properly treat a disease. Disease diagnostics plays a key role to initiate proper treatment of diseases.[1] Mankind is still lacking behind in devising a groundbreaking, rapid, sensitive and portable disease diagnostics method for detection of pathogens or gene-based mutation. The existing diagnostic methods or tools (PCR, immunoassay, lateral-flow, nano-sensor, etc.) still lagging behind to outreach to every nook and corner of the world, especially in developing or underdeveloped countries or economically benighted population.

A constant need arises to develop an ideal diagnostic platform that is rapid, sensitive, specific and cost-effective for routine monitoring of diseases. The platform requires to be user-friendly and should be able to give instrument-free readouts for early detection of causative agents. Nowadays, the advent of number of new diagnostic methods or tools including CRISPR-based rapid diagnostic tools[2–6] for detection of number of pathogens and diseases has changed the perspective of modern diagnostic systems. A rapid, sensitive and high-throughput detection has been made possible with recent developments in microfluidics[7] or lab-on-chip-based point-of-care (POC) disease diagnostics which has recently gained huge scientific and public attention due to ease of usage and eco-friendly features. It has been designed in a way that has readily simplified sample preparation, extraction of nucleic acid and analysis post the reaction.[8] Bringing in automation and integrating it with electronics and optics has made it more convenient and handy. These modern diagnostic tools are now being implemented in healthcare and smart sensors for non-invasive disease diagnosis for accurate treatment.[9] In this chapter, we highlight microfluidics for POC diagnostic for managing and curbing progression of diseases.

2. Microfluidics based point-of-care for infectious disease diagnosis

Infectious diseases are responsible for higher morbidity and mortality especially in a large population. The global infectious disease market size is expected to reach USD 86.2 billion in 2025.[10] Growth of this market is driven by POC diagnostics and research funding in the area associated with infectious diseases. The global burden of infectious disease stresses upon the need to develop neoteric and advanced POC diagnostics. While molecular biology based PCR method is widely used for detection of pathogens, it is hampered by it need for infrastructure, cost and trained-manpower. Microfluidics technology can be designed by integration of miniaturized PCR process, sample preparation, amplification and sample analysis in a hassle-free manner to aid infection disease detection.[11]

Romao et al.[12] developed a portable microfluidics device for early detection of pathogens including Ebola virus, Coronavirus and Zika virus (Fig. 1). Lab-on-chip devices are affordable and have been successful in drawing attention of the scientific community. It has been used as a rapid diagnostic tool to test, track and isolate patient's samples to ensure better treatment.[12] Foudeh et al.[32] reported about the advantage that microfluidics lab-on-chip offers in pathogen detection in fully automated procedure including sample preparation, small volume processing, extraction and subsequent detection of nucleic acid.

The novel coronavirus (SARS-CoV-2 virus) causes severe pulmonary COVID-19 disease which can be prevented if diagnosed at an early stage.[13] Conventional immunoassays and quantitative real-time polymerase chain reaction (qRT-PCR) are commonly used for diagnosing coronavirus which is time-consuming, require skilled manpower and difficult to extend to remote areas. Therefore, POC biosensors-based on chip or paper-based sensors can be a preferred choice which is normally low-cost and user-friendly,

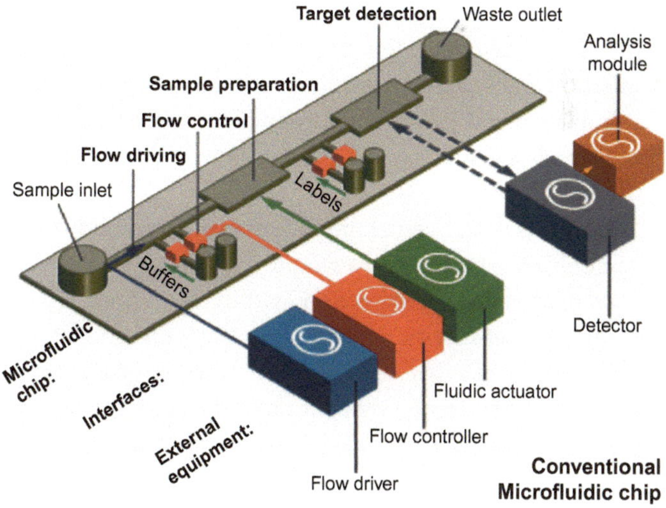

Fig. 1 Design and fabrication of fully automation microfluidic chip for detection of biomarkers. *Figure reproduced with permission from Romao VC, Martins SAM, Germano J, Cardoso FA, Cardoso S, Freitas PP. Lab-on-chip devices: gaining ground losing size.* ACS Nano. *2017;11:10659–10664. © ACS.*

and also offers rapid and sensitive diagnosis. In order to detect coronavirus, Wang et al.[14] reported about integrating lab-on-chip, lateral-flow, lab-on-a-disc (LOAD) devices, microfluidic paper-based analytical devices (μPADs), and isothermal nucleic acid amplification (INAA) can be integrated and using it for rapid and sensitive detection of coronavirus for controlling its spread.

POC diagnostics allow rapid and sensitive diagnosis of infectious diseases which is crucial for improving public health. POC platform is affordable, portable, automated and efficient for diagnostic applications.[15,16] Some of these POC tools are additionally integrated with immunoassay for detection of pathogens, over and above the commonly assembled different chambers including sample preparation, extraction and separation of nucleic acid.

3. Microfluidics-based point-of-care for disease diagnostics

Rapid, sensitive and specific diagnostic options are important from clinical point of view when using samples of bodily fluid including saliva, blood and urine collected from patients and analyzed for presence of various biomarkers. This could be really helpful for monitoring disease progression, response to treatment and patient's prognosis.[17–20] A conventional microfluidics lab-on-chip device has been modified for point-of-care diagnosis of diseases[20,21] based on continuous flow in microchannels which is restricted in reconfiguration and scalability. In that work, the tool was designed as fully integrated droplets based "digital" microfluidics lab-on-chip for diagnosing human body fluids including saliva, urine, blood, sweat, tear, serum and plasma.[20]

In addition, Herr et al.[22] developed a portable microfluidics POC device for detection of biomarkers in saliva sample of patients. It has been tested for chip-based immunoassay for detection within 3–10 min with just 10 μL sample volume and sensitivity limit ranging from nM to pM. This devise was integrated with the facility of sample pre-treatment with electrophoretic immunoassay and it was noted that it required less pre-treatment of saliva samples. Further integration with electronics, optical elements and software helped to expedite the detection of biomarkers in saliva and in future the same can be extended for other body fluids, too. In a subsequent study, Herr et al.[23] developed POC diagnostic for rapid quantification of oral disease biomarkers from patient's saliva. It facilitates touch-free saliva analysis which is integrated with sample preparation and electrophoretic immunoassay. Initially, 20 μL of saliva sample was used to measure collagen-cleaving enzyme matrix metalloproteinase-8 (MMP-8) and was detected in <10 min. This study allows rapid and reliable detection of proteinaceous disease markers from body fluids.

In order to improve treatment and clinical outcomes, microfluidics technology has proved to be useful technology for rapid and sensitive detection of proteins, DNA and RNA present at even low concentrations in samples. Currently, with advancements in microfluidics, and its integrated with automation have greatly reduced the chances of errors. The small quantity of sample that is used for POC diagnostic applications increases the rapidity and accuracy of the technique.[18] An automated microchip has been designed and used for rapid extraction and processing of samples at very small scale for diagnostic applications.[17]

According to Wang et al.[24] the use of paper and polymers as substrates in POC diagnostics makes it highly affordable, addresses disposability issues and facilitates its large production. To date, conventional techniques including ELISA and PCR are the first to be used for diagnosis of diseases. Microfluidics is user friendly and offers non-invasive diagnosis not just from saliva, but also from urine and blood samples.[25,26] Certainly, microfluidics technology has the potential to reduce time and reagent consumption price for detection towards accurate disease treatment.[27] Siegrist et al.[19] used microfluidics for medical diagnostics based on nucleic acids. The sample preparation, amplification and detection were all integrated in one device. In that work, a centrifugal microfluidics platform was used for mechanical lysis of samples and homogenization, following which the supernatant was transferred for extraction using siphon. This tool has been used for detection of *Bacillus subtilis* and respiratory virus. The samples were ready for amplification within 6 min which is indeed rapid for POC diagnosis of clinical samples.

Kim et al.[28] designed microfluidics-based microsystems fabricated for cell lysis using mechanical, chemical, thermal and electrical disruption of cells. It was integrated with nucleic acid separation methods using silica-based surface affinity and nano-porous membrane filtration. This facilitates the post-processing of samples. Microfluidics offers and enables miniaturization, integration and automation for measuring biomolecules. Some may also be designed with capillary test strips which is called as "microfluidic large scale integration" that is pressure-driven droplet-based and electrowetting based microfluidics. This allows sample transportation, mixing, incubation, separation, droplet formation, splitting, dispensing and detection of biomarkers all in one go[29] and can be used to meet several biomedical applications.

Cardiovascular diseases (CVDs) rank foremost in causing morbidity and mortality worldwide. CVDs are preventable and manageable with early diagnosis, prognosis and treatment.[30] Currently, due to advancements in microfluidics integrated with nanotechnology, a fully automated and affordable device has been designed for early detection of CVDs. The technology is further connected with smartphones in order to receive information about measuring and analyzing physiological indexes.[30] Li et al.[31] developed a POC testing for multiplex biomarkers of non-small cell lung cancer (NSCLC). It allowed to perform timely surgery and to initiate treatment for saving lives. A nanoporous glass (NPG) integrated with volumetric bar-chart chip (V-Chip) was used for detection of CEA, CYFRA 21-1, and SCCA from NSCLC by naked eye. The NPG-V-chip enabled detection within 50 pg/mL (10-fold higher than detection on V-chip). The total detection time also reduced from 4 to 0.5 h. It can be further extended for diagnosis of other cancer cells.

4. Conclusion and future remarks

Microfluidics and lab-on-chip technology is rapidly evolving and expanding exponentially for many applications from basic biology to biotechnology, biomedical and translational sciences. Rapid, sensitive and user-friendly diagnosis of diseases plays a vital role for detection of biomarkers toward better treatment of diseases. POC rapid diagnostics aims at diagnosing a diseased state accurately at an early stage before it advances to a stage that is hard to be reversed. It can be extended and designed to offer low-cost, rapid, more sensitive and reliable kit for diagnosis of diseases to economically deprived or remote areas. Therefore, once a disease is sensitively detected, it becomes easier to treat and save lives. Integrating microfluidics with lab-on-chip or paper-based lateral flow makes it simple by design and the automation reduces errors of experiments, thus endowing highly sensitive and specific mode of detection systems.

Acknowledgments

The financial support from Gujarat State Biotechnology Mission (GSBTM) (Project ID: 5LY45F), Gujarat, India to G.B. and V.S. is duly acknowledged. N.G. acknowledges the Indian Council of Medical Research, Government of India for financial assistance as Senior Research Fellowship (File No. 5/3/8/63/ITR-F/2020). G.B., R.M., N.G. and V.S. thank Indrashil University, Rajpur, Mehsana, India for providing the infrastructure facility.

References

1. Pandey CM, Augustine S, Kumar S, et al. Microfluidics based point-of-care diagnostics. *Biotechnol J.* 2017;13:1700047. https://doi.org/10.1002/biot.201700047.
2. Bhattacharjee G, Gohil N, Lam NL, Singh V. CRISPR-based diagnostics for detection of pathogens. *Prog Mol Biol Transl Sci.* 2021;181:45–57.
3. Gootenberg JS, Abudayyeh OO, Kellner MJ, Joung J, Collins JJ, Zhang F. Multiplexed and portable nucleic acid detection platform with Cas13, Cas12a, and Csm6. *Science.* 2018;360:439–444.
4. Gootenberg JS, Abudayyeh OO, Lee JW, et al. Nucleic acid detection with CRISPR-Cas13a/C2c2. *Science.* 2017;356:438–442.
5. Khambhati K, Bhattacharjee G, Singh V. Current progress in CRISPR-based diagnostic platforms. *J Cell Biochem.* 2019;120:2721–2725.
6. Myhrvold C, Freije CA, Gootenberg JS, et al. Field-deployable viral diagnostics using CRISPR-Cas13. *Science.* 2018;360:444–448.
7. Carrera J, Rodrigo G, Singh V, Kirov B, Jaramillo A. Empirical model and *in vivo* characterization of the bacterial response to synthetic gene expression show that ribosome allocation limits growth rate. *Biotechnol J.* 2011;6:773–783.
8. Kim JA, Lee JY, Seong S, et al. Fabrication and characterization of a PDMS–glass hybrid continuous-flow PCR chip. *Biochem Eng J.* 2006;29:91–97.
9. Cui P, Wang S. Application of microfluidic chip technology in pharmaceutical analysis: a review. *J Pharm Anal.* 2019;9:238–247.
10. Infectious Disease Therapeutics Market. *Infectious Disease Therapeutics Market Worth $86.2 Billion by 2025.* Grand View Research; 2017. https://www.grandviewresearch.com/press-release/global-infectious-disease-therapeutics-market. Accessed on 15.05.2021.
11. Park S, Zhang Y, Lin S, Wang TH, Yang S. Advances in microfluidic PCR for point-of-care infectious disease diagnostics. *Biotechnol Adv.* 2011;29:830–839.
12. Romao VC, Martins SAM, Germano J, Cardoso FA, Cardoso S, Freitas PP. Lab-on-chip devices: gaining ground losing size. *ACS Nano.* 2017;11:10659–10664.
13. Foudeh AM, Didar TF, Veres T, Tabrizian M. Microfluidic designs and techniques using lab-on-a-chip devices for pathogen detection for point-of-care diagnostics. *Lab Chip.* 2012;12:3249–3266.
14. Choi JR. Development of point-of-care biosensors for COVID-19. *Front Chem.* 2020;8:517.
15. Wang C, Liu M, Wang Z, Li S, Deng Y, He N. Point-of-care diagnostics for infectious diseases: from methods to devices. *Nano Today.* 2021;37, 101092.
16. Su W, Gao X, Jiang L, Qin J. Microfluidic platform towards point-of-care diagnostics in infectious diseases. *J Chromatogr A.* 2015;1377:13–26.
17. Zarei M. Infectious pathogens meet point-of-care diagnostics. *Biosens Bioelectron.* 2018;106:193–203.
18. Cui F, Rhee M, Singh A, Tripathi A. Microfluidic sample preparation for medical diagnostics. *Annu Rev Biomed Eng.* 2015;17:267–286.
19. McCalla SE, Tripathi A. Microfluidic reactors for diagnostics applications. *Annu Rev Biomed Eng.* 2011;13:321–343.
20. Siegrist J, Gorkin R, Bastien M, et al. Validation of a centrifugal microfluidic sample lysis and homogenization platform for nucleic acid extraction with clinical samples. *Lab Chip.* 2010;10:363–371.
21. Srinivasan V, Pamula VK, Fair RB. An integrated digital microfluidic lab-on-a-chip for clinical diagnostics on human physiological fluids. *Lab Chip.* 2004;4:310–315.
22. Hart RW, Mauk MG, Liu C, et al. Point-of-care oral-based diagnostics. *Oral Dis.* 2011;17:745–752.
23. Herr AE, Hatch AV, Giannobile WV, et al. Integrated microfluidic platform for oral diagnostics. *Ann N Y Acad Sci.* 2007;1098:362–374.

24. Herr AE, Hatch AV, Throckmorton DJ, et al. Microfluidic immunoassays as rapid saliva-based clinical diagnostics. *Proc Natl Acad Sci U S A*. 2007;104:5268–5273.
25. Wang S, Chinnasamy T, Lifson MA, Inci F, Demirci U. Flexible substrate-based devices for point-of-care diagnostics. *Trends Biotechnol*. 2016;34:909–921.
26. Giannobile WV, McDevitt JT, Niedbala RS, Malamud D. Translational and clinical applications of salivary diagnostics. *Adv Dent Res*. 2011;23:375–380.
27. Gorkin R, Park J, Siegrist J, et al. Centrifugal microfluidics for biomedical applications. *Lab Chip*. 2010;10:1758–1773.
28. Sackmann EK, Fulton AL, Beebe DJ. The present and future role of microfluidics in biomedical research. *Nature*. 2014;507:181–189.
29. Kim J, Johnson M, Hill P, Gale BK. Microfluidic sample preparation: cell lysis and nucleic acid purification. *Integr Biol (Camb)*. 2009;1:574–586.
30. Haeberle S, Zengerle R. Microfluidic platforms for lab-on-a-chip applications. *Lab Chip*. 2007;7:1094–1110.
31. Hu J, Cui X, Gong Y, et al. Portable microfluidic and smartphone-based devices for monitoring of cardiovascular diseases at the point of care. *Biotechnol Adv*. 2016;34: 305–320.
32. Li Y, Xuan J, Song Y, et al. Nanoporous glass integrated in volumetric bar-chart chip for point-of-care diagnostics of non-small cell lung cancer. *ACS Nano*. 2016;10: 1640–1647.

CHAPTER NINE

Lab on a body for biomedical electrochemical sensing applications: The next generation of microfluidic devices

Itthipon Jeerapan[a,b,*], Chochanon Moonla[c], Panote Thavarungkul[a,b], and Proespichaya Kanatharana[a,b]

[a]Division of Physical Science, Faculty of Science, Prince of Songkla University, Hat Yai, Songkhla, Thailand
[b]Center of Excellence for Trace Analysis and Biosensor, Prince of Songkla University, Hat Yai, Songkhla, Thailand
[c]School of Chemistry, Institute of Science, Suranaree University of Technology, Nakhon Ratchasima, Thailand
*Corresponding author: e-mail address: itthipon.j@psu.ac.th

Contents

Progress in Molecular Biology and Translational Science, Volume 187
ISSN 1877-1173
https://doi.org/10.1016/bs.pmbts.2021.07.025
Copyright © 2022 Elsevier Inc.
All rights reserved.

Abstract

This chapter highlights applications of microfluidic devices toward on-body biosensors. The emerging application of microfluidics to on-body bioanalysis is a new strategy to establish systems for the continuous, real-time, and on-site determination of informative markers present in biofluids, such as sweat, interstitial fluid, blood, saliva, and tear. Electrochemical sensors are attractive to integrate with such microfluidics due to the possibility to be miniaturized. Moreover, on-body microfluidics coupled with bioelectronics enable smart integration with modern information and communication technology. This chapter discusses requirements and several challenges when developing on-body microfluidics such as difficulties in manipulating small sample volumes while maintaining mechanical flexibility, power-consumption efficiency, and simplicity of total automated systems. We describe key components, e.g., microchannels, microvalves, and electrochemical detectors, used in microfluidics. We also introduce representatives of advanced lab-on-a-body microfluidics combined with electrochemical sensors for biomedical applications. The chapter ends with a discussion of the potential trends of research in this field and opportunities. On-body microfluidics as modern total analysis devices will continue to bring several fascinating opportunities to the field of biomedical and translational research applications.

1. Introduction

"Lab on a body" is described as a new class of miniaturized total analysis systems in a platform that can operate in vivo, such as a transdermal sensing patch for biochemical analysis in sweat.[1,2] Such a system ideally transforms several units of a full analytical laboratory into a miniaturized and integrated platform.[3] The careful design combines numerous functional building blocks into a single patch of bioelectronics. The functions of such microfluidics range from biofluid collection,[4] storage,[5] separation,[6,7] elimination of interferences,[8] reactions,[9,10] amplification,[11] and chemical analysis.[12] Unlike the diagnostic device without microfluidics, the unique feature of microfluidics is manipulating the fluid flow. Emerging diagnostic technology aims to obtain comprehensive insight into how concentrations of chemicals (such as glucose,[1,13] lactate,[1,14] electrolytes,[13,15] and hormones[16]) rise up or drop in a continuous manner. Therefore, the development of microfluidics is essential to enable the continuous replacement of biofluids (such as sweat,[17,18] tear,[19,20] and saliva[19,20]) on the electrochemical sensor throughout the detection. This chapter summarizes the advance of on-body microfluidic biosensing devices that open special opportunities to track the dynamic change of chemical parameters.

Miniaturization of chemical sensors is a key in the design of micro total analysis devices; electrochemical sensors are among the best candidates of many detectors.[21–23] Even though the fabricated microfluidic channel is small, the total size of the full system may be large when considering other integrated components, including detectors. The large detector will limit practical applications for on-body analysis. For example, downscaling other techniques, such as spectroscopy, is complex. In fact, other than electrochemical detectors, it is impossible to shrink many chemical analyzers down to a few mm scale. Hence, it is required to address the key challenge involving the fabrication of tiny detectors that can monitor multiple analytes. Advantageously, electrochemical sensors satisfy many requirements for lab-on-a-body devices. We can downscale sensing electrodes to locate at the detection zone of microfluidics, while not deteriorating the performances of sensors. Moreover, microelectrodes enhance the quality of analytical signals and require small sample volumes.[24] In addition to the opportunity of using small electrodes, the recent advance of stretchable and flexible electrodes paves a crucial supporting path for the seamless integration with on-body microfluidic devices.[25,26] Another desirable feature of electrochemical sensors is their responses, e.g., voltage,[27] current,[28] impedance.[29] These electrical quantities are ready to be integrated with smart electronics and information and communication technology (ICT).

This chapter discusses one of the most updated aspects of microfluidics applied to on-body biosensors (Fig. 1). Some advanced microfluidic components toward the development of on-body microfluidic devices are elaborated. Expanding from recent review articles and book chapters on microfluidics,[2,12,30–32] novel technologies are updated in this chapter. Importantly, the acceleration in the development of microfluidics and miniaturized electrochemical sensors allows the incorporation of wireless and automatic systems, pushing the technology further toward smart personalized healthcare monitoring and digital interactive networks. The convergence of research areas, including soft electronics, wireless communication, and on-body sensors paves the way for the fabrication of total analysis systems for in vivo electrochemical sensings. Representative devices for many analytes, such as glucose, lactate, and electrolytes, are also presented. The discussion will emphasize perspiration obtained from the skin. However, principles can be extended for other biofluids. We also point out future opportunities in this translational research development.

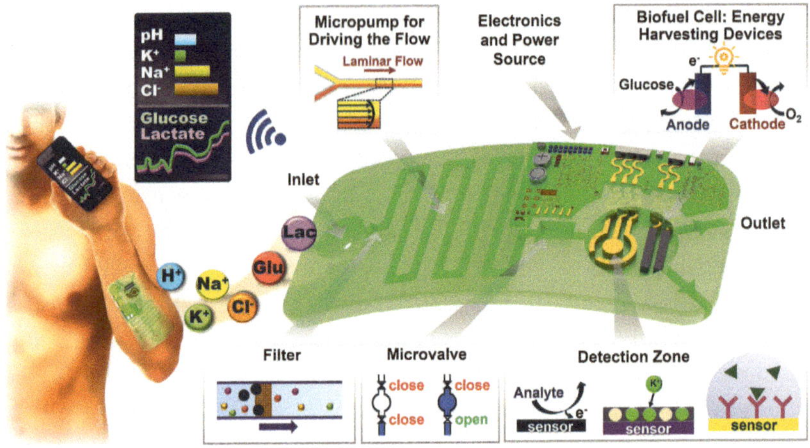

Fig. 1 On-body microfluidic systems that open opportunities for biomedical electrochemical sensing applications. The device captures and analyzes biomarkers in biofluids to reveal informative profiles. The integrated bioelectronics enables smart personalized healthcare applications.

2. Concepts of on-body microfluidic sensing devices

The term "microfluidics" can be called when a dimension of enclosed channels is in the range of a few μm or smaller than mm.[33] Typically, a small volume of biofluids, such as sweat around μL or less than nL ranges, is manipulated precisely on a microscale device.[34,35] As shown in Fig. 1, the on-body microfluidic sensing device mainly includes microfluidics (microchannels, microreservoirs, microvalves, or micropumps), detectors, power sources, and electronics, integrated within a single device.

In addition to the traditional puncture to the skin, an alternative approach of collecting biofluids toward the inlet of the microfluidic network can be done passively or actively (by applying electrical[36,37] or thermal[38,39] energy across the skin). Microfluidics serve to sample the biofluid, then store it in reservoirs, or transport it through microchannels to detection zones. One of the most significant building blocks of microfluidics is the microvalve controlling the passage of samples or reagents through the microchannel to micromixing units or detectors with programmable actuation. Apart from the active pumping of the fluid by external controllers, the combination of the capillary system (small fluidic channels or porous fabric/paper substrates) and the draining system at the waste reservoir (e.g., due to natural evaporation) induces natural micropumping, allowing the design of fluid-flow systems without the need of external pumps. When the fluid reaches the detector, the chemical information of the biofluid is analyzed. The integration of state-of-the-art electronics interfaced with microfluidics offers the miniaturization of sensing and data-acquisition systems. Electronics, such as potentiostats and wireless-communication modules, are essential to control the sensor, process data, and transmit signals to users. Inevitably, electronics and microfluidics require efficient energy systems. Strategies to realize self-sustainable devices should thus be developed to meet the requirements of on-body microfluidics (such as low-power consumption). One strategy is integrating the assembled energy-harvesting/conversion[40,41] (e.g., biofuel cells and supercapacitors) and energy-efficient electronics[1,42] to communicate between sensing microfluidics and data receivers (e.g., smartphones). Finally, at the outlet (waste zone), the excessive fluid drains out, allowing the replenishment of fresh biofluid samples. Not limit to sweat, this example of the concept can be applied for other in vivo microfluidic devices, such as the diagnosis in blood,[22,31,43] tear,[17,18] and saliva.[19,20]

3. Challenges when developing on-body microfluidic sensing devices

When transforming common microfluidics into on-body microfluidic devices coupled with chemical sensors, several challenges deserve special attention:

1. Collecting and handling the small volume of bio-samples represent key challenges. The normal resting state generates only a small volume of sweat. The volume of tears is also small. Hence, on-body microfluidic devices coupled with sensors require efficient strategies to collect enough biofluids for continuously tracking human physiological information. Due to the ultralow sample volume, sample evaporation and transportation need to be considered and well-handled.
2. Another grand challenge is the low concentration of analytes available in untreated biofluid matrixes comprising of a variety of species. For example, when detecting glucose in sweat (10–200 μM range[30]), possible interfering agents can be lactate, uric acid, ascorbic acid, hormones, therapeutic drugs, electrolytes, or even microorganisms. When directly analyzing biofluid samples in the microfluidic device, electroactive compounds, such as uric acid and ascorbic acid, can cause matrix effects, representing a source of inaccuracy in the chemical analysis of glucose.

Moreover, materials used for microfluidic fabrication should not absorb the analyte. For instance, poly(dimethylsiloxane) (PDMS) has low interaction with glucose and pyruvate, while it may trap vitamins and hormones.[44,45]

3. The traditional system of microfluidic lab-on-chip devices relies on external instrument setups such as microsyringe pumps and valve systems with external power sources. It is needed to be simplified and minimized for true on-body sensing applications. A self-sustainable system holds the answer to this challenge. For example, downscaling passive integrated designs and autonomous self-powered sensing devices while maintaining standards should be considered.
4. The architectural design of the on-body microfluidic devices is one of the most important parameters. The detailed understanding of complex geometries and materials of sensors is a fundamental guide when designing microfluidics to meet requirements (e.g., stability and wearability).
5. We need to explore substrate materials and components which provide satisfactory stretchability, flexibility, and biocompatibility to acquire a body-conformal platform since mechanically compliant features play key roles in improving the wearability of sensing microfluidics. Materials with low modulus and elastic mechanical properties (e.g., rubbers, PDMS, and hydrogels) are suitable. The use of hybrid materials is an alternative approach that can offer reliable measurement. The high modulus part of the microfluidic integrated device provides a robust platform with minimal deformations, while the soft area (low modulus materials) facilitates wearability.[46]
6. In addition to provide reliable information, on-body microfluidic sensing devices are expected to be simple and user-friendly for personalized healthcare applications by employing a smart system. This holds another key challenge to manage big data, effectively regulate and simplify data collection, storage, and processing.
7. Microfluidics in practical applications requires viable energy to supply several functions, such as flow management (e.g., pumps and valve controllers), electrochemical detection, and data transfer between devices and receivers (e.g., smartphones) via a connectionless fashion. To invent self-sustainable systems, the use of intelligent technology such as near field communication (NFC) technology and energy-harvesting modules (such as biofuel cells to harvest electricity from biofluids themselves) holds great significance to overcome the limitation of practical portability and miniaturization.[47,48]

8. Among various current fabrication techniques, it is challenging to develop a universal strategy that meets the requirement of simple, efficient, and cost-effective manufacturing for mass production of sophisticate miniaturized microfluidic sensing devices.
9. Inherently, unlike the fluid mechanism in a big duct, the fluid behavior in microfluidics is unique, related to the interface and surface.[35,49] Therefore, fluid mechanics should be kept in mind when developing on-body microfluidics.

4. Advances of microfluidic components toward the development of on-body microfluidic devices

On-body microfluidics require several components to control the flow of fluids that are geometrically confined to a miniaturized scale. Microfluidic systems include functional units for processing fluids, e.g., transportation, separation, and mixing. Note that the heart of microfluidic is manipulating fluids. To govern the fluids, it is essential to develop fluid chambers (such as microreservoirs), flow resistors (such as valves or chromatographic columns for separation or extraction), flow accelerators (such as pumps), and flow turbulation (such as mixers). This section describes some designs, potentially used as key components in lab-on-a-body devices.

4.1 Sampling systems

A sampling system allows biofluid samples to enter the microfluidic device. It should provide capabilities to reliably collect fluid samples, even at a very small volume. An example design is Archimedean spiral shape (Fig. 2A).[4] The biofluid is first collected at the sweat collection chamber, then the pressure will push the sweat further to the fluid channel. It is also essential to minimize volume in the chamber zone by designing a spacer. Moreover, the filtration system should be integrated at the inlet zone to decrease contamination.[50]

Ideally, researchers aim to analyze biofluids qualitatively and quantitatively with a time-stamping profile. It is crucial to design the sampling system to track time-dependent information. An example was demonstrated by designing transdermal microfluidic systems with passive galvanic stopwatches, as shown in Fig. 2B.[51] Five chambers (capturing 50 μL of biofluid per chamber) are connected with passive valves. The galvanic stopwatch, consisting of a tetrathiafulvalene-based anode and a Pt-based cathode, is integrated at the entrance of each chamber. The anode displays a potential

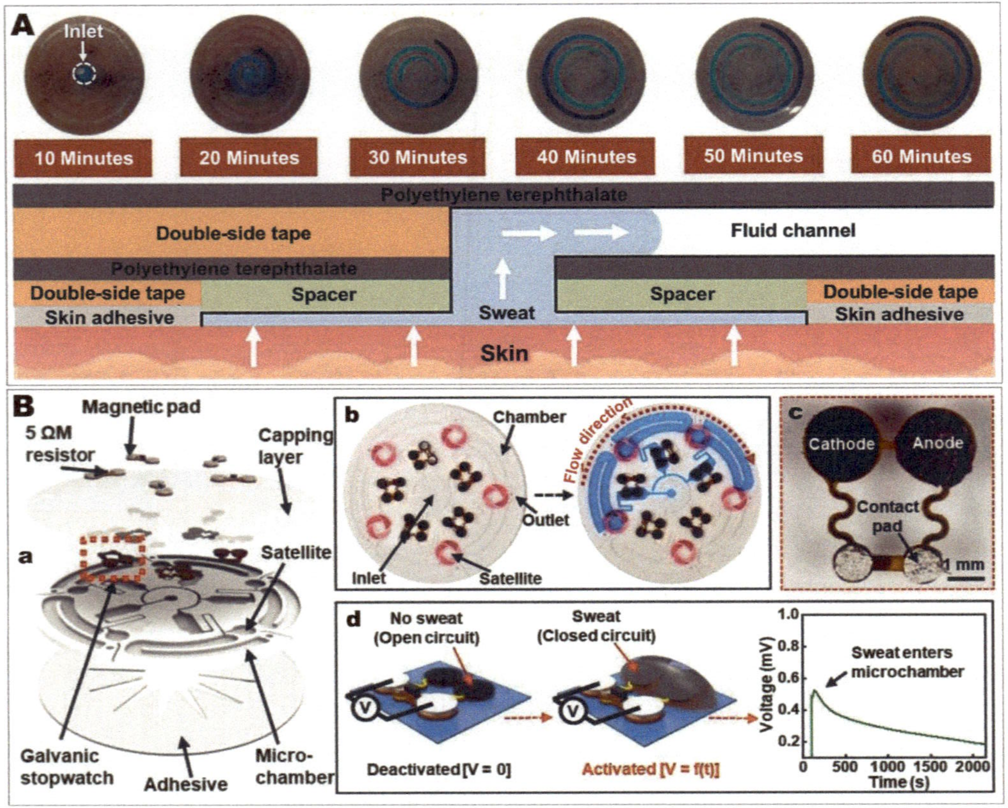

Fig. 2 Microfluidic biofluid sampling systems. (A) Wearable microfluidic biofluid collectors: (top) in situ collection in the sweat collector; (bottom) the cross-sectional view of the collector. (B) Passive galvanic stopwatches for chronological sampling: (a) the schematic of the chronosampling soft microfluidic system; (b) photographs of the device before and during filling with the blue-dyed solution; (c) the photograph of a galvanic cell; (d) the operation of the stopwatch and a voltage–time curve. *Panel (A): Adapted with permission from Lin H, Zhao Y, Lin S, et al. A rapid and low-cost fabrication and integration scheme to render 3D microfluidic architectures for wearable biofluid sampling, manipulation, and sensing.* Lab Chip. *2019;19(17):2844–2853, Copyright 2019, Royal Society of Chemistry. Panel (B): Adapted with permission from Bandodkar AJ, Choi J, Lee SP, et al. Soft, skin-interfaced microfluidic systems with passive galvanic stopwatches for precise chronometric sampling of sweat.* Adv Mater. *2019;31(32):1902109, Copyright 2019, Wiley-VCH.*

of 0 V vs Ag/AgCl, while the cathode shows a high cathodic potential of 0.6 V vs Ag/AgCl due to the oxygen reduction reaction. The net voltage of the cell, therefore, reaches 0.6 V when exposing to electrolyte-containing biofluids. When the biofluid reaches the galvanic system, the cell displays the spontaneous voltage, indicating chronological signals of voltage. Such time-dependent voltage signals reveal the time-stamped biofluid collection.

4.2 Microchannels: Soft microfluidic platforms

Common strategies to fabricate on-body microfluidics rely on the use of soft materials, such as PDMS[52] and Ecoflex.[11] The high elasticity is preferable (moduli of kPa or few MPa) as microfluidics need mechanical compliance to be compatible with daily activities involving omnidirectional motions. In addition to the commonly used PDMS, a poly(styrene-isoprene-styrene)-based device can retain bio-samples for 4 h with less than 20% loss.[53] Moreover, epidermal fluidics would require water barrier properties when using in real applications, e.g., for swimmers.

4.3 Flow controlling systems

The way to sample and control the flow of biofluids is the key to develop microfluidics. In general, we can use a simple and passive way due to the pressure of the natural biofluid flow (e.g., a sweat gland produces the pressure up to 70 kPa[54]) and the capillary force of the microchannel. The passive design holds a great promise to minimize power consumption. Therefore, the flow of transdermal microfluidic devices can be inherently and continuously pushed for real-time biofluid analysis. Apart from passive approaches, several strategies (relying on piezoelectrics,[55] electrokinetics,[56] and surface chemistry,[57] etc.) can be applied to govern the flow in microfluidics.

4.4 Fabric-based microfluidic devices: Microchannels to enhance flow

Fiber-based platforms provide advantages because sewing, weaving, and embroidering techniques can be integrated. In addition, they allow seamless integrations; for example, sensing electrodes in the form of threads[58,59] suggest potentials toward the wearable integration. Importantly, the flow speed in microfluidics can be enhanced by using the wicking principle that helps to absorb the fluid.[60,61] The hydrophilicity of the fluidic track and the fluid viscosity affect the flow rate. The hydrophobicity can be tailored by material modification, such as ultraviolet[62] and corona discharge[63] treatments. In general, the flow rate also depends on the hydrophilic channel width.

The fluid velocity is low when using a large channel width. This phenomenon can be estimated by Young–Laplace equation.[64] The hydrophilic thread is an example that can be attached with hydrophobic substrates (including textiles).[65] This idea offers passive sampling and attractive wearability for microfluidics that can wick the biofluid.

4.5 Microvalves

A microvalve is another key component that opens and closes to manage the flow of biofluids or reagents. It is necessary to include the valve when flow rate variability can influence the detection. For example, the variation in flow rate directly affects the amperometric signal in flow analysis due to hydrodynamic conditions at the electrode sensor.[66] Moreover, many analytical protocols (such as immunoassays) involve multistep and many chambers that require the precise control of time and fluid flow. Large mechanical tools or optical actuation systems used for traditional bench-top microfluidics may not be convenient to regulate the valve for true on-body applications.

Valves can be categorized into passive and active valves. The former can eliminate the extra unit to actuate the valve; therefore, it is extensively used in wearable microfluidics.[67] Considerations to design the valve depend on applications, which will be elaborated in this section. Hydrophobic valves[68] and capillary bursting valves[39] are among the most attractive passive systems, representing the great potential for lab-on-a-body microfluidics. The mechanism relies on microfluidic networks pushing fluids via capillary forces generated by microchannels or osmotic pressure produced by hydrogel materials. The key advantage is to eliminate external actuators, energy sources, or supplementary modules to trigger the valve, allowing the miniaturization and simplification of the system. In addition to passive valves, active valves can be programmed to manage the fluid. For example, an electronically controllable valve is an attractive alternative as mechanical parts in mechanical valves are susceptible to dysfunction and hard to control in small fluidics. Some representatives are discussed.

4.5.1 *Capillary bursting valves*

The passive capillary bursting valves can be integrated with soft microfluidics for chronological capture of biofluids (Fig. 3A).[39] Microreservoirs are connected to form interconnected series. Capillary bursting valves are engineered to open at different pressures; for example, the bursting pressure (BP) of the valve #3 is very high (3 kPa). The estimated BPs for valves #1

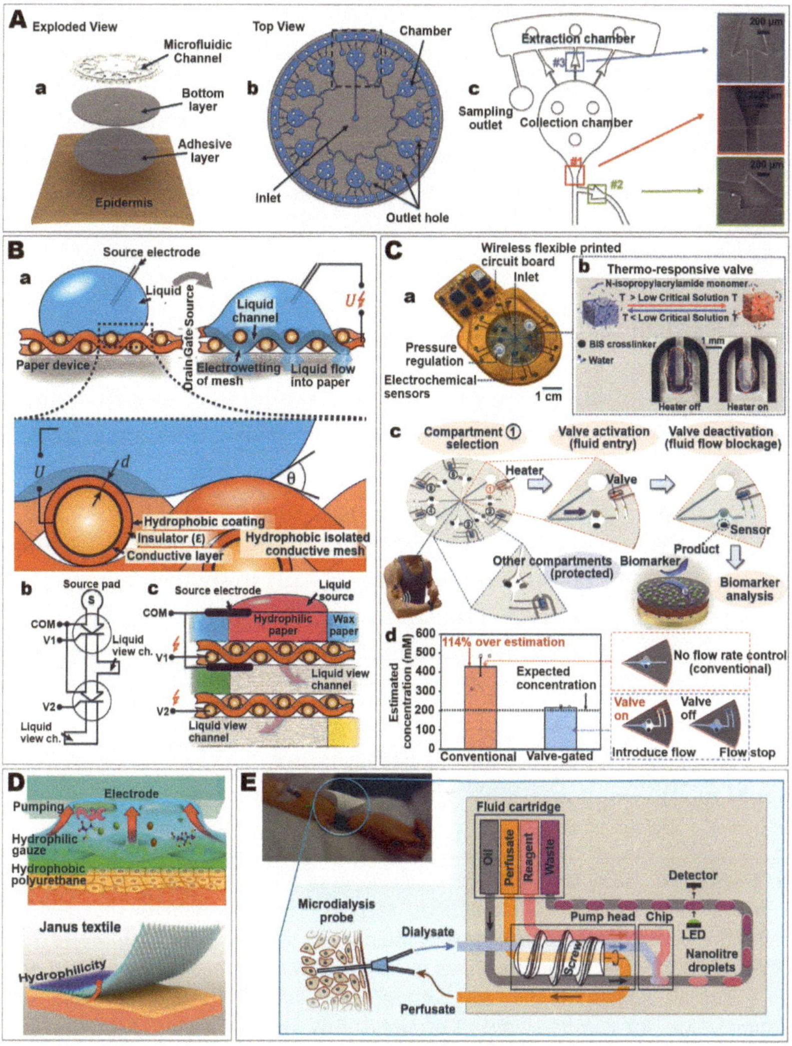

Fig. 3 Examples of microvalving, micropumping, and micromixing systems. (A) Capillary bursting valves for chrono-sampling of sweat: (a and b) schematic illustrations of epidermal microfluidic channels; (c) the illustration of a unit cell in a device, including a collection chamber, extraction chamber, sampling outlet, and three capillary bursting valves and microscopic images of these valves. (B) Electrical textile valves: (a) the operation of the valve by the application of a voltage between the liquid and the electrotextile, permitting the liquid to flow; (b) the diagram of two electrofluidic valves; and (c) the device containing microfluidic layers separated by two layers of valves. (C) Thermo-responsive valves: (a) the wearable bioanalytical microfluidic device; (b) shrinkage/expansion of hydrogel; (c) the programmable microfluidic valving system; (d) the comparison of the glucose concentration evaluated at 5 μL min^{-1} (no valve) and 0 μL min^{-1} (corresponding to valve-gated system). (D) Janus textiles for self-pumping

(Continued)

and #2 are designed to be only 16% and 29%, compared with BP #3. When the pressure of the fluid exceeds BP #1, it will only open the valve #1 (while valves #2 and #3 are still closed), allowing the fluid to be collected in the chamber. Once this collection chamber is full, the fluid pressure will burst the valve #2, allowing the fluid to move to the next chamber. In a temporal fashion, this example design allows all chambers to be filled sequentially. Therefore, chrono-sampling can be obtained.

4.5.2 Electrical textile valves

Another strategy to fabricate a valve is using the electrical textile (mesh opening size of 60–160 μm).[69] Acknowledging the electrowetting on dielectric, the insulated, hydrophobic, electrically conductive fabric can be applied because the normal state of the hydrophobic material blocks the fluid flow ("off" state). Applying a high voltage via the valve will turn this textile on to allow the fluid penetration ("on" state). The actuation of the activated valve consumes low energy (27 μJ) to switch the valve within 1 s, and the actuation can be programmed electronically, offering individual or multiplex controls, satisfying multiple solutions or multistep reactions (as shown in Fig. 3B). Additionally, the successful application can be extended to integrate with paper-based microfluidics. The low-surface-tension fluid, such as 50:50 (v/v) ethanol/water mixture (surface tension around 30 mN m^{-1}), requires 80 V of actuation voltage, while blood or

Fig. 3—Cont'd sampling. The hydrophilic layer of the Janus textile faced on the skin. (E) Wearable droplet microfluidic-based sensors. The operation of the device that applies the pump to feed perfusate into the microdialysis loop and withdraws the dialysate toward the sensing device. *Panel (A): Adapted with permission from Choi J, Kang D, Han S, Kim SB, Rogers JA. Thin, soft, skin-mounted microfluidic networks with capillary bursting valves for chrono-sampling of sweat.* Adv Healthc Mater. *2017;6(5):1601355, Copyright 2017, Wiley-VCH. Panel (B): Adapted with permission from Ainla A, Hamedi MM, Güder F, Whitesides GM. Electrical textile valves for paper microfluidics.* Adv Mater. *2017;29(38):1702894, Copyright 2017, Wiley-VCH. Panel (C): Adapted under the terms and conditions of the CC-BY, Lin H, Tan J, Zhu J, et al. A programmable epidermal microfluidic valving system for wearable biofluid management and contextual biomarker analysis.* Nat Commun. *2020;11(1):4405, Copyright 2020, Lin et al., published by* Nat. Commun. *Panel (D): Adapted with permission from He X, Yang S, Pei Q, et al. Integrated smart janus textile bands for self-pumping sweat sampling and analysis.* ACS Sens. *2020;5 (6):1548–1554, Copyright 2020, American Chemical Society. Panel (E): Adapted under the terms and conditions of the CC-BY, Nightingale AM, Leong CL, Burnish RA, et al. Monitoring biomolecule concentrations in tissue using a wearable droplet microfluidic-based sensor.* Nat Commun. *2019;10(1):2741. Copyright 2020, Nightingale et al., published by* Nat. Commun.

protein-containing fluids (i.e., 50 g L^{-1} bovine serum albumin in phosphate buffer solution) need high voltages of 240–480 V to allow the fluids to flow through the porous textile. It is also challenging that extra energy (e.g., heat or electricity) may affect targets or fluid characteristics.

4.5.3 *Thermo-responsive valves*

Another example of an electronically programmable valve which can be applied for on-skin microfluidic biosensing devices is shown in Fig. 3C.[70] The mechanism of this valve relies on thermo-responsive materials, namely poly(N-isopropylacrylamide). This hydrogel can be actuated by using the local heat, which is controlled by the integrated microheater adjacent to the valve. The thermal energy can tailor hydrogen-bonding interactions in the hydrogel network. When activating the responsive hydrogel at the temperature of 45 °C, the volume of the hydrogel decreases, allowing the flow of fluids. Note that the responsive hydrogel should be engineered carefully as the volumetric transition temperature cannot be close to the body temperature because natural heat transfer from the skin can also confound the valve actuation. Remarkably, programmable valve is important to regulate the flow rate. Without the controllable valve and biofluid compartmentalization, the fluctuation of flow rate directly affects the detection as the electrochemical signal also involves the flux of fluids. When designing the microfluidic system by integrating the valve and the electrochemical sensor located upstream of the channel, the sample can be retained in the zero-flow-rate compartmentalized chamber for the detection step. This design is crucial to avoid the overestimation of the resulting signal as the flow rate of the sample is systemically controlled by the microvalve.

4.6 Micropumps

The self-pumping concept is suitable to couple with on-body devices. Self-pumping becomes a particularly significant requirement in situations when adding complex pumps, controllers, and energy sources into a single microfluidic device is impractical or impossible. A representative strategy is developing stand-alone microfluidics by leveraging Janus structures to generate the net force to drive the fluid in one direction.[71,72] One practical alternative is using the flow induction by hydrophilic/-phobic characteristics. Electrospinning hydrophobic polyurethane fibers onto the superhydrophilic fabric to form the textile with two different faces: hydrophobic and hydrophilic faces (Fig. 3D).[71] This Janus design enables a passive system that can pump the biofluid toward the hydrophilic zone integrated with electrochemical sensors.

Additionally, programmable pumping using an external control is essential to dispense the liquid when the capillary force alone and physiological pressure from the body (such as sweat glands) are not enough to regulate the transport of the liquid. An example of active micropumps is developing droplet-flow systems. A screw-driven push-pull peristaltic micropump can be integrated with the wearable microfluidic device to enable nanoliter-sized droplets (Fig. 3E).[73] As this system employs the microdialysis process to extract the analyte (dissolved in dialysate and delivered to the detection zone), it is required to keep low perfusion flow rates (0.1–5 $\mu L\ min^{-1}$); the backflow can cause a problem. Therefore, to address this challenge, unidirectional screw turns in the pumping system are designed. Importantly, the fluid (aqueous-based sample) changes to form droplets by introducing an immiscible oil. In contrast to the continuous flow system, the droplet-flow system can reduce Taylor dispersion,[74] thus improving the temporal resolution for the analysis step. The use of the droplet-flow system is preferable because it is capable to shorten the response time, preserve the sample concentration profile, and capture fast transient concentration changes.[73,75]

4.7 Micromixers: Systems to enhance mass transport

The micromixer is of great importance for enhancing mass transport in microfluidics.[76–78] Laminar flows cause another challenge as flow velocities in microfluidics are not high enough to allow turbulent flows. Moreover, the driving inertial force is trivial because of the low Reynolds number when scaling down; natural mixing in microchannels involves only diffusion. Therefore, careful designs of structural micromixers (e.g., serpentine[79] and staggered grooves[80]) are needed. Moreover, micromotors[81,82] as stirrers can be alternatives because local fluid mixing and hydrodynamics can be enhanced by engineering micromotors to overcome challenges when mixing in the microworld.[83]

4.8 Electrochemical detection units

Apart from input units for sample- and reagent-handling systems, the detection unit is crucial to provide analytical signal output. For on-body applications, the electrode for the detection system should also be mechanically compliant.[25,84] Flexible and stretchable electrodes can be fabricated by utilizing elastomeric and conducting material composites[85] or structural designs.[86] Strategies using island-bridge designs,[87] serpentine shapes,[85] or wavy structures[88] have been applied to accommodate the strain when using the device in vivo. Various electrochemical detection approaches have been designed to display analytical electric signals, such as potentiometry[89] (recording potential), voltammetry[90]/amperometry[91] (recording current as a function of applied potential), conductometry[92] (recording conductivity), and electrochemical impedance spectroscopy[93] (recording impedance). Moreover, stand-alone detectors, such as self-powered electrodes based on biofuel cells, are attractive in miniaturized systems.[94] Importantly, microfluidic flow helps to enhance biofuel-cell performances.[95] Challengingly, traditional electrochemical sensors are susceptible to environments, such as fluctuating pH, temperature, and oxygen. Unlike bench-top microfluidics, perfect controlled conditions cannot be done when dealing with on-body operations, representing a grand challenge. For instance, varying oxygen can cause inaccuracy of biosensing signals.[96,97] The electrode should be designed by employing a redox mediator as a second-generation biosensor[98] or developing oxygen-rich nanocomposites to supply internal oxygen to mitigate the fluctuating condition of environments.[99] These are only representative considerations when developing electrochemical detectors toward real on-body microfluidics.

5. Examples of on-body microfluidic devices integrated electrochemical sensors for diagnosis

In addition to the analysis of non-chemical parameters such as sweat rate[100] and thermography,[53] on-body microfluidic devices coupled with electrochemical sensors can expand the analysis to realize comprehensive information of chemical compositions. Sections 5.1–5.5 provide examples of the advancement made in applications of microfluidic devices integrated with electrochemical sensors in biosensing fields.

5.1 The detection of glucose

Among several biomarkers, glucose is a primary example. It is required to monitor blood glucose levels to guide the therapy to effectively manage glucose levels.[101] Alternative biofluids, such as sweat, saliva, and tear, are interesting as we can access them noninvasively. Electrochemical glucose sensors can be categorized into two groups: enzymatic and non-enzymatic sensors.[102,103] The former relies on the enzyme immobilization on the working electrode surface, while the latter employs electrocatalysts (e.g., noble metal composites) coated on the working electrode. Amperometric sensors are among the most utilized techniques. The reaction of glucose with glucose oxidase (GOx) generates electrons or important by-products (such as peroxide) that can further provide measurable electrochemical current signals by using the suitable electrode transducer. The use of enzymes allows specific detection of glucose available in complex matrixes which include many other metabolites and ions. However, the nonenzymatic system is another route to address the challenge of instability of biocomponents. The functionalization of electrocatalysts (either enzymes or non-enzymes) while maintaining electrocatalytic activity and selectivity is a grand challenge. The on-body application also involves very dynamic parameters; therefore, pH, ions, and temperature may affect the resulting sensitivity and stability for long-term monitoring.

A paper-based microfluidic device for sweat-glucose detection was demonstrated (Fig. 4A).[104] The passive collection and fluid flow are driven by the capillary effect. Geometrical shape of the homogeneous porous substrate governs the fluid motion.[105] The hydrophobic pattern as a wall is formed by screen-printing wax. After folding the five-layer paper, the three-dimensional (3D) microfluidic system is realized. The obtained hydrophilic chambers act as a sweat collector, and the cellulose vertical and transverse microchannels act as the fluid track, delivering the sample to the electrochemical detector and finally to the evaporation zone. The GOx/Prussian Blue (PB) electrode is used as the electrode transducer. The glucose target is oxidized by GOx immobilized on the electrode, generating H_2O_2. At a low potential of −0.1 V, the H_2O_2 can then be measured by the reduction reaction on PB material. The low supplied potential is important to minimize the interference effect from electroactive compounds, such as ascorbic acid and uric acid, that can be oxidized at a high potential.[106] This epidermal paper-based device can real-time monitor glucose in sweat for ~30 min. Similarly, another 3D paper-based microfluidic glucose biosensor was demonstrated by using PB deposited reduced graphene oxide-tetraethylene pentamine modified paper working electrode, allowing a low detection potential of −0.3 V.[107] The aldehyde group on the hydrophilic cellulose substrate can be generated via periodate oxidation to offer the firm functionalization of GOx. This paper-based microfluidic device could be promising for the detection of glucose in sweat and whole blood.

Besides enzymatic biosensing approaches, another representative is developing a non-enzymatic electrochemical sensor coupled with a stretchable microfluidic patch (Fig. 4B).[64] The nanoporous gold electrode is

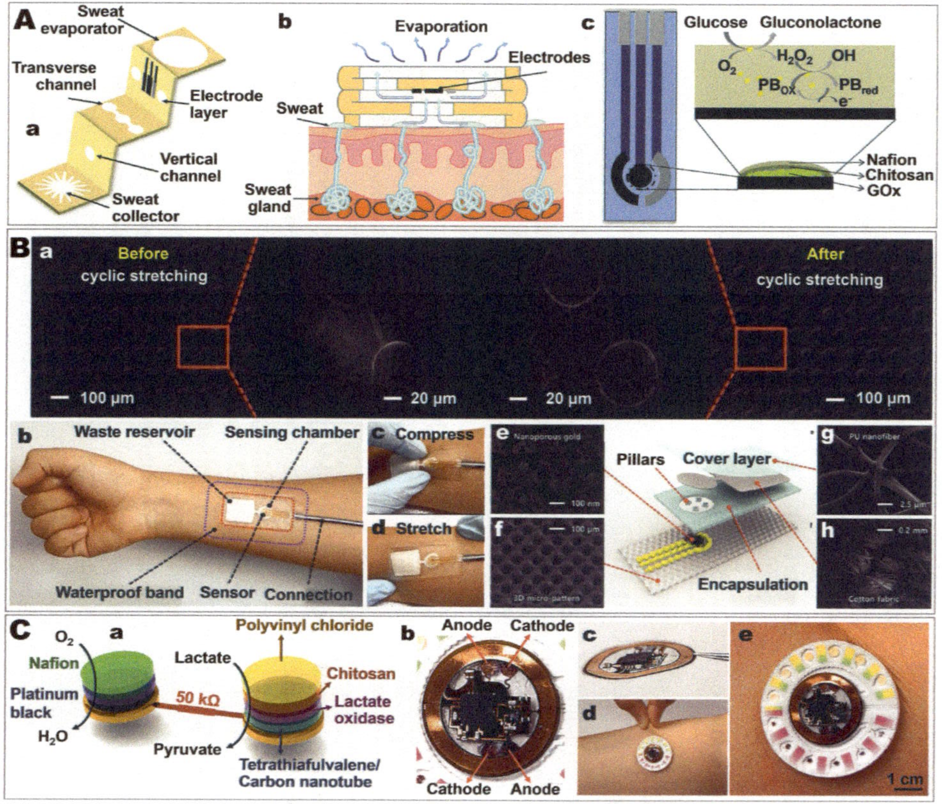

Fig. 4 See figure legend on next page.

designed to be stretchable by anchoring it on mogul-patterned PDMS substrate. While this electrode is multidirectionally stretchable, it also acts as the electrocatalyst for glucose detection at a low potential of around −0.2 V. Integrating stretchable hydrophilic textile (including 10-μm cotton fiber) with a polyurethane nanofiber-reinforced PDMS channel offers the passive sampling and delivery of the biofluid from the skin to the detection zone. The fluid flow can be controlled by the channel width. In a channel-width range of 1–3 mm, volumetric flow rates up ~20 $\mu m^3\,s^{-1}$ and linear velocities of 20–30 $\mu m\,s^{-1}$ can be obtained. Note that when the fluidic channel is stretched, the flow will be faster due to the narrower channel during the stretching. When this microfluidic glucose sensor is used to monitor sweat glucose, the combinatorial use with pH monitoring is necessary to compensate for the pH effect on the glucose sensor.

5.2 The detection of lactate

The principle of glucose sensors, e.g., relying on the oxidation of analytes or the detection of peroxide, can be extended to several biosensors, including lactate,[108,109] cholesterol,[110] and ethanol[36] sensors. In addition to design

Fig. 4 On-body microfluidic devices for detecting metabolites. (A) Three-dimensional paper-based microfluidic devices for wearable electrochemical glucose detection: (a) layers of the device; (b and c) schematic images of the device applied on the human skin and the transduction event. (B) Stretchable microfluidics integrated with nanoporous gold sensors for wearable glucose monitoring: (a) images of working-electrode surfaces before and after stretching; (b) the photograph of the sensing patch mounted on human body, demonstrating the conformality under (c) compression or (d) stretching. Images showing (e) the nanoporous gold electrode, (f) three-dimensional micropatterned substrate, (g) polyurethane nanofiber, and (h) stretchable cotton fabric. (C) Battery-free microfluidic system: (a) layers of the biofuel-cell-based lactate sensor; (b) the photograph of the sensor; (c) the photograph of the near-field communication (NFC) electronics; (d–e) the photographs of the complete integrated sensing system attached on the human skin. *Panel (A): Adapted under the terms and conditions of the CC-BY-NC, Cao Q, Liang B, Tu T, Wei J, Fang L, Ye X. Three-dimensional paper-based microfluidic electrochemical integrated devices (3D-PMED) for wearable electrochemical glucose detection.* RSC Adv. *2019;9(10):5674–5681. Copyright 2019, Cao et al., published by* RSC Adv. *Panel (B): Adapted with permission from Bae CW, Toi PT, Kim BY, et al. Fully stretchable capillary microfluidics-integrated nanoporous gold electrochemical sensor for wearable continuous glucose monitoring.* ACS Appl Mater Interfaces. *2019;11(16):14567–14575, Copyright 2019, American Chemical Society. Panel (C): Adapted with permission from the photographs kindly provided by Bandodkar AJ, Gutruf P, Choi J, et al. Battery-free, skin-interfaced microfluidic/electronic systems for simultaneous electrochemical, colorimetric, and volumetric analysis of sweat.* Sci Adv. *2019;5(1):eaav3294.*

energy-efficient devices, a critical challenge of on-body bioelectronics is a power-supply unit. Small and soft power sources are ideal. Acknowledging the viable energy that can be harvest from biomolecules (such as lactate), biofuel cells are highly attractive for addressing energy challenges.[47] After the first successful demonstration of applying biofuel cells for on-body self-powered biosensors,[40] researchers have continued to leverage such ideas of battery-free bioelectronics.[42] Another example of battery-free microfluidics were demonstrated by utilizing lactate and glucose biofuel cells (Fig. 4C).[1] The pressure from the sweat gland forces the fluid to enter the soft microfluidic device, then the sample goes through the biofuel-cell-based sensor. The self-powered sensor relies on: (1) the oxidation of metabolites (i.e., lactate or glucose) via the oxidase biocatalytic system on the anode, and (2) the reduction of oxygen on the cathode. The resistor is loaded across two electrodes, giving the signal that can be transmitted wirelessly via near-field communication electronics. This self-powered idea is a promising solution to eliminate the requirement of potentiostats and external energy-involving units, simplifying on-body microfluidic sensors.

5.3 The detection of peptides

The advancement of on-body microfluidics for immunosensing analysis faces grand challenges. For example, sandwich-based systems are complicated when designing automated on-body microfluidics. Apart from simplifying steps in analysis protocols, designing microfluidic systems to effectively sampling the biofluid and manage the accurate flow of fluids and reagents are also challenging. Therefore, label-free and reagent-free approaches are possible directions to be integrated with microfluidics. Without the requirement of any signal label (e.g., enzymes or metal nanoparticles), they eliminate marking steps, hence shortening the time and simplifying microfluidic designs. By using impedometric sensors, various analytes can be detected, such as environmental contaminants[111] and pathogens.[112] An example of a stretchable microfluidic system was demonstrated for immunosensing neuropeptide Y, which is a stress indicator present in human sweat (Fig. 5).[113] The system combined a stretchable track to manipulate fluids with an electrochemical impedimetric detector. First, sweat is collected at the middle circle of the absorption pad, allowing the analyte to bind with the sensing electrode surface. After the incubation (forming the antigen–antibody complex), the user pushes a button on the washing buffer chamber to wash the sensing electrode. Only the bonded target will be attached to the electrode

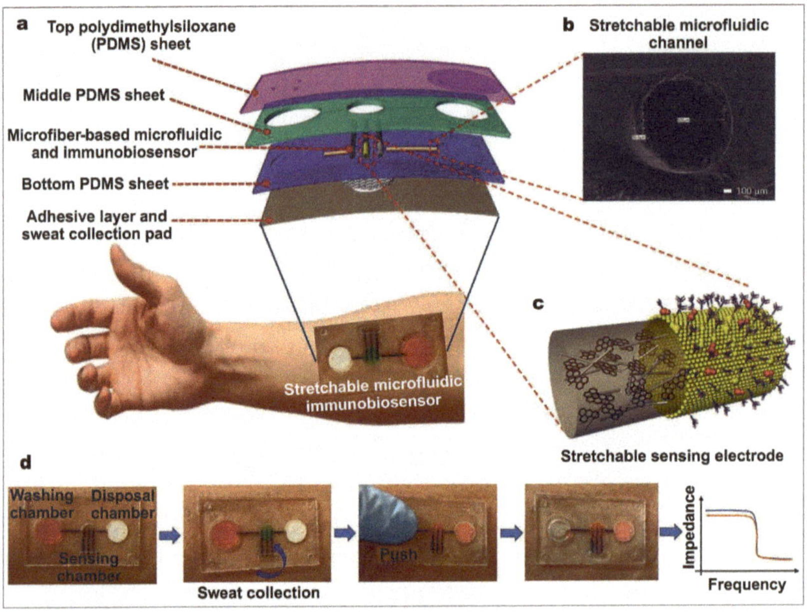

Fig. 5 On-body microfluidic immunosensing devices: (A) the illustration of the sensing patch attached on the human arm; (B) the cross-sectional image of a hollow microfiber; (C) the conductive microfiber electrode; and (D) the operation concept of the microfluidic immunobiosensor patch. *Adapted with permission from Huynh L, Trung Q, Meeseepong M, et al. Hollow microfibers of elastomeric nanocomposites for fully stretchable and highly sensitive microfluidic immunobiosensor patch.* Adv Funct Mater. *2021;30(46):2004684. Copyright 2020, Wiley-VCH.*

surface for the impedance measurement. The impedance increases with the increasing concentration of the analyte. While this represents an interesting start, much special attention must be given to develop other immunosening microfluidics.

5.4 The detection of hormones

Another example of a soft microfluidic device was combined with a label-free impedimetric cortisol biosensor as a wearable patch (Fig. 6).[16] This passive microfluidic device consists of two PDMS layers, i.e., the fluid passage layer (FPL) and fluid connection layer (FCL). These sheets form a sweat inlet chamber (*I*), sensing chamber (*S*), waste disposal chamber (*D*), reagent reservoir (*R*), and microchannel connections with one another. The generated sweat is accumulated at the *I* chamber and then passively traveled toward the *S* chamber. The 200-μm-width zig-zag pattern is composed

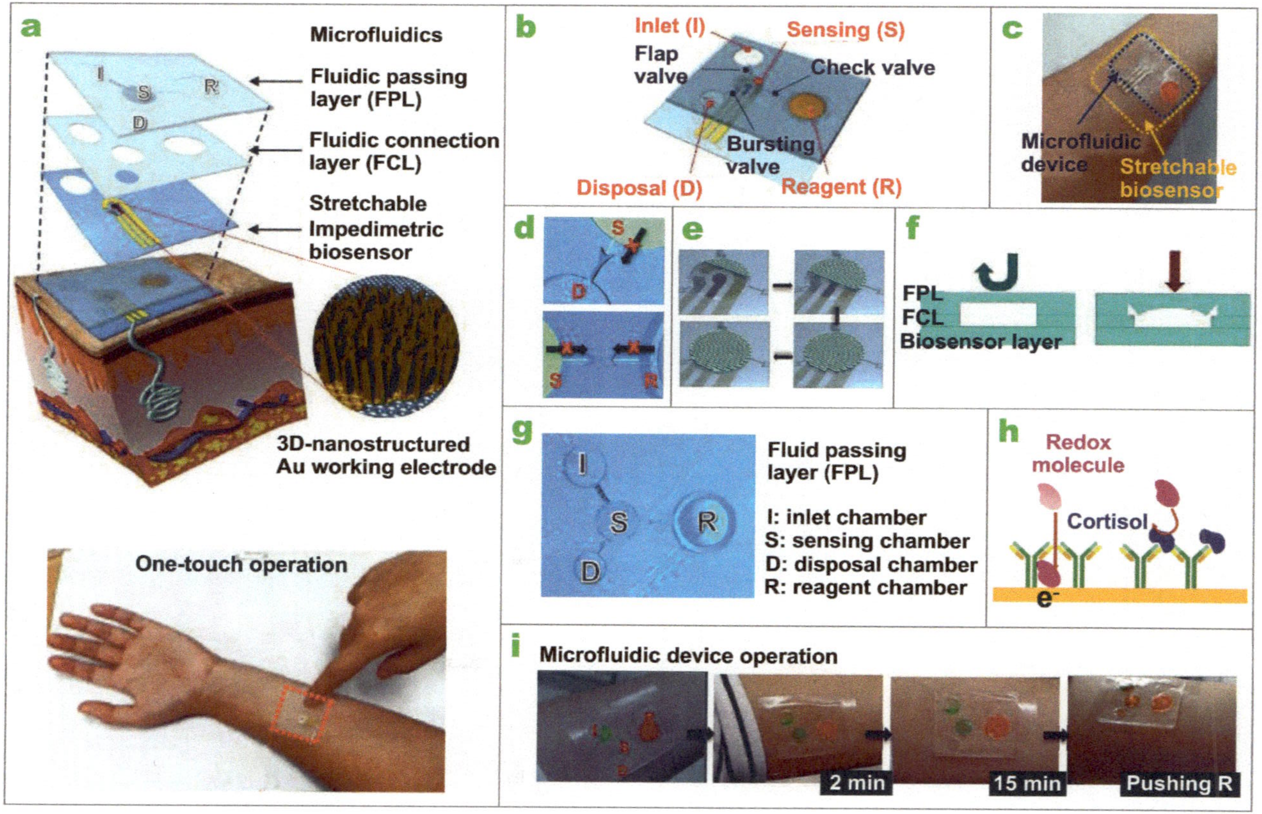

Fig. 6 See figure legend on next page.

inside the *S* to enhance the capillary effect, enabling the liquid to entirely fill the chamber. The arched flap valve, acting like venous valves in human veins, between the *I* and the *S* and the burst valve between the *S* and the *D* are installed to prevent the backflow during reagent injection and hold the fluid in the *S* during the incubation time, respectively. After the incubation that allows the cortisol target to bind with the antibody probe, the *R* with the larger volume than the *S* is pushed by the user's finger to deliver the reagent ($K_3Fe(CN)_6$ redox mediator) toward the *S* and effectively wash the *S* (all the liquid pushed toward the *D*), simultaneously. The tapered shape of the *R* reservoir eliminates the backflow while the busting and check valves between the *R* and the *S* are installed to control the incubation and reagent injection because the busting valves are opened only when applying sufficient pressure. With the presence of cortisol, the target bound by the specific recognition will hinder the redox probe, increasing the electron transfer resistance. Therefore, electrochemical impedance spectroscopy is performed to observe the response with respect to cortisol concentration. This example can be extended to develop lab-on-a-body devices for monitoring various targets, including steroid hormones that control a wide range of physiological processes.[114–116]

5.5 The detection of electrolytes

Electrochemical and impedance-based sweat-rate sensors, merged inside a microfluidic channel, offer real-time studies of sweat-secretion and chemical profiles. The detector to record electrical conductivity or impedance can be integrated with microfluidics, offering possibilities for in-vivo studies of sweat conductivity and sweat rate.[117,118] The microfluidic platform with

Fig. 6 On-body microfluidic devices for detecting cortisol: (A) a wearable patch along with the demonstration of one-touch operation; (B) the main components of a stretchable microfluidic device and a stretchable impedimetric biosensor; (C) a photo image of a wearable patch attached to a forearm; (D) (top) bursting valve between the sensing chamber and the disposal chamber and (bottom) check valve between the sensing chamber and reagent chamber; (E) sequence of the fluid flow through the zig-zag patterned capillaries in the sensing chamber; (F) cross-sectional view of the rectangular and taper-shaped push button on the reagent chamber; (G) top-view image of the fluidic passing layer; (H) mechanism for detecting cortisol; and (I) the demonstration of microfluidic device operation. *Adapted with permission from Lee H-B, Meeseepong M, Trung TQ, Kim B-Y, Lee N-E. A wearable lab-on-a-patch platform with stretchable nanostructured biosensor for non-invasive immunodetection of biomarker in sweat.* Biosens Bioelectron. *2020;156:112133, Copyright 2020, Elsevier.*

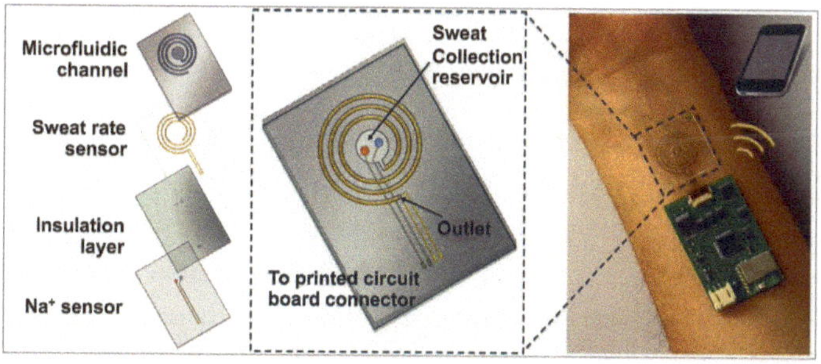

Fig. 7 On-body microfluidic devices for detecting electrolytes. Wearable microfluidic sensing patches for dynamic sweat-secretion analysis. The platform consists of four layers: a spiral-patterned microfluidic channel, a pair of parallel Au electrodes for sweat-rate sensing, an insulator layer, and Na^+-sensing electrodes. *Adapted with permission from Nyein HYY, Tai L-C, Ngo QP, et al. A wearable microfluidic sensing patch for dynamic sweat secretion analysis.* ACS Sens. *2018;3(5):944–952, Copyright 2018, American Chemical Society.*

flexible plastic substrate allows a conformal contact with the skin, thus minimizing the time lag between sweat secretion and arrival to the sensing site. As shown in Fig. 7, a 0.4-mm depth and 5-mm diameter opening of the microfluidic patch as a collection reservoir attached with sensors enables detection of multi-analytes, such as H^+, Na^+, K^+, and Cl^-, while minimizing evaporation and contamination of sweat samples. By spontaneous pumping due to perspiration pressure and capillary action, sweat flow is then directed through a 600-μm-width and 200-μm-depth microchannel attached the sweat-rate sensor to monitor sweat-secretion rate along the way to the end of a 0.5-mm outlet. Change in admittance increasing in polynomial behavior with respect to both increasing concentration and distance of the fluid allows sweat-rate monitoring. This integrated platform facilitates physiological and clinical investigations by closely monitoring interrelated sweat parameters.[100] Another example relies on microfluidics coupled with ion-selective field-effective transistors (ISFETs) for pH level and Na^+ and K^+ analysis.[119] The passive arborescent microfluidic system holding a volume capacity of 170 nL is patterned on an epoxy-based negative photoresist layer with a low contact angle of 64° to passively drive the sample through a 50-μm-width and 25-μm-depth microchannels. Additionally, the pH-ISFET integrated with temperature sensor enables to track sweat pH along with temperature[120] and the use of fin FET merged with microfluidics offers a low-power pH sensing.[121]

6. Conclusion and prospects

The recent research shows a paradigm shift to translate traditional microfluidic devices to fully automatic and even smaller microfluidic systems. In this chapter, we describe an overview of on-body microfluidics. New on-body applications can be realized due to advanced manipulation of biofluids, directly sampled from the body, with miniaturized channels (even few micrometers). The research on microfluidics coupled with electrochemical sensors on new platforms, such as soft devices, represents an emerging and attractive discipline. On-body or soft microfluidic devices coupled with electrochemical sensors are expected to play important roles in modern personalized healthcare.

New microfluidics help to advance on-body biosensors and bioelectronics for trace analysis by integrating the amplification concept. For example, integrating DNA amplification and analysis with microfluidics opens opportunities for convenient and on-body screening.[11,122] Besides the integration of microfluidics with electrochemical sensors, microfluidics can be equipped with other techniques, such as spectrophotometric techniques. The microfluidics coupled with fluorometric sensors offers a noninvasive disposable sensing device for monitoring sweat biomarkers.[123,124] Importantly, the research in microfluidics would greatly benefit from engineering materials to achieve multifunctional and high-performance sensors and microfluidic building blocks. Tailoring effects of materials (e.g., chemical compositions, surface, quantum size, etc.) would promote smart reactive or responsive materials that will facilitate the development of electrode sensors, microchannels, valves, mixers, and autonomous or programmable transportation of fluids. Furthermore, it is essential to consider small-scale fluidic behaviors, particularly when significant characteristics in the microworld involve surface tension and frictional forces. Much endeavor is needed to carefully design the geometry of the microchannel and engineer surface properties of the microchannel. Moreover, the addition of wireless flexible printed circuit board should be coupled with microfluidics to enable full integration, offering the state-of-the-art internet of things (IoTs) and ICT for personalized healthcare management. Microfluidic sensors can monitor the target in user's biofluids, revealing the wellness status in a smart way. This microfluidic device integrated with biosensing systems can be extended for screening tests of diseases.[125–127] The database obtained from microfluidics should link to the feedback system and allow timely treatment. This would also require efficient power sources or energy-harvesting devices that can sustainably supply on-body bioelectronics. Ultimately, it is preferable to miniaturize and simplify all units while maximizing sensitivity and speed of analysis. Shrinking more functional units in a single device would rely on microfabrication. Printing microfluidics is an interesting example of bottom-up approaches that allows mass and versatile fabrication.[128] We envision that a wide range of research fields can keep shifting the existing microfluidic technology to practical and reliable devices that will play an important role in the modern lifestyle in the near future.

References

1. Bandodkar AJ, Gutruf P, Choi J, et al. Battery-free, skin-interfaced microfluidic/electronic systems for simultaneous electrochemical, colorimetric, and volumetric analysis of sweat. *Sci Adv*. 2019;5(1):eaav3294.
2. Bohr A, Colombo S, Jensen H. Chapter 15—Future of microfluidics in research and in the market. In: Santos HA, Liu D, Zhang H, eds. *Microfluidics for Pharmaceutical Applications*. William Andrew Publishing; 2019:425–465.
3. Patabadige DEW, Jia S, Sibbitts J, Sadeghi J, Sellens K, Culbertson CT. Micro total analysis systems: fundamental advances and applications. *Anal Chem*. 2016;88(1):320–338.
4. Lin H, Zhao Y, Lin S, et al. A rapid and low-cost fabrication and integration scheme to render 3D microfluidic architectures for wearable biofluid sampling, manipulation, and sensing. *Lab Chip*. 2019;19(17):2844–2853.
5. Koh A, Kang D, Xue Y, et al. A soft, wearable microfluidic device for the capture, storage, and colorimetric sensing of sweat. *Sci Transl Med*. 2016;8(366):366ra165.
6. Antfolk M, Laurell T. Continuous flow microfluidic separation and processing of rare cells and bioparticles found in blood—a review. *Anal Chim Acta*. 2017;965:9–35.
7. Sonker M, Sahore V, Woolley AT. Recent advances in microfluidic sample preparation and separation techniques for molecular biomarker analysis: a critical review. *Anal Chim Acta*. 2017;986:1–11.
8. Wei X, Gao X, Zhao L, et al. Fast and interference-free determination of glyphosate and glufosinate residues through electrophoresis in disposable microfluidic chips. *J Chromatogr A*. 2013;1281:148–154.
9. Zhu Y, Huang Z, Chen Q, et al. Continuous artificial synthesis of glucose precursor using enzyme-immobilized microfluidic reactors. *Nat Commun*. 2019;10(1):4049.
10. Zhang Y, Sesen M, de Marco A, Neild A. Capacitive sensing for monitoring of microfluidic protocols using nanoliter dispensing and acoustic mixing. *Anal Chem*. 2020;92(15):10725–10732.
11. Yang B, Kong J, Fang X. Bandage-like wearable flexible microfluidic recombinase polymerase amplification sensor for the rapid visual detection of nucleic acids. *Talanta*. 2019;204:685–692.
12. Chen G, Zheng J, Liu L, Xu L. Application of microfluidics in wearable devices. *Small Methods*. 2019;3(12):1900688.
13. Nyein HYY, Bariya M, Kivimäki L, et al. Regional and correlative sweat analysis using high-throughput microfluidic sensing patches toward decoding sweat. *Sci Adv*. 2019;5(8):eaaw9906.
14. Ji X, Lau HY, Ren X, et al. Highly sensitive metabolite biosensor based on organic electrochemical transistor integrated with microfluidic channel and poly(N-vinyl-2-pyrrolidone)-capped platinum nanoparticles. *Adv Mater Technol*. 2016;1(5):1600042.
15. McCaul M, Porter A, Barrett R, et al. Wearable platform for real-time monitoring of sodium in sweat. *ChemPhysChem*. 2018;19(12):1531–1536.

16. Lee H-B, Meeseepong M, Trung TQ, Kim B-Y, Lee N-E. A wearable lab-on-a-patch platform with stretchable nanostructured biosensor for non-invasive immunodetection of biomarker in sweat. *Biosens Bioelectron*. 2020;156:112133.
17. Yetisen AK, Jiang N, Tamayol A, et al. Paper-based microfluidic system for tear electrolyte analysis. *Lab Chip*. 2017;17(6):1137–1148.
18. Agustini D, Bergamini MF, Marcolino-Junior LH. Tear glucose detection combining microfluidic thread based device, amperometric biosensor and microflow injection analysis. *Biosens Bioelectron*. 2017;98:161–167.
19. Pappa A-M, Curto VF, Braendlein M, et al. Organic transistor arrays integrated with finger-powered microfluidics for multianalyte saliva testing. *Adv Healthc Mater*. 2016; 5(17):2295–2302.
20. de Castro LF, de Freitas SV, Duarte LC, de Souza JAC, Paixão TRLC, Coltro WKT. Salivary diagnostics on paper microfluidic devices and their use as wearable sensors for glucose monitoring. *Anal Bioanal Chem*. 2019;411(19):4919–4928.
21. Zhang W, Wang R, Luo F, Wang P, Lin Z. Miniaturized electrochemical sensors and their point-of-care applications. *Chin Chem Lett*. 2020;31(3):589–600.
22. Senel M, Dervisevic E, Alhassen S, et al. Microfluidic electrochemical sensor for cerebrospinal fluid and blood dopamine detection in a mouse model of Parkinson's disease. *Anal Chem*. 2020;92(18):12347–12355.
23. Nesakumar N, Kesavan S, Li C-Z, Alwarappan S. Microfluidic electrochemical devices for biosensing. *J Anal Test*. 2019;3(1):3–18.
24. Pletcher D. Why microelectrodes? In: Montenegro MI, Queirós MA, Daschbach JL, eds. *Microelectrodes: Theory and Applications*. Dordrecht: Springer Netherlands; 1991: 3–16.
25. Jeerapan I, Poorahong S. Review—flexible and stretchable electrochemical sensing systems: materials, energy sources, and integrations. *J Electrochem Soc*. 2020;167(3): 037573.
26. Huang S, Liu Y, Zhao Y, Ren Z, Guo CF. Flexible electronics: stretchable electrodes and their future. *Adv Funct Mater*. 2019;29(6):1805924.
27. Parrilla M, Cánovas R, Jeerapan I, Andrade FJ, Wang J. A textile-based stretchable multi-ion potentiometric sensor. *Adv Healthc Mater*. 2016;5(9):996–1001.
28. Anastasova S, Crewther B, Bembnowicz P, et al. A wearable multisensing patch for continuous sweat monitoring. *Biosens Bioelectron*. 2017;93:139–145.
29. Furniturewalla A, Chan M, Sui J, Ahuja K, Javanmard M. Fully integrated wearable impedance cytometry platform on flexible circuit board with online smartphone readout. *Microsyst Nanoeng*. 2018;4(1):20.
30. Bariya M, Nyein HYY, Javey A. Wearable sweat sensors. *Nat Electron*. 2018;1(3): 160–171.
31. Kang YJ, Lee S-J. In vitro and ex vivo measurement of the biophysical properties of blood using microfluidic platforms and animal models. *Analyst*. 2018;143(12): 2723–2749.
32. Yeo JC, Kenry LCT. Emergence of microfluidic wearable technologies. *Lab Chip*. 2016;16(21):4082–4090.
33. Chen S, Shamsi MH. Biosensors-on-chip: a topical review. *J Micromech Microeng*. 2017;27(8):083001.
34. Whitesides GM. The origins and the future of microfluidics. *Nature*. 2006;442 (7101):368–373.
35. Song Y, Zhao X, Tian Q, Liang H. Fundamental concepts and physics in microfluidics. In: Song Y, Cheng D, Zhao L, eds. *Microfluidics: Fundamental, Devices and Applications*. Wiley-VCH Verlag GmbH & Co. KGaA; 2018:19–111.
36. Kim J, Jeerapan I, Imani S, et al. Noninvasive alcohol monitoring using a wearable tattoo-based iontophoretic-biosensing system. *ACS Sens*. 2016;1(8):1011–1019.

37. Qiao L, Benzigar MR, Subramony JA, Lovell NH, Liu G. Advances in sweat wearables: sample extraction, real-time biosensing, and flexible platforms. *ACS Appl Mater Interfaces*. 2020;12(30):34337–34361.
38. Zhang Y, Guo H, Kim SB, et al. Passive sweat collection and colorimetric analysis of biomarkers relevant to kidney disorders using a soft microfluidic system. *Lab Chip*. 2019;19(9):1545–1555.
39. Choi J, Kang D, Han S, Kim SB, Rogers JA. Thin, soft, skin-mounted microfluidic networks with capillary bursting valves for chrono-sampling of sweat. *Adv Healthc Mater*. 2017;6(5):1601355.
40. Jeerapan I, Sempionatto JR, Pavinatto A, You J-M, Wang J. Stretchable biofuel cells as wearable textile-based self-powered sensors. *J Mater Chem A*. 2016;4(47):18342–18353.
41. Lv J, Jeerapan I, Tehrani F, et al. Sweat-based wearable energy harvesting-storage hybrid textile devices. *Energy Environ Sci*. 2018;11(12):3431–3442.
42. Yeknami AF, Wang X, Jeerapan I, et al. A 0.3-V CMOS biofuel-cell-powered wireless glucose/lactate biosensing system. *IEEE J Solid State Circuits*. 2018;53(11):3126–3139.
43. Phillips EA, Moehling TJ, Ejendal KFK, et al. Microfluidic rapid and autonomous analytical device (microRAAD) to detect HIV from whole blood samples. *Lab Chip*. 2019;19(20):3375–3386.
44. Halldorsson S, Lucumi E, Gómez-Sjöberg R, Fleming RMT. Advantages and challenges of microfluidic cell culture in polydimethylsiloxane devices. *Biosens Bioelectron*. 2015;63:218–231.
45. Wang JD, Douville NJ, Takayama S, ElSayed M. Quantitative analysis of molecular absorption into pdms microfluidic channels. *Ann Biomed Eng*. 2012;40(9):1862–1873.
46. Choi J, Chen S, Deng Y, et al. Skin-interfaced microfluidic systems that combine hard and soft materials for demanding applications in sweat capture and analysis. *Adv Healthc Mater*. 2021;10(4):2000722.
47. Jeerapan I, Sempionatto JR, Wang J. On-body bioelectronics: wearable biofuel cells for bioenergy harvesting and self-powered biosensing. *Adv Funct Mater*. 2019;30(29):1906243.
48. Cansiz M, Altinel D, Kurt GK. Efficiency in RF energy harvesting systems: a comprehensive review. *Energy*. 2019;174:292–309.
49. Kirby BJ. *Micro-and Nanoscale Fluid Mechanics: Transport in Microfluidic Devices*. Cambridge University Press; 2010.
50. Javanmard M, Emaminejad S, Gupta C, Provine J, Davis RW, Howe RT. Depletion of cells and abundant proteins from biological samples by enhanced dielectrophoresis. *Sens Actuators B Chem*. 2014;193:918–924.
51. Bandodkar AJ, Choi J, Lee SP, et al. Soft, skin-interfaced microfluidic systems with passive galvanic stopwatches for precise chronometric sampling of sweat. *Adv Mater*. 2019;31(32):1902109.
52. Yeo JC, Yu J, Koh ZM, Wang Z, Lim CT. Wearable tactile sensor based on flexible microfluidics. *Lab Chip*. 2016;16(17):3244–3250.
53. Reeder JT, Choi J, Xue Y, et al. Waterproof, electronics-enabled, epidermal microfluidic devices for sweat collection, biomarker analysis, and thermography in aquatic settings. *Sci Adv*. 2019;5(1):eaau6356.
54. Sonner Z, Wilder E, Heikenfeld J, et al. The microfluidics of the eccrine sweat gland, including biomarker partitioning, transport, and biosensing implications. *Biomicrofluidics*. 2015;9(3):031301.
55. Ma T, Sun S, Li B, Chu J. Piezoelectric peristaltic micropump integrated on a microfluidic chip. *Sens Actuators A Phys*. 2019;292:90–96.
56. Ren Q, Wang Y, Lin X, Chan CL. AC electrokinetic induced non-Newtonian electrothermal blood flow in 3D microfluidic biosensor with ring electrodes for point-of-care diagnostics. *J Appl Phys*. 2019;126(8):084501.

57. Zhao B, Moore JS, Beebe DJ. Principles of surface-directed liquid flow in microfluidic channels. *Anal Chem*. 2002;74(16):4259–4268.
58. Wang L, Wang L, Zhang Y, et al. Weaving sensing fibers into electrochemical fabric for real-time health monitoring. *Adv Funct Mater*. 2018;28(42):1804456.
59. Liu X, Lillehoj PB. Embroidered electrochemical sensors on gauze for rapid quantification of wound biomarkers. *Biosens Bioelectron*. 2017;98:189–194.
60. Choi JR, Nilghaz A, Chen L, Chou KC, Lu X. Modification of thread-based microfluidic device with polysiloxanes for the development of a sensitive and selective immunoassay. *Sens Actuators B Chem*. 2018;260:1043–1051.
61. Naeimirad M, Abuzade R, Babaahmadi V, Dabirian F. Microfluidic through fibrous structures: recent developments and future trends. *Mater Design Process Commun*. 2019;1(4):e78.
62. Hu S, Ren X, Bachman M, Sims CE, Li GP, Allbritton N. Surface modification of poly(dimethylsiloxane) microfluidic devices by ultraviolet polymer grafting. *Anal Chem*. 2002;74(16):4117–4123.
63. Jiang Y, Hao Z, He Q, Chen H. A simple method for fabrication of microfluidic paper-based analytical devices and on-device fluid control with a portable corona generator. *RSC Adv*. 2016;6(4):2888–2894.
64. Bae CW, Toi PT, Kim BY, et al. Fully stretchable capillary microfluidics-integrated nanoporous gold electrochemical sensor for wearable continuous glucose monitoring. *ACS Appl Mater Interfaces*. 2019;11(16):14567–14575.
65. Nilghaz A, Ballerini DR, Shen W. Exploration of microfluidic devices based on multi-filament threads and textiles: a review. *Biomicrofluidics*. 2013;7(5):051501.
66. MacKoul D, Johnson DC, Schick KG. Effect of variation in flow rate on amperometric detection in flow injection analysis. *Anal Chem*. 1984;56(3):436–439.
67. Ishida T, McLaughlin D, Tanaka Y, Omata T. First-come-first-store microfluidic device of droplets using hydrophobic passive microvalves. *Sens Actuators B Chem*. 2018;254:1005–1010.
68. Zhang Y, Chen Y, Huang J, et al. Skin-interfaced microfluidic devices with one-opening chambers and hydrophobic valves for sweat collection and analysis. *Lab Chip*. 2020;20(15):2635–2645.
69. Ainla A, Hamedi MM, Güder F, Whitesides GM. Electrical textile valves for paper microfluidics. *Adv Mater*. 2017;29(38):1702894.
70. Lin H, Tan J, Zhu J, et al. A programmable epidermal microfluidic valving system for wearable biofluid management and contextual biomarker analysis. *Nat Commun*. 2020;11(1):4405.
71. He X, Yang S, Pei Q, et al. Integrated smart janus textile bands for self-pumping sweat sampling and analysis. *ACS Sens*. 2020;5(6):1548–1554.
72. Yu T, Athanassiadis AG, Popescu MN, et al. Microchannels with self-pumping walls. *ACS Nano*. 2020;14(10):13673–13680.
73. Nightingale AM, Leong CL, Burnish RA, et al. Monitoring biomolecule concentrations in tissue using a wearable droplet microfluidic-based sensor. *Nat Commun*. 2019;10(1):2741.
74. Feng S, Shirani E, Inglis DW. Droplets for sampling and transport of chemical signals in biosensing: a review. *Biosensors*. 2019;9(2):80.
75. Leong CL, Coleman S, Nightingale AM, et al. Lactate monitoring in droplet microfluidics: a cautionary tale in assay miniaturisation. *Anal Methods*. 2019;11(48):6119–6123.
76. Lee C-Y, Chang C-L, Wang Y-N, Fu L-M. Microfluidic mixing: a review. *Int J Mol Sci*. 2011;12(5):3263–3287.
77. Lee C-Y, Wang W-T, Liu C-C, Fu L-M. Passive mixers in microfluidic systems: a review. *Chem Eng J*. 2016;288:146–160.

78. Xia HM, Wan SYM, Shu C, Chew YT. Chaotic micromixers using two-layer crossing channels to exhibit fast mixing at low Reynolds numbers. *Lab Chip*. 2005; 5(7):748–755.
79. Clark J, Kaufman M, Fodor PS. Mixing enhancement in serpentine micromixers with a non-rectangular cross-section. *Micromachines*. 2018;9(3):107.
80. Hama B, Mahajan G, Fodor PS, Kaufman M, Kothapalli CR. Evolution of mixing in a microfluidic reverse-staggered herringbone micromixer. *Microfluid Nanofluid*. 2018; 22(5):54.
81. Soto F, Lopez-Ramirez MA, Jeerapan I, et al. Rotibot: use of rotifers as self-propelling biohybrid microcleaners. *Adv Funct Mater*. 2019;29(22):1900658.
82. Cinti S, Valdés-Ramírez G, Gao W, Li J, Palleschi G, Wang J. Microengine-assisted electrochemical measurements at printable sensor strips. *Chem Commun*. 2015; 51(41):8668–8671.
83. Cohen N, Boyle JH. Swimming at low Reynolds number: a beginners guide to undulatory locomotion. *Contemp Phys*. 2010;51(2):103–123.
84. Economou A, Kokkinos C, Prodromidis M. Flexible plastic, paper and textile lab-on-a chip platforms for electrochemical biosensing. *Lab Chip*. 2018;18(13):1812–1830.
85. Bandodkar AJ, Jeerapan I, You J-M, Nuñez-Flores R, Wang J. Highly stretchable fully-printed CNT-based electrochemical sensors and biofuel cells: combining intrinsic and design-induced stretchability. *Nano Lett*. 2016;16(1):721–727.
86. Liu Y, Pharr M, Salvatore GA. Lab-on-skin: a review of flexible and stretchable electronics for wearable health monitoring. *ACS Nano*. 2017;11(10):9614–9635.
87. Silva CA, lv J, Yin L, et al. Liquid metal based island-bridge architectures for all printed stretchable electrochemical devices. *Adv Funct Mater*. 2020;30(30):2002041.
88. Jin Z-H, Liu Y-L, Chen J-J, Cai S-L, Xu J-Q, Huang W-H. Conductive polymer-coated carbon nanotubes to construct stretchable and transparent electrochemical sensors. *Anal Chem*. 2017;89(3):2032–2038.
89. Ding R, Lisak G. Sponge-based microfluidic sampling for potentiometric ion sensing. *Anal Chim Acta*. 2019;1091:103–111.
90. Liu Z, Jin M, Cao J, et al. Electrochemical sensor integrated microfluidic device for sensitive and simultaneous quantification of dopamine and 5-hydroxytryptamine. *Sens Actuators B Chem*. 2018;273:873–883.
91. Hayter EA, Castiaux AD, Martin RS. 3D-printed microfluidic device with in-line amperometric detection that also enables multi-modal detection. *Anal Methods*. 2020;12(15):2046–2051.
92. Koman VB, Lew TTS, Wong MH, Kwak S-Y, Giraldo JP, Strano MS. Persistent drought monitoring using a microfluidic-printed electro-mechanical sensor of stomata in planta. *Lab Chip*. 2017;17(23):4015–4024.
93. Kinnamon D, Ghanta R, Lin K-C, Muthukumar S, Prasad S. Portable biosensor for monitoring cortisol in low-volume perspired human sweat. *Sci Rep*. 2017;7(1):13312.
94. Jeerapan I, Ma N. Challenges and opportunities of carbon nanomaterials for biofuel cells and supercapacitors: personalized energy for futuristic self-sustainable devices. *C—J Carbon Res*. 2019;5(4).
95. Escalona-Villalpando RA, Reid RC, Milton RD, Arriaga LG, Minteer SD, Ledesma-García J. Improving the performance of lactate/oxygen biofuel cells using a microfluidic design. *J Power Sources*. 2017;342:546–552.
96. Prévoteau A, Mano N. Oxygen reduction on redox mediators may affect glucose biosensors based on "wired" enzymes. *Electrochim Acta*. 2012;68:128–133.
97. Bedekar AS, Feng JJ, Krishnamoorthy S, Lim KG, Palmore GTR, Sundaram S. Oxygen limitation in microfluidic biofuel cells. *Chem Eng Commun*. 2007; 195(3):256–266.
98. Wang J. Electrochemical glucose biosensors. *Chem Rev*. 2008;108(2):814–825.

99. Jeerapan I, Sempionatto JR, You J-M, Wang J. Enzymatic glucose/oxygen biofuel cells: use of oxygen-rich cathodes for operation under severe oxygen-deficit conditions. *Biosens Bioelectron.* 2018;122:284–289.
100. Nyein HYY, Tai L-C, Ngo QP, et al. A wearable microfluidic sensing patch for dynamic sweat secretion analysis. *ACS Sens.* 2018;3(5):944–952.
101. Teymourian H, Barfidokht A, Wang J. Electrochemical glucose sensors in diabetes management: an updated review (2010–2020). *Chem Soc Rev.* 2020;49(21):7671–7709.
102. Hwang D-W, Lee S, Seo M, Chung TD. Recent advances in electrochemical non-enzymatic glucose sensors—a review. *Anal Chim Acta.* 2018;1033:1–34.
103. Mohapatra J, Ananthoju B, Nair V, et al. Enzymatic and non-enzymatic electrochemical glucose sensor based on carbon nano-onions. *Appl Surf Sci.* 2018;442:332–341.
104. Cao Q, Liang B, Tu T, Wei J, Fang L, Ye X. Three-dimensional paper-based microfluidic electrochemical integrated devices (3D-PMED) for wearable electrochemical glucose detection. *RSC Adv.* 2019;9(10):5674–5681.
105. Elizalde E, Urteaga R, Berli CLA. Rational design of capillary-driven flows for paper-based microfluidics. *Lab Chip.* 2015;15(10):2173–2180.
106. Kim J, Jeerapan I, Sempionatto JR, et al. Wearable bioelectronics: enzyme-based body-worn electronic devices. *Acc Chem Res.* 2018;51(11):2820–2828.
107. Cao L, Han G-C, Xiao H, Chen Z, Fang C. A novel 3D paper-based microfluidic electrochemical glucose biosensor based on rGO-TEPA/PB sensitive film. *Anal Chim Acta.* 2020;1096:34–43.
108. Hiraka K, Kojima K, Tsugawa W, Asano R, Ikebukuro K, Sode K. Rational engineering of Aerococcus viridans l-lactate oxidase for the mediator modification to achieve quasi-direct electron transfer type lactate sensor. *Biosens Bioelectron.* 2020;151, 111974.
109. Zaryanov NV, Nikitina VN, Karpova EV, Karyakina EE, Karyakin AA. Nonenzymatic sensor for lactate detection in human sweat. *Anal Chem.* 2017;89(21):11198–11202.
110. Salazar P, Martín M, González-Mora JL. In situ electrodeposition of cholesterol oxidase-modified polydopamine thin film on nanostructured screen printed electrodes for free cholesterol determination. *J Electroanal Chem.* 2019;837:191–199.
111. Cheng YH, Barpaga D, Soltis JA, et al. Metal–organic framework-based microfluidic impedance sensor platform for ultrasensitive detection of perfluorooctanesulfonate. *ACS Appl Mater Interfaces.* 2020;12(9):10503–10514.
112. Jeerapan I, Sonsa-ard T, Nacapricha D. Applying nanomaterials to modern biomedical electrochemical detection of metabolites, electrolytes, and pathogens. *Chemosensors.* 2020;8(3):71.
113. Huynh L, Trung Q, Meeseepong M, et al. Hollow microfibers of elastomeric nanocomposites for fully stretchable and highly sensitive microfluidic immunobiosensor patch. *Adv Funct Mater.* 2021;30(46):2004684.
114. Torrente-Rodríguez RM, Tu J, Yang Y, et al. Investigation of cortisol dynamics in human sweat using a graphene-based wireless mhealth system. *Matter.* 2020;2(4):921–937.
115. Shrivastava S, Trung TQ, Lee N-E. Recent progress, challenges, and prospects of fully integrated mobile and wearable point-of-care testing systems for self-testing. *Chem Soc Rev.* 2020;49(6):1812–1866.
116. Nah JS, Barman SC, Zahed MA, et al. A wearable microfluidics-integrated impedimetric immunosensor based on Ti3C2Tx MXene incorporated laser-burned graphene for noninvasive sweat cortisol detection. *Sens Actuators B Chem.* 2021;329:129206.
117. Francis J, Stamper I, Heikenfeld J, Gomez EF. Digital nanoliter to milliliter flow rate sensor with in vivo demonstration for continuous sweat rate measurement. *Lab Chip.* 2019;19(1):178–185.

118. Kim SB, Lee K, Raj MS, et al. Soft, skin-interfaced microfluidic systems with wireless, battery-free electronics for digital, real-time tracking of sweat loss and electrolyte composition. *Small.* 2018;14(45):1802876.
119. Garcia-Cordero E, Bellando F, Zhang J, et al. Three-dimensional integrated ultra-low-volume passive microfluidics with ion-sensitive field-effect transistors for multiparameter wearable sweat analyzers. *ACS Nano.* 2018;12(12):12646–12656.
120. Nakata S, Arie T, Akita S, Takei K. Wearable, flexible, and multifunctional healthcare device with an ISFET chemical sensor for simultaneous sweat pH and skin temperature monitoring. *ACS Sens.* 2017;2(3):443–448.
121. Garcia-Cordero E, Guerin H, Muhech A, Bellando F, Ionescu AM. Heterogeneous integration of low power pH FinFET sensors with passive capillary microfluidics and miniaturized Ag/AgCl quasi-reference electrode. In: *Paper Presented at: 2016 46th European Solid-State Device Research Conference (ESSDERC); 12–15 Sept. 2016*; 2016.
122. Kong M, Li Z, Wu J, et al. A wearable microfluidic device for rapid detection of HIV-1 DNA using recombinase polymerase amplification. *Talanta.* 2019;205:120155.
123. Ardalan S, Hosseinifard M, Vosough M, Golmohammadi H. Towards smart personalized perspiration analysis: an IoT-integrated cellulose-based microfluidic wearable patch for smartphone fluorimetric multi-sensing of sweat biomarkers. *Biosens Bioelectron.* 2020;168:112450.
124. Kim S, Lee B, Reeder JT, et al. Soft, skin-interfaced microfluidic systems with integrated immunoassays, fluorometric sensors, and impedance measurement capabilities. *Proc Natl Acad Sci.* 2020;2020:12700.
125. Limsakul P, Charupanit K, Moonla C, Jeerapan I. Advances in emergent biological recognition elements and bioelectronics for diagnosing COVID-19. *Emerg Mater.* 2021;4 (1):231–247.
126. Berkenbrock JA, Grecco-Machado R, Achenbach S. Microfluidic devices for the detection of viruses: aspects of emergency fabrication during the COVID-19 pandemic and other outbreaks. *Proc Math Phys Eng Sci P Roy Soc A.* 2020;476(2243):20200398.
127. Tayyab M, Sami MA, Raji H, Mushnoori S, Javanmard M. Potential microfluidic devices for COVID-19 antibody detection at point-of-care (POC): a review. *IEEE Sens J.* 2021;21(4):4007–4017.
128. Loo JFC, Ho AHP, Turner APF, Mak WC. Integrated printed microfluidic biosensors. *Trends Biotechnol.* 2019;37(10):1104–1120.

CHAPTER TEN

Microfluidic tools for veterinary and zoonotic disease diagnostics

Nidhi Verma[a], Rutuparna Kulkarni[b], and Alok Pandya[a,*]
[a]Department of Engineering and Physical Sciences, Institute of Advanced Research, Gandhinagar, Gujarat, India
[b]Department of Biological Sciences and Biotechnology, Institute of Advanced Research, Gandhinagar, Gujarat, India
*Corresponding author: e-mail address: alokpandya20@gmail.com

Contents

Abstract

Animal disease diagnostics has linked as the cause and cure of any disease. It also plays a vital role in disease management and prevention. A small outbreak of disease can pose a threat to the entire animal community as we realized in corona pandemic. Thus, to ensure the overall welfare of animals and disease spread monitoring, the development of detection tools for veterinary diagnosis becomes essential. Currently, the animal disease diagnosis is relied on laboratory-based testing. There is a parallel necessity for rapid, reliable and low-cost diagnostic tests to be done by intervention of growing area such as microfluidic platform. Therefore, in this chapter, we have discussed about various microfluidic platform and their application for early diagnosis of veterinary disease. Followed by, we also lightened on future perspective of role of microfluidic in animal disease diagnostics.

Progress in Molecular Biology and Translational Science, Volume 187
ISSN 1877-1173
https://doi.org/10.1016/bs.pmbts.2021.07.026
Copyright © 2022 Elsevier Inc.
All rights reserved.

1. Introduction

Point-of-care (POC) or point of need diagnostics is a subject of attention which has drawn significant attention over the past years. Detection at Point-of-care means that analysis events are done in close proximity to the patients hence also denoted as "bed-side" testing. Due to numerous advantages of point-of-care like reduction in the cost of diagnosis, easy availability and early detection as well as reducing the time between sampling and a treatment choice. Ideally, POC tests must adhere to World Health Organization (WHO) benchmark criteria for the ideal rapid test, under the acronym "ASSURED (Affordable, Sensitive, Specific, User-friendly, Rapid and robust, Equipment-free and Deliverable to end-users)." Point of care based microfluidic based platform has emerged as a great field having and attributing to have all the requirements to be accredited as an ASSURED device.[1]

In the area of veterinary science, there is a parallel necessity for rapid, reliable and low-cost diagnostic tests to be done at the point of need microfluidic platform. In case of veterinary diagnosis, it is generally referred to as animal-side testing and several advantages over laboratory-based testing, which typically comprises laborious and costly laboratory set up as well as requirement of technical personnel.[2] All of the processes concerned with testing, to collection of the sample to analysis the samples could possibly be achieved in a single step by significantly decreasing the time between testing and treatment. The microfluidic system has capability translate into more affordable veterinary care, reduced handling of animals, targeted treatments and rapid testing in more remote geographic areas.[3] The need for more inexpensive, accessible and rapid tests is a need of the hour for veterinary field. In many cases cattle healthy is highly neglected causing losses to dairy as well as agriculture.[4] Furthermore, the overall globalization of trade of animal products and animals has significantly amplified the risk of rapid and wide-ranging spread of emerging and exotic diseases which might in certain cases cause a public health concern. Henceforth, it is imperative to have timely and efficient ways of dealing with diseases related to animals that could have disastrous consequences for the individual farmer, as well as to the whole economy and highly affect the international trade. In circumstances regarding disease outbreaks, where rapid proliferation of infectious agents causing high mortality as in the case of the highly pathogenic COVID 19 which the current world is surrounding with as well as H5N1 strain of avian influenza virus, a rapid "animal-side" test would represent a critical tool for both collecting investigation data and for supporting in the regulation of outbreaks.[2,5] After the current outbreak, it becomes evident that animal health monitoring should be at utmost importance. Presently available veterinary POC tests are limited and under developed. Recent developments in context to the microfluidic technologies for POC testing in the field of disease diagnosis for human has revolutionized it implementation in veterinary medicine could overcome these difficulties. This chapter aims to evaluate the current status of POC testing in veterinary medicine, describing their applications and limitations, whereas also assessing the potential of microfluidic technologies to progress current POC tests and solve some of their fundamental limitations.

2. Types of microfluidic tools

2.1 Microfabrication of paper based devices for application in veterinary detection

Microfabrication techniques enables to learn and produce micro structures down to the micrometer scale and smaller, to be assembled into microfluidic devices there are ranges of microfabrication techniques used on various substrates ranging from paper to silica in addition to PDMS as well as glass. It is used in a wide range of applications to exactly functions by creating adapted microstructures.[6]

Paper based devices are now often referred to as Microfluidic paper-based analytical devices (μPADs) conceptualized brought to attention by the Whiteside group at Harvard University, it was initially used for determination of pH. The devices are used ever since its discovery as it is low cost and have easy multiplexing properties. While preserving the benefits of traditional microfluidic technology, such as speed, size and less volumes requirement both for the sample as well as reagents. Although, paper is relatively a new substrate but have significant advantages over other substrate which have been used in microfluidics over time including silica, glass or PDMS. Paper as a substrate have proved to be an inert easy to use, biodegradable and cheap, as well as a naturally available substrate unlike conventional substrates. Point of care devices made from paper always have a dominance as they do not need any external power or pump for its functioning, as well the fabrication techniques utilized and equipment for production are generally low cost than those for other materials, with minimal technical expertise required. Paper signifies an exceptional medium for diagnostic testing not only for human but for in veterinary application as it offers properties of being easily available and rapid reaction times. In terms of their footmark in veterinary applications first falls the production costs of the device followed by easy production as well as easy detection at point of care. It is of great importance where curative facilities for animals are limited.

Yang et al. established a multiplexed and easy to use and paper-origami tests in semen samples from the test site for the detection of three bovine infectious reproductive diseases. Bovine herpes virus-1 (BoHV-1) DNA was detected along with extraction of Brucella and Leptospira, followed by amplified by means of loop-mediated isothermal amplification and fluorescently, detected with a detection of limit of <1 pg. The paper based sensors offers as an showed exceptional for the diagnosis multiple infectious diseases from semen samples.[7] Another study revealed the development of paper-based enzyme-linked immunoassays and was named as Multi-Pad Paper Plate (MP^3). The proposed hybrid paper-based model microfluidic platform. The hybrid model made use of laser micromachining of paper along with packaging through thermal lamination. The method proved to be simple, quick and universal as well as comparable with a standard 96-well microplate format. The hybrid method was used for detection inflammation marker known as bovine haptoglobin (Hp) having sensitivity of 0.73 μg/mL. The proposed method had potential to transform similar assays into P-ELISA and could be a beginning point for accomplishing animal-side testing.[8] Another field to be explored is in regards with the detection of the antibiotics which is administered to the cattle's as this in certain case can possess as a risk of contaminate for the milk during the process of milking. These residues have potential to cause stern allergic reactions creating public health concern leading to severe monetary losses to the farmer. Hence, in order to address this a paper device was fabricated by using aptamer having a limit of detection of 300 nM, with a time of reaction of around 2 min.[9] A microfluidic paper-based analytical device (μPAD) was fabricated for detection of aflatoxin M1. unmodified gold nanoparticles was used as a probe and aptamer as a recognition molecule. Under optimal conditions, the linear range was 1 μM to 1 pM having detection limit up to 3 pM in standard buffer and 10 nM in spiked milk samples, respectively.[10]

2.2 Microchip and microcapillary fabrication for detection

Microfluidics technology proposes a stage for growth and expansion of point-of-care diagnostic devices for various diseases related to veterinary diagnostics. Wadhwa et al. developed a microfluidic assay system for diagnosis of Johne's disease using a bead-based system. The antigens of Johne's disease were coated on the beads and incubated with samples and detected through fluorescence. Under optimized conditions were performed by magnetically immobilizing a number of beads within based glass microchip consisting of microchannel was immobilized with a number of magnetic beads and detected using laser-induced fluorescence. The proposed bead-based microfluidic system can act as an great alternative for on-site sero diagnosis of JD.[11] Not only is it necessary to detect disease of animals but also important to discriminate between them as in certain cases few disease might have over lapping symptoms making it difficult to detect.

Hence specific diagnosis of diseases having common clinical signs requires multiple as well as independent diagnostic tests specific for each disease to confirm it. Microfluidic platform offers wide range of application to not only detect disease but to effectively discriminate between them. Livestock hemorrhagic diseases including of African swine fever (ASF), classical swine fever (CSF) and vesicular diseases, such as vesicular stomatitis (VS), foot-and-mouth disease (FMD) and swine vesicular disease (SVD) can have a huge influence on the livestock industry. Lung et al. described the development user friendly assay for instantaneous detection and differentiation of several viruses of veterinary importance. The proposed system detects 24 samples within 6 h wherein just sample loading and initiation of assay rest being automated. The system was checked with livestock samples and it was observed that no cross-reactivity was detected with 15 non target viruses that affect cattle. The developed could be a boon in veterinary diagnostics.[12] A very recent approach in veterinary based detection and diagnosis of diseases is integration of electrochemical sensing into microfluidic platforms. One such system was developed by Heinze et al. for detection of Bovine Viral Diarrhea in dairy and beef industry. A microfluidic cartridge based electrochemical sensors for on ground sample analysis which is an economic microfluidic system. As soon as the sample is positioned into the cartridge it is then introduced in a handheld console that will transmit results via Bluetooth to a phone app. A desktop vinyl cutter is used for fabrication of the devices having six electrochemical sensors. Due to ease in fabrication the entire process from testing to materials can be finished quickly and leak-free environment on the microfluidic platform utilizing equipment.[13]

3. Applications of microfluidics in veterinary diagnostics

Microfluidics platform is by now extensively used in veterinary medicine, as new and evolving technologies bring considerable developments to

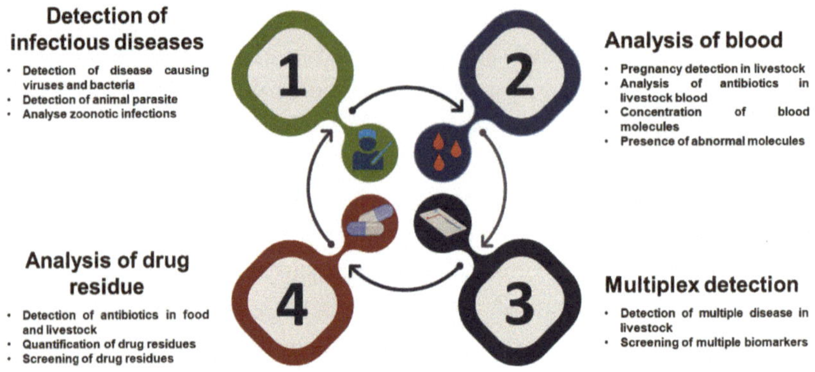

Fig. 1 Applications of microfluidics in veterinary diagnostics.

both the range of tests and their characteristic implementation. The low cost and time effective as one-step functions make these devices attractive the testing of animals. Although microfluidics is currently, mainly useful in human diagnostics there have been applications in areas of veterinary interest as well which are listed below and as shown in Fig. 1.

3.1 Detection of infectious diseases

The infectious diseases in animals are caused by mainly pathogens such as bacteria and viruses. These diseases severely affect the health of livestock, wildlife and also cause decline in productivity and economy. Viral infections also pose a threat of human zoonotic infections. For these purposes rapid and early detection of infectious diseases is very important. Over the past few years, there have been many developments in the devices and techniques being utilized for detection of infectious pathogens.

A silicon-based microfluidic chip along with a sensor was developed for the analysis of pathogenic DNA sequences by Chen et al. The system, which uses loop-mediated isothermal amplification (LAMP) was shown to sense as little as 15 μL test sample within 30 min.[14] It gave positive or negative result about presence of respiratory pathogen, based upon fluorescence emitted by LAMP reagents and easy detection is possible.

Sexually transmitted infections in animals are difficult to detect due to limited availability of infrastructure. Yang et al. addressed this problem by developing paper-origami based tests for detection of one viral (bovine herpes virus-1) and two bacterial (*Leptospira* and *Brucella*) infections. The method, used to test semen samples of bulls, showed high sensitivity and specificity and is considered to be easy to use. Highly sensitive method was developed by Yu et al. to detect avian influenza virus. They constructed a microfluidic immunoassay having ZnO nanorod, which is shown to have ability to detect multiple viruses simultaneously.[7,15]

Bovine mastitis is a disease in cattle which causes inflammation of mammary glands. There have been many advances in developing rapid and accurate tools for diagnosis of mastitis pathogens. This disease is associated with multiple bacterial infections and thus it is necessary to detect and characterize all organisms simultaneously. Viveiros et al. developed a microchip consisting of arrays of magneto-resistive sensors for analysis of nucleic acids. This system was shown to be quite sensitive and could detect five mastitis pathogens namely *E. coli*, Klebsiella sp., *S. aureus*, *Streptococcus uberis* and *Streptococcus agalactiae*.[16]

A study by Yeh et al. focused on constructing rapid and accurate testing device for detection of viral infections. They named this device as VIRRION, that is virus capture with Raman spectroscopy detection and identification. The method coupled with Raman identification, enabled real-time detection of multiple viral strains with high accuracy and within few minutes.[17]

A novel method called PRESCIENT (Platform for the Rapid Evaluation of antibody SucCess using Integrated microfluidics ENabled Technology) was developed by Wippold et al. to detect the antibodies elicited in response to viral infections. This droplet-based microfluidic system has shown to overcome the problems associated with conventional assays by enabling detection with high-sensitivity and high-reliability assays.[18]

Ticks are associated with many infectious diseases of animals such as tick-borne encephalitis virus infection. To detect these tick-borne viruses (TBV), Gondard et al. developed a microfluidics-based real-time PCR assay system having the ability to detect presence or absence of many TBVs. This method thus enabled detection of a number of TBVs from many ticks and from previously undiscovered regions.[19]

In poultry, respiratory syndrome is commonly associated with coinfections by pathogens like metapneumovirus, bronchitis virus, laryngotracheitis virus, avian influenza A virus, Mycoplasma spp., *E. coli*, etc. To detect these multiple pathogens simultaneously, Croville et al. made use of quantitative PCR by incorporating nano and microfluidic technology. This method could detect as many as 96 co-infectious pathogens simultaneously and has potential for diagnostics of poultry respiratory diseases.[20]

Harmful diseases commonly seen in pigs are the infections by porcine epidemic diarrhea virus (PEDV), transmissible gastroenteritis virus (TGEV) and porcine delta-coronavirus (PDCoV). The microfluidic device constructed by El-Tholoth et al. allowed co-detection of these three viruses along with semi-quantitative analysis by real-time, reverse transcription LAMP.[21] The on-site, rapid and specific detection of these pathogens can be possible with help of this disease. Another common disease in pigs is infection by influenza A virus (IAV). Goecke et al. developed a technique to improve detection and monitoring of swine influenza virus which made use of gene specific qPCR assays. The method provides real time and high-throughput analysis along with reducing the cost of virus subtyping.[22]

3.2 Analysis of blood

Blood tests are often performed for obtaining accurate diagnosis of various diseases. These tests analyze various parameters like blood cells, concentration of blood molecules, presence of abnormal molecules and so on.

Detection of innate and adaptive immune responses in chickens requires high-throughput systems which are highly sensitive. For this purpose, Borowska et al. developed a microfluidic PCR-based system which could enable detection and analysis of 96 samples in 6 nano-liter reaction mixtures.[22] The specificity of these assays was determined using amplicon sequencing and the assays showed that blood cells from various tissues could be reliably used to analyze immune responses in birds. The assay can have application in selective breeding programs and screening for vaccine responses.

For analysis of salmon blood, Lulijwa et al. developed a microvolume-based method to analyze peripheral blood cellular composition. They compared the results of their microfluidic system with that of the traditional hemocytometer and showed that the new system gave more accurate cell viability analysis. Apart from accuracy the system is also rapid, easy to use and can be applied for on-site analysis of samples.[23]

The investigation conducted by Ulum et al. focused on a microfluidic thread based analytical device based on cotton thread matrix. This device was characterized for blood plasma separation of healthy sheep using EDTA (ethylene diamine tetra-acetic acid) as anticoagulant. The blood treated with EDTA showed better separation from plasma and also depends on cotton fibers.[24] Such microfluidic thread-based device can be useful for blood plasma separation and its assay for detection.

Ghosh et al. developed a microfluidic paper-based assays with help of laser printer. The conventional methods used to develop such assays have the drawback of being expensive and complex.[25] The use of laser printer made it possible to reduce cost and complexity of microfluidic paper-based assay construction. This may stand application in point-of-care diagnosis of veterinary diseases.

In-vivo screening of blood coagulation can be done by the microfluidic assay known as total thrombus analysis system (T-TAS) fabricated by Iwanaga et al. This novel method comprises of two chips which mimic in-vivo conditions and can give the results for canine blood coagulation.[26]

3.3 Analysis of drug residue

The drugs used in veterinary disease treatments have different retention times which can affect the animal health as well as productivity. For this purpose, the screening and quantification of drug residues is important.

Zhang et al. made use of microfluidic chip-based electrophoresis for the analysis of kanamycin with help of a specific aptamer probe, labeled with gold.[27] The signal was amplified by combining this process with hybridized chain reaction and detected with fluorescence detection. The procedure takes very less time, gives accurate results and is sensitive. Wang et al. employed a similar microfluidic chip-based technique, except that it was combined with target catalyzed hairpin assemble method for signal amplification.[28] This method was also shown to have high efficiency and sensitivity. Another study which made use of kanamycin as a model analyte by He et al., was based on a microfluidic chip-based aptasensor. Here, the amplification was done by rolling circle mechanism with help of stir bar probes which triggered the amplification only in presence of the antibiotic. The circular DNA template was separated and detected with the microfluidic chip.[29] This technique has been claimed to have very high sensitivity and selectivity and can be used for field-based antibiotic analysis. The aptamer probes in these methods can be easily synthesized and all reactions can be performed in a single vessel, which gives them an edge over other conventional detection method.

A novel microfluidic chip-based strategy was developed by Chen et al. to simultaneously screen antibiotic, estrogen residues and heavy metal ions in samples. Target-specific aptamers labeled with magnetic bead-encoded probes were used along with primers to activate hybridized chain reaction.[30] The method is highly selective due to magnetic separation, reusable and has high efficiency. Similarly, He et al. combined magnetic tripartite DNA structure-functionalized nanocomposites with the microfluidic aptasensor for screening kanamycin, aflatoxin and estrogen residues all at once. Incorporation of DNA structures resulted in increased sensitivity of the method along with high stability.[31]

The microfluidic assay known as total thrombus analysis system (T-TAS) enables analysis of various anticoagulation drugs in-vivo. Similarly, the effect of acetyl-salicylic acid can also be studied with help of platelet function and fibrin clot formation chips, which mimic in-vivo conditions.

3.4 Multiplex detection

Often, diseases are associated with multiple pathogens, which need to be diagnosed early. Multiplex assays aid in the rapid screening of multiple analytes for different pathogens simultaneously using microfluidic devices. Multiplex assays have importance in clinical diagnosis as they can be utilized for analysis of various biomarkers.

Chen et al. developed a microfluidic chip-based device for screening of target sequences from equine respiratory pathogens. The loop-mediated isothermal amplification reagents present in wells of microfluidic chip give rise to fluorescent products when target sequences are present in sample.[14] The signals can be detected by smartphone camera and analyzed by software installed in smartphone giving accurate test results.

A major problem in livestock industry are the hemorrhagic diseases of pigs such as classical swine fever, vesicular disease and so on. To alleviate this problem, Erickson et al. combined automated microarray assay with multiplex RT-PCR which could accurately detect seven different swine viruses with 68 different strains.[32] Similarly, Lung et al.[12] developed an automated multiplex assay based on RT-PCR and reverse dot-blot detection. This has been shown to detect 37 different strains of target viruses from various samples with high accuracy.

For early detection of various re-emerging coronaviruses in pigs, Zhou et al. developed a novel technique involving microfluidic RT-LAMP chip. This system could detect three different coronaviruses simultaneously with high sensitivity and accuracy.[32]

The cost involved in development and use of such assays is very high and there is requirement of wax-printer for constructing microfluidic channels. To overcome this, Fava et al. suggested a disposable electrochemical paper-based microfluidic device with multiplexing capacity of up to 16 channels.

This strategy was used to develop a biosensor to screen glucose in urine and has been shown to have applicability in clinical diagnostics.[32]

Jiang et al. developed a rapid real-time quantitative PCR system for on-site detection of swine viruses. This technique could specifically and accurately analyze up to eight samples at a time.[33] Another microfluidic chip-based PCR system called One start was developed by Huang et al. for detection of respiratory tract infection pathogens. This method is rapid, can be used to analyze multiple pathogens and showed little batch-to-batch variation.[34]

4. Conclusion and future perspective

In view of the wide array of veterinary conditions and the nature of veterinary diagnostics, POC testing offers discrete benefits above traditional testing which are laboratory-based. The dawn of microfluidic technologies has further improved the prospects for wider and more treasured use of point of care testing. Even though these technologies have not yet been applied as extensively to veterinary medicine as they have in human medicine, they still offer great capacity. Many of the obstacles faced both in human and animal medicine are quite common, hence development in one field will therefore offer assistances to both fields, given that detailed needs confronted from an animal health point of view are acknowledged. Significantly, an intimate association among engineers and veterinary healthcare workers for developing new and current technologies for on ground application where diagnosis at point of care is limited is of paramount importance. Also integrating microfluidics with a more advanced form of sensing platform like an electrochemical sensor may prove to beneficial. As well as integrating microfluidic platforms with mobile interface can help in easy and continues monitoring of animal health.

Acknowledgments

We gratefully acknowledge Gujarat State Biotechnology Mission (GSBTM Project ID: AJJV48) and SHODH: Scheme of Developing High Quality Research Scholarship (Ref No: 201901440009) for financial support.

References

1. Smith S, Korvink JG, Mager D, Land K. The potential of paper-based diagnostics to meet the ASSURED criteria. *RSC Adv.* 2018;8(59):34012–34034.
2. Kumar S, Henrickson KJ. Update on influenza diagnostics: lessons from the novel H1N1 influenza A pandemic. *Clin Microbiol Rev.* 2012;25(2):344–361.

3. Busin V, Wells B, Kersaudy-Kerhoas M, Shu W, Burgess ST. Opportunities and challenges for the application of microfluidic technologies in point-of-care veterinary diagnostics. *Mol Cell Probes.* 2016;30(5):331–341.
4. Huxley J, Archer S, Biggs A, et al. An expert review of the diagnosis, prognosis and treatment of recumbency in adult cattle. *Cattle Pract.* 2010;18(1):53–60.
5. Verma N, Patel D, Pandya A. Emerging diagnostic tools for detection of COVID-19 and perspective. *Biomed Microdevices.* 2020;22(4):1–18.
6. Martinez AW, Phillips ST, Whitesides GM. Devices (uPADs)-are a new platform designed for ASSURED. *Anal Chem.* 2010;82:3–10.
7. Yang Z, Xu G, Reboud J, et al. Rapid veterinary diagnosis of bovine reproductive infectious diseases from semen using paper-origami DNA microfluidics. *ACS Sens.* 2018; 3(2):403–409.
8. Busin V, Burgess S, Shu W. A hybrid paper-based microfluidic platform toward veterinary P-ELISA. *Sensors Actuators B Chem.* 2018;273:536–542.
9. Ramalingam S, Collier CM, Singh A. A paper-based colorimetric aptasensor for the detection of gentamicin. *Biosensors.* 2021;11(2):29.
10. Kasoju A, Shahdeo D, Khan AA, et al. Fabrication of microfluidic device for aflatoxin M1 detection in milk samples with specific aptamers. *Sci Rep.* 2020;10(1):1–8.
11. Wadhwa A, Foote RS, Shaw RW, Eda S. Bead-based microfluidic immunoassay for diagnosis of Johne's disease. *J Immunol Methods.* 2012;382(1–2):196–202.
12. Lung O, Fisher M, Erickson A, Nfon C, Ambagala A. Fully automated and integrated multiplex detection of high consequence livestock viral genomes on a microfluidic platform. *Transbound Emerg Dis.* 2019;66(1):144–155.
13. Agnew M, Creedon AN, O'Connell I. Low-Cost, rapid prototyping of a microfluidic biosensor cartridge for on-farm diagnostics. In: *Paper Presented at: 2020 IEEE International Symposium on Circuits and Systems (ISCAS)*; 2020.
14. Chen W, Yu H, Sun F, et al. Mobile platform for multiplexed detection and differentiation of disease-specific nucleic acid sequences, using microfluidic loop-mediated isothermal amplification and smartphone detection. *Anal Chem.* 2017;89(21): 11219–11226.
15. Yu X, Xia Y, Tang Y, et al. A nanostructured microfluidic immunoassay platform for highly sensitive infectious pathogen detection. *Small.* 2017;13(24):1700425.
16. Viveiros S, Rodrigues M, Albuquerque D, Martins SA, Cardoso S, Martins VC. Multiple Bacteria identification in the point-of-care: an old method serving a new approach. *Sensors.* 2020;20(12):3351.
17. Yeh Y-T, Gulino K, Zhang Y, et al. A rapid and label-free platform for virus capture and identification from clinical samples. *Proc Natl Acad Sci.* 2020;117(2):895–901.
18. Wippold JA, Wang H, Tingling J, Leibowitz JL, de Figueiredo P, Han A. PRESCIENT: platform for the rapid evaluation of antibody success using integrated microfluidics enabled technology. *Lab Chip.* 2020;20(9):1628–1638.
19. Gondard M, Michelet L, Nisavanh A, et al. Prevalence of tick-borne viruses in Ixodes ricinus assessed by high-throughput real-time PCR. *Pathog Dis.* 2018;76(8):fty083.
20. Croville G, Foret C, Heuillard P, et al. Disclosing respiratory co-infections: a broad-range panel assay for avian respiratory pathogens on a nanofluidic PCR platform. *Avian Pathol.* 2018;47(3):253–260.
21. El-Tholoth M, Bai H, Mauk MG, Saif L, Bau HH. A portable, 3D printed, microfluidic device for multiplexed, real time, molecular detection of the porcine epidemic diarrhea virus, transmissible gastroenteritis virus, and porcine deltacoronavirus at the point of need. *Lab Chip.* 2021;21(6):1118–1130.
22. Borowska D, Kuo R, Bailey RA, et al. Highly multiplexed quantitative PCR-based platform for evaluation of chicken immune responses. *PLoS One.* 2019;14(12): e0225658.

23. Lulijwa R, Alfaro AC, Merien F, et al. Characterisation of Chinook salmon (Oncorhynchus tshawytscha) blood and validation of flow cytometry cell count and viability assay kit. *Fish Shellfish Immunol.* 2019;88:179–188.
24. Ulum MF, Maylina L, Noviana D, Wicaksono DHB. EDTA-treated cotton-thread microfluidic device used for one-step whole blood plasma separation and assay. *Lab Chip.* 2016;16(8):1492–1504.
25. Ghosh R, Gopalakrishnan S, Savitha R, Renganathan T, Pushpavanam S. Fabrication of laser printed microfluidic paper-based analytical devices (LP-μPADs) for point-of-care applications. *Sci Rep.* 2019;9(1):1–11.
26. Iwanaga T, Fukushima R, Nagasato T, Maruyama I, Miura N. Analysis of blood clotting with the total thrombus analysis system in healthy dogs. *J Vet Diagn Investig.* 2021;33 (2):357–361.
27. Zhang K, Gan N, Hu F, Chen X, Li T, Cao J. Microfluidic electrophoretic non-enzymatic kanamycin assay making use of a stirring bar functionalized with gold-labeled aptamer, of a fluorescent DNA probe, and of signal amplification via hybridization chain reaction. *Microchim Acta.* 2018;185(3):1–9.
28. Wang Y, Gan N, Zhou Y, et al. Novel label-free and high-throughput microchip electrophoresis platform for multiplex antibiotic residues detection based on aptamer probes and target catalyzed hairpin assembly for signal amplification. *Biosens Bioelectron.* 2017;97:100–106.
29. He L, Shen Z, Cao Y, et al. A microfluidic chip based ratiometric aptasensor for antibiotic detection in foods using stir bar assisted sorptive extraction and rolling circle amplification. *Analyst.* 2019;144(8):2755–2764.
30. Chen X, Wang J, Shen H-Y, et al. Microfluidic chip for multiplex detection of trace chemical contaminants based on magnetic encoded aptamer probes and multibranched DNA nanostructures as signal tags. *ACS Sens.* 2019;4(8):2131–2139.
31. He L, Shen Z, Wang J, et al. Simultaneously responsive microfluidic chip aptasensor for determination of kanamycin, aflatoxin M1, and 17β-estradiol based on magnetic tripartite DNA assembly nanostructure probes. *Microchim Acta.* 2020;187(3):1–11.
32. Erickson A, Fisher M, Furukawa-Stoffer T, et al. A multiplex reverse transcription PCR and automated electronic microarray assay for detection and differentiation of seven viruses affecting swine. *Transbound Emerg Dis.* 2018;65(2):e272–e283.
33. Jiang Y, Jiang S, Wu Y, et al. Multiplex and on-site PCR detection of swine diseases based on the microfluidic chip system. *BMC Vet Res.* 2021;17(1):1–8.
34. Huang E, Wang Y, Yang N, Shu B, Zhang G, Liu D. A fully automated microfluidic PCR-array system for rapid detection of multiple respiratory tract infection pathogens. *Anal Bioanal Chem.* 2021;413(7):1787–1798.

CHAPTER ELEVEN

Continuous monitoring of molecular biomarkers in microfluidic devices

Andrea Idili[a,b], Helena Montón[a], Mariana Medina-Sánchez[c], Bergoi Ibarlucea[d,e], Gianaurelio Cuniberti[d,e], Oliver G. Schmidt[c,f,g], Kevin W. Plaxco[a,h], and Claudio Parolo[a,i,*]

[a]Department of Chemistry and Biochemistry, University of California, Santa Barbara, CA, United States
[b]Department of Chemical Science and Technologies, University of Rome, Tor Vergata, Rome, Italy
[c]Institute for Integrative Nanosciences, Leibniz IFW Dresden, Dresden, Germany
[d]Institute for Materials Science and Max Bergmann Center for Biomaterials, Technische Universität Dresden, Dresden, Germany
[e]Center for Advancing Electronics Dresden (CFAED), Technische Universität Dresden, Dresden, Germany
[f]Research Center for Materials, Architectures and Integration of Nanomembranes (MAIN), Chemnitz, Germany
[g]School of Science, TU Dresden, Dresden, Germany
[h]Interdepartmental Program in Biomolecular Science and Engineering University of California, Santa Barbara, CA, United States
[i]Barcelona Institute for Global Health (ISGlobal) Hospital Clínic, Barcelona, Spain
*Corresponding author: e-mail address: claudio.parolo@gmail.com

Contents

Progress in Molecular Biology and Translational Science, Volume 187
ISSN 1877-1173
https://doi.org/10.1016/bs.pmbts.2021.07.027
Copyright © 2022 Elsevier Inc. All rights reserved.

Abstract

The ability to monitor molecular targets is crucial in fields ranging from healthcare to industrial processing to environmental protection. Devices employing biomolecules to achieve this goal are called biosensors. Over the last half century researchers have developed dozens of different biosensor approaches. In this chapter we analyze recent advances in the biosensing field aiming at adapting these to the problem of continuous molecular monitoring in complex sample streams, and how the merging of these sensors with lab-on-a-chip technologies would be beneficial to both. To do so we discuss (1) the components that comprise a biosensor, (2) the challenges associated with continuous molecular monitoring in complex sample streams, (3) how different sensing strategies deal with (or fail to deal with) these challenges, and (4) the implementation of these technologies into lab-on-a-chip architectures.

1. Introduction

1.1 Biosensors and society

Molecules and microorganisms are, obviously, not visible to the naked human eye, and thus for centuries we have instead employed tools that allow us to indirectly screen for their presence.[1] During the early 1900s, for example, miners used canaries to detect the presence of harmful gases (Fig. 1A).[2] Similarly, we humans have long used trained dogs to find things ranging from missing persons and delicious fungi to landmines and contraband goods.[3] More recently, however, technological advancements are leading to the creation of sensing strategies that, while not yet fully supplanting the approach, are far more convenient than carting around live birds and dogs.[4,5] Today, for example, we can easily find at the pharmacy small, user-friendly devices able to accurately detect pregnancy in matter of minutes (Fig. 1B).[6] Implantable sensors that monitor glucose levels in real time have likewise been available for some years[7] and, more recently, have been adapted to closed-loop feedback control over the release of insulin by an artificial pump (Fig. 1C).[8–10] Birds, dogs, pregnancy tests, continuous glucose monitors: the thing these widely separate technologies have in common is that they are biosensors, approaches to the detection of specific molecular targets using a biological recognition component, such as a protein or a nucleic acid. Here we review the state of the art of biosensors specifically enabling the continuous monitoring of molecular targets and how their integration into lab-on-a-chip (LoC) devices has the potential to revolutionize the biosensing field.

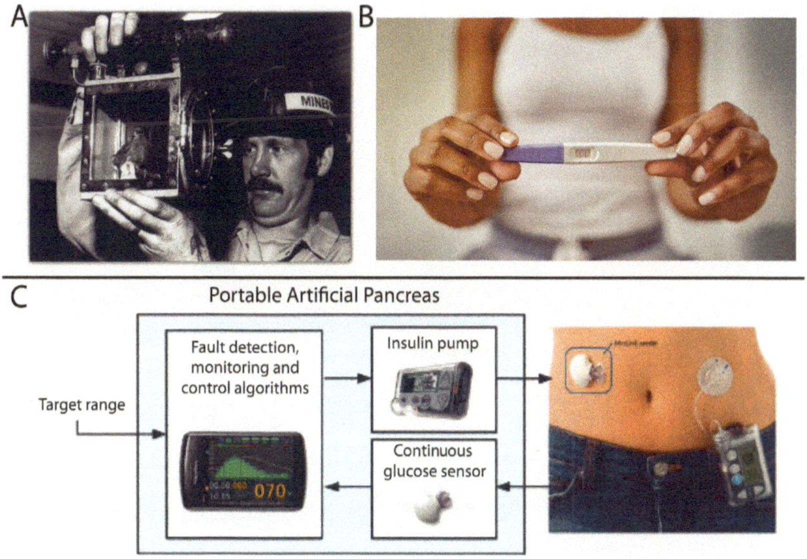

Fig. 1 Examples of biosensors. (A) For centuries miners employed canaries to detect the presence of harmful gases. (B) Since 1970s, commercial lateral flow assays have been available for the qualitative naked-eye detection of many molecular cues, including the biomarkers indicative of pregnancy. (C) More recently insulin pumps use the continuous, real-time concentration information provided by wearable glucose sensors to autonomously deliver insulin in the treatment of diabetes. *Figure adapted from* Canaries As Poisonous Gas Detectors. *Amusing Planethttps://www.amusingplanet.com/2018/05/canaries-as-poisonous-gas-detectors.html; n.d. Accessed May 20, 2021,* When is the Best Time to Take Pregnancy Test?. *Ohio State Medical Center https://wexnermedical.osu.edu/blog/best-time-to-take-pregnancy-test; n.d. Accessed May 20, 2021, Resalat N, Hilts W, El Youssef J, Tyler N, Castle JR, Jacobs PG. Adaptive control of an artificial pancreas using model identification, adaptive postprandial insulin delivery, and heart rate and accelerometry as control inputs.* J Diabetes Sci Technol. *2019;13:1044–1053. https://doi.org/10.1177/1932296819881467.*

1.2 What is a biosensor

Biosensors share a generic structure comprised of two, class-defining components: a bioreceptor, that is a biomolecule (such as an antibody or a DNA strand) that tightly and specifically recognizes the target, and a mechanism (whose defined elements are together defined as transducer) of converting this binding event into an easily detectable output (Fig. 2). In the continuous glucose monitor, for example, the bioreceptor is the enzyme glucose oxidase and the transducer is an electrochemical platform able to measure the enzyme's electroactive product, hydrogen peroxide.[11] As a result, higher concentrations of glucose generate higher currents, the intensity of which is easily measured and is directly related to the amount of glucose present.[11]

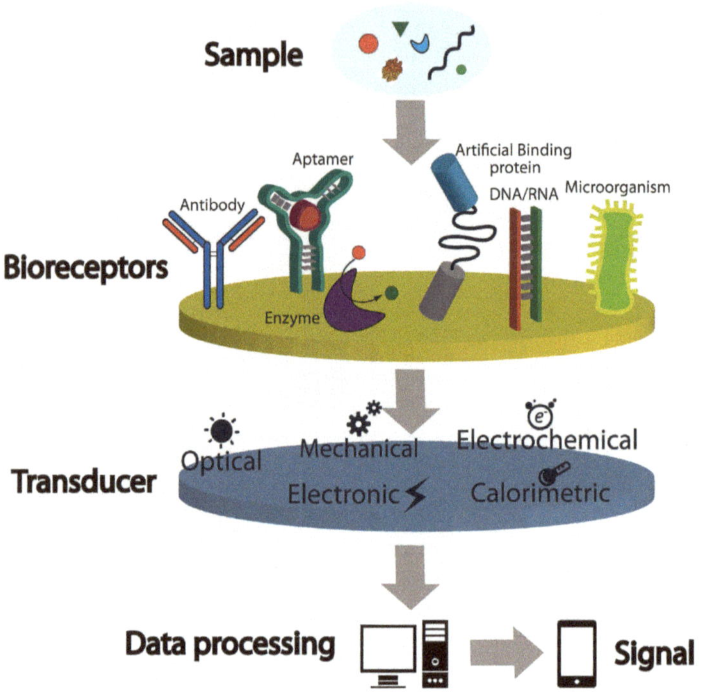

Fig. 2 Biosensors are comprised of two class-defining components: a bioreceptor and a transducer. The former is a biomolecule, such as a protein or a nucleic acid, that binds specifically to the target molecule. The latter is the component that transduces the binding event into a measurable output.

1.3 Bioreceptors

Due to the broad range of targets they are able to recognize, and the high affinity and specificity with which they generally do so, proteins and nucleic acids are the most commonly used bioreceptors. Proteins are chains of amino acids that carry out countless of functions within an organism (e.g., catalysis, structural integrity).[12] Their versatility relies on the wide range of different chemical properties of the side chains of the 20 different amino acids from which they are comprised. Looking in detail at how proteins are used as bioreceptors, we find two broad classes are employed: enzymes[13] and antibodies.[5] The applicability of the former, however, is limited to the specific substrate of the enzyme, rendering enzymatic sensors poorly generalizable. In contrast, antibodies are selected by the immune system of an animal in response to the presence of an exogenous molecule, rendering it possible to generate receptors that bind any of a wide range of targets. Due to the resulting generalizability, sensitivity and specificity, antibodies are the most widely used protein-based bioreceptors.[14]

Nucleic acid-based bioreceptors are extremely versatile allowing for the designing of biosensors able to detect different types of molecular targets.[15] In fact, although the primary biological functions of the nucleic acids RNA and DNA are the storing, replication, and transmission of genetic information, since 1990,[16] it has been possible to select nucleic acid sequences in vitro, called "aptamers," able to bind specific molecular targets other than DNA or RNA complements.[17,18] The methodology to do so, which is termed systematic evolution of ligands by exponential enrichment (SELEX), consists of challenging a diverse pool of nucleic acids oligomers (often of order 10^{15} unique sequences) with the biomarker of interest to select those rare molecules that bind to it. Advantages of aptamers over antibodies include the use of negative selection during SELEX (removing any sequence that binds to potential interferent molecules), quicker development (weeks vs months), and improved stability, resulting in a longer shelf-life.[19,20] To date aptamers have been reported that bind a range of clinically-relevant-biomarkers, from protein to small molecules (such as drugs and amino acids), and achieving analytical performance comparable to antibodies.[21,22] Moreover, the use of SELEX has allowed to expand the functions of DNA strands beyond target binding, obtaining molecules (DNAzymes, ribozymes) that, in addition to recognition properties, also catalyze reactions. Using these, developers have created biosensors in which the aptamer acts as both bioreceptor and transducer.[23]

1.4 Transducers and labels

The large majority of bioreceptors do not respond in any easily measurable way to target binding. Antibodies, for example, do not change their shape, emit photons or electrons in response to binding their target antigens. For this reason, biosensors require a second component (or set of components), the "transducer," whose function is to give an output that the user can interpret to identify and quantify the target molecule. Some transduction mechanisms rely on the intrinsic physical properties of the target molecule, such as its mass (e.g., quart crystal microbalances[24]), charge (e.g., field effect transistors[25]), steric bulk (e.g., microcantilvers[26]), or index of refraction (e.g., surface plasmon resonance[27]). Other transduction mechanisms require the presence of a "reporter" (i.e., a molecule or nanoparticle that has an easily detectable physical-chemical characteristic, such as redox,[28] optical,[29] or catalytic activity[30]), that is attached to the receptor during sensor fabrication and whose output is modulated in an easily measurable way by target binding. Still other approaches add a reporter to the target molecule through covalent modification or via the binding of a reporter-modified secondary receptor to generate a signal when the target binds the sensor's primary receptor.[31] In these circumstances, which includes sandwich assays, such as ELISAs[32] and Lateral Flow Assays,[6] the reporter is referred to as a "label." i.e., when the reporter is used to directly or indirectly modify the target itself. We believe, however, that such approaches are better thought of as bio-assays rather than biosensors, as they require multiple steps and thus, for example, do not easily support continuous monitoring.

1.5 Single-point vs continuous measurements

In this chapter we make a significant distinction between measurement approaches that can measure the analyte once during a single operation, and biosensors that can measure variations of the target over time.[33] For instance, the miners of the previous example required the constant monitoring for the presence of harmful gases, because they could not know where and when their life was at risk. There is no need, in contrast, to continuously monitor human chorionic gonadotropin, as a single time-point measurement of this target is sufficient to detect pregnancy. Currently, the vast majority of available bioanalytical measurements support only single-point analysis, even when, in many cases, continuous monitoring would be preferred (Fig. 3). For example, the current gold standard for the monitoring of kidney failure is performing single point measurements of biomarkers, such as creatinine and neutrophil gelatinase associated lipocalin.[34] Imagine how much more valuable the information would be if clinicians instead had available continuously recorded data, which would not only indicate the current levels of these molecules, but also their well-resolved trends, which would likely be far more indicative of a patient's health status and response to treatment.

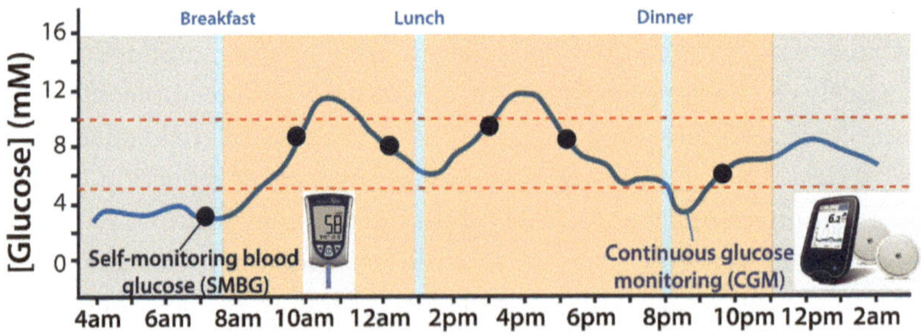

Fig. 3 Single-point measurements provide far less information than continuous measurements. Specifically, the latter allows for the identification of trends, and thus provides for more information on, for example, a patient's health status. Shown is the continuous monitoring of glucose, which permits the identification of glycemic peaks (following food intake) that would otherwise go undetected.

1.6 The challenges of continuous measurements

To achieve continuous molecular monitoring a number of technical hurdles must be surmounted simultaneously. Specifically, to support this a biosensor must:

- Exhibit sufficiently rapidly reversible binding and signaling to monitor the rising and falling target concentrations with time resolution relevant to the application at hand (Fig. 4A). Antibodies, for example, are generally not suitable for continuous monitoring because the rate with which they dissociate from their target is generally slow relative to the time resolution required to achieve clinically relevant, continuous monitoring (typical dissociation rate constants are of order $10^{-5}\ s^{-1}$).[35,36] Instead, more recently, aptamers showed excellent binding kinetics and reversibility (with average k_{off} often two orders of magnitude more rapid than those of antibodies), improving the time resolution.[37] Similarly, the transducer mechanism must support measurements at a frequency that is high relative to the necessary time resolution.
- Second, the biosensor must operate under clinically-relevant conditions (e.g., in bodily fluids) for long enough to provide a meaningful time window to study the event of interest (Fig. 4B).[38,39] That is the final device must ensure that the sample stream is in constant contact with the sensing area and that all components are not affected by the environment (for example, the blood flow inside a vein).
- Third, the overall device must be fully automated. Although it may sound obvious, this is a tricky engineering problem to solve. In fact, this implies the use of a software and a hardware that potentially support thousands of measurements per hour (Fig. 4C and D). Based on this, sensors that do not require batch processing steps, such as target labeling, separations, incubations, have a clear advantage over those that do.

1.7 Lab on a chips and continuous molecular monitoring

A platform capable of providing fully automated, continuous analysis would expand the possible applications of molecular monitoring to a variety of fields, including at the point of care and in the absence of specialized-personnel. In this context, LoCs appear promising.[40,41] During the last two decades LoCs have passed from being single and straight channels mounted on top of a rigid glass slide, to the current complex systems integrating multiple, often 3D interconnected functionalities (e.g., lab-in-a-tube or lab-in-a-tip)[42–44] that can even mimic organ functions (e.g., organ-on-a-chip[45]) (Fig. 5). The possibility of integrating continuous

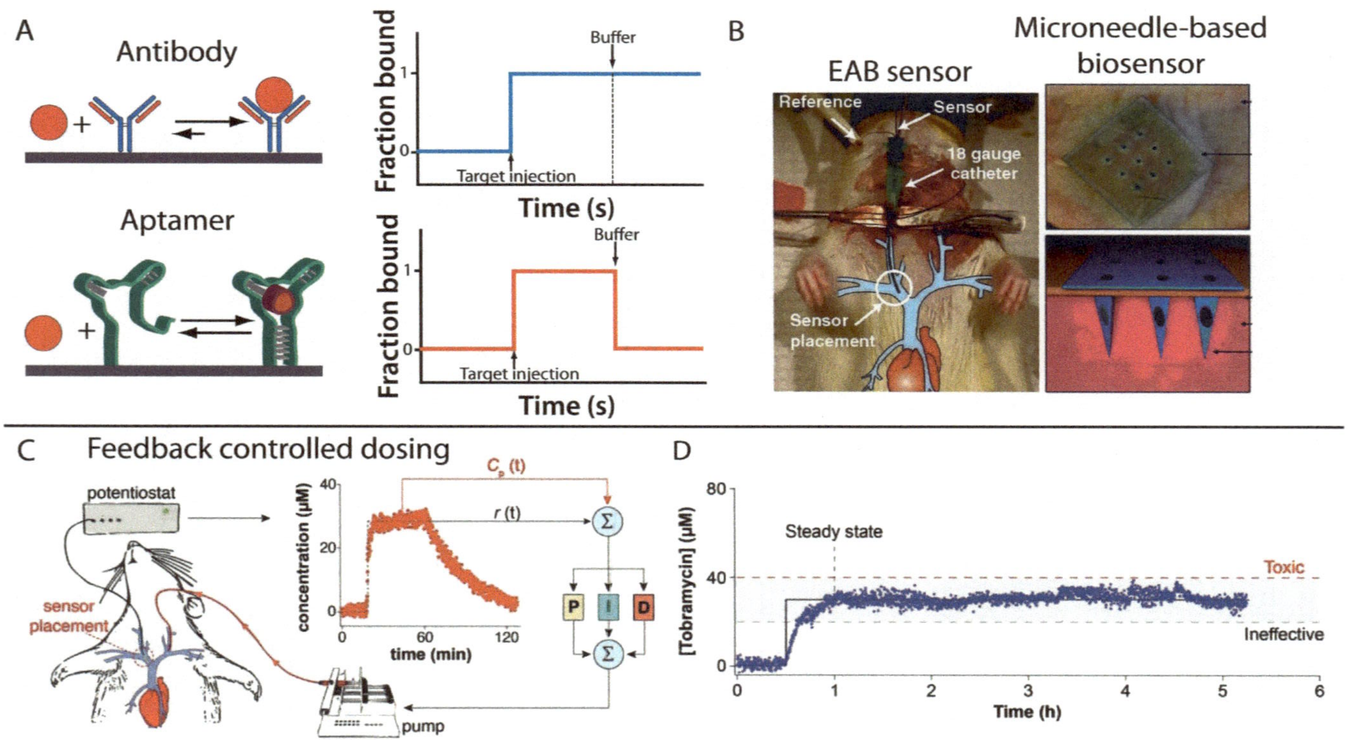

Fig. 4 Multiple challenges are associated with performing continuous measurements. (A) The receptor must bind its target both rapidly and reversibly. (B) The sensor must remain in continuous contact with the sample stream. (C) The system must be automated. (D) The sensor must be stable as long as the application requires, generally in the order of hours to days. *Reproduced from Mishra RK, Vinu Mohan AM, Soto F, Chrostowski R, Wang J. A microneedle biosensor for minimally-invasive transdermal detection of nerve agents.* Analyst. *2017;142:918–924. https://doi.org/10.1039/c6an02625g with permission from The Royal Society of Chemistry.*

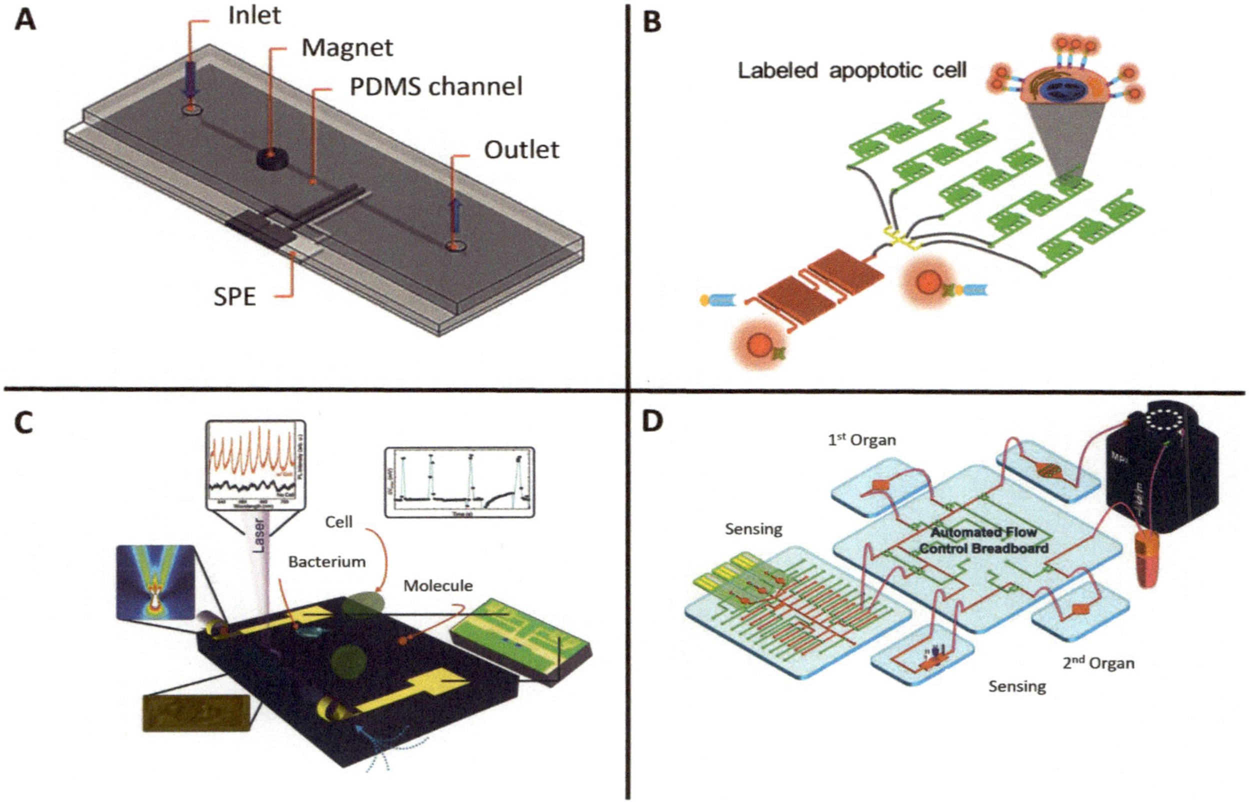

Fig. 5 See figure legend on next page.

-and even real time- molecular monitoring into LoCs would be mutually beneficial: on one hand, when necessary sample preparation can be supported by the automation provided by LoC devices[46]; on the other, the ability to monitor specific molecules continuously and in real time would advance LoCs by supporting the development of new sense-and-respond functionalities. For example, real-time molecular concentration information can support closed-loop feedback control (Fig. 4C and D),[47] such that the concentration of a specific molecule is maintained at constant levels. To date, however, examples of LoCs integrated with continuous molecular monitoring -much less LoCs harnessing those measurements to actuate- remain limited. In the next section we will analyze which analytical techniques can potentially provide continuous and real time measurements and how and why they have been integrated into LoCs.

2. Techniques

In this section we explore the sensing strategies supporting the continuous monitoring of specific molecular targets that have been integrated into LoCs.

Fig. 5 Lab-on-a-chip devices have evolved from simple, single channel structures to, for example, complex organ-on-a-chip architectures. (A) Early lab-on-a-chip devices consisted of simple, straight microfluidic channels. (B) Increases in complexity allowed the integration of functionalities such as mixing and separations. More recently (C) lab-in-a-tube and (D) organ-on-a-chip devices support the study of complex biological events in a miniaturized, often highly parallelized format. *Panel (A) Reproduced from Medina-Sánchez M, Miserere S, Morales-Narváez E, Merkoçi A. On-chip magneto-immunoassay for Alzheimer's biomarker electrochemical detection by using quantum dots as labels.* Biosens Bioelectron. *2014;54:279–284. https://doi.org/10.1016/j.bios.2013.10.069 with permission from Elsevier. Panel (B) Reproduced from Montón H, Medina-Sánchez M, Soler JA, Chałupniak A, Nogués C, Merkoçi A, Rapid on-chip apoptosis assay on human carcinoma cells based on annexin-V/quantum dot probes.* Biosens Bioelectron. *2017;94:408–414. https://doi.org/10.1016/j.bios.2017.03.034 with permission from Elsevier. Panels (C and D) Reproduced from Qiu Z, Shu J, He Y, Lin Z, Zhang K, Lv S, Tang D. CdTe/CdSe quantum dot-based fluorescent aptasensor with hemin/G-quadruplex DNzyme for sensitive detection of lysozyme using rolling circle amplification and strand hybridization.* Biosens Bioelectron.*2017;87:18–24. https://doi.org/10.1016/j.bios.2016.08.003 with permission from National Academy of Sciences and from Smith EJ, Xi W, Makarov D, Mönch I, Harazim S, Bolaños Quiñones VA, Schmidt CK, Mei Y, Sanchez S, Schmidt OG. Lab-in-a-tube: ultracompact components for on-chip capture and detection of individual micro-/nanoorganisms, in:* Lab Chip. *Royal Society of Chemistry; 2012:1917–1931. https://doi.org/10.1039/c2lc21175k The Royal Society of Chemistry.*

2.1 Optical sensors

Optical biosensors use colorimetric or luminescent (fluorescent or phosphorescent) methods for detection and are, to date, the most widely used class of biosensors in clinical applications. They rely on producing a measurable optical signal which intensity is proportional to the analyte concentration.[48]

2.1.1 Colorimetric biosensors

Colorimetric biosensors employ chromophores -strongly absorbing molecules or nanoparticles- as reporters to produce a measurable color signal whose intensity depends on the analyte concentration (Fig. 6A).[49,50] Their commercial success to carry out single point measurements (e.g., ELISA and LFA) depends on the nature of their signal that can be easily read by naked eye or cameras or scanners.[4,51,52] However, as we discuss below in Section 2.1.3, their use in continuous monitoring remains limited.

2.1.2 Fluorescence-based biosensors

Fluorescence is the property of some substances (fluorophores) to absorb light of one wavelength and emit light at a longer wavelength during a brief interval -known as the fluorescence lifetime- typically in the order of few nanoseconds (Fig. 6B).[53,54] It is by far the most exploited optical property in biosensors due to a significant number of related techniques that provide measurable parameters such as intensity, lifetime, energy transfer, and quenching efficiency. Fluorescence-based biosensors require a source of excitation for the fluorophore, typically a UV lamp or LEDs. Their transducer component often consists in the use of dedicated filters and detectors such as cameras, photomultiplier detectors and photon counters. They generally achieve lower limits of detection than their colorimetric counterparts, albeit at a higher cost due to the need for more expensive optical components as signal transduction elements (e.g., laser excitation, optical filters).[51]

2.1.3 Continuous sensing using optical sensors

The use of optical biosensors for continuous molecular sensing in clinical settings remains limited. There are two main reasons for this: the need for an optical reporter[55] and the presence of optical interferents in many important samples matrixes.[56,57] Regarding the former, the vast majority of fluorescent-based biosensors rely on the labeling of the target molecule with a fluorescent reporter, making them unable to support continuous monitoring.[55] The latter instead refers to the strong absorbance, scattering and auto-fluorescence of most biological matrixes, including blood, urine

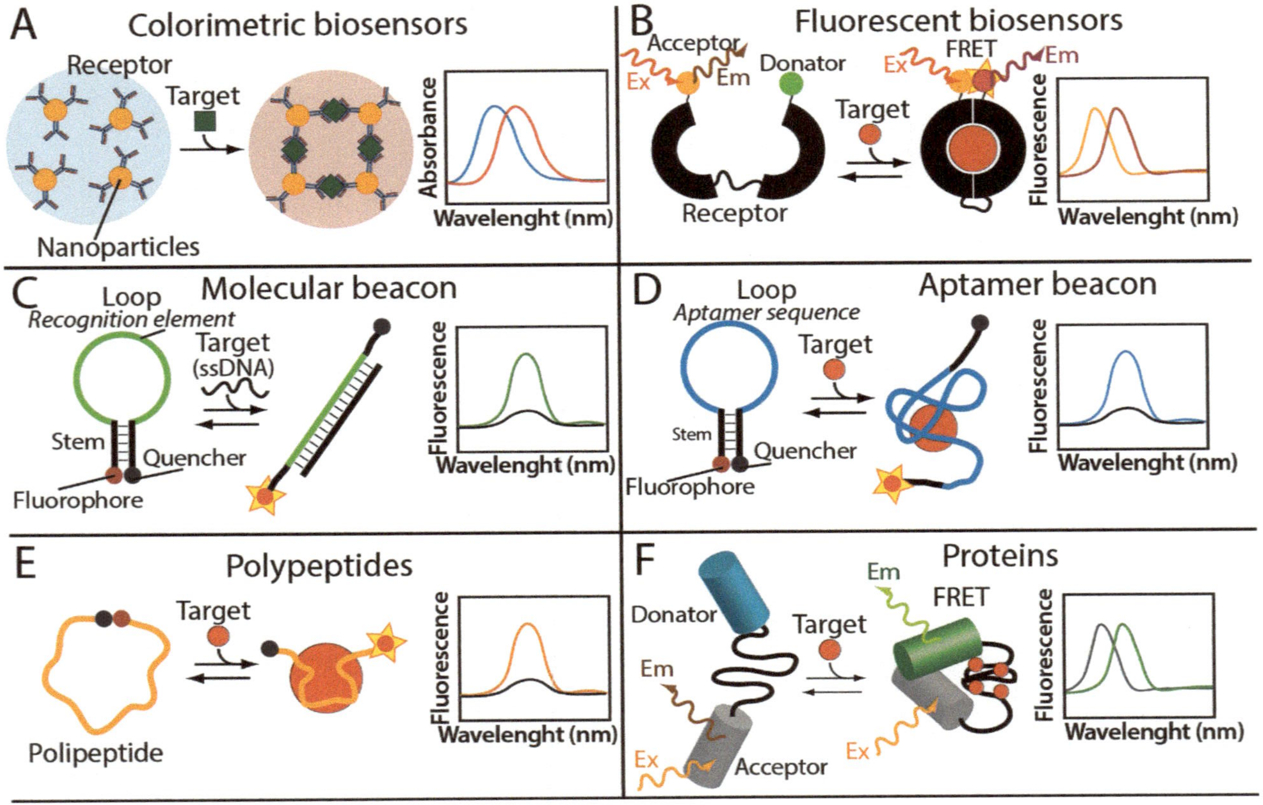

Fig. 6 See figure legend on opposite page.

and foodstuffs. Imagine, for example, how difficult it is to see an optical signal against the strong red color of undiluted whole blood. Because of this, optical biosensors have been used predominantly for single-point measurements that employ washing or dilution steps to reduce this background. For continuous sensing, however, these are "deal-breakers."[56,57]

If we exclude non-generalizable sensors detecting chromogenic or fluorogenic targets, promising optical sensing approaches that succeed in continuous monitoring are those employing a bioreceptor undergoing a binding-induced conformational change.[58] Although it has not generally been used for continuous monitoring, among the first examples of this type of sensors, which illustrates the principles of the class, is the molecular beacon. First described by Tyagi and Kramer in 1996,[59] these oligonucleotide-detecting sensors are comprised of a nucleic acid with complementary ends that thus forms a stem-loop structure bringing the two termini into proximity (Fig. 6C). The hybridization of a target oligonucleotide to the loop sequence opens the stem, separating its two ends. A fluorophore quencher pair placed at the termini converts this conformational change into an easily measurable optical output. More recently, researchers have adapted aptamers to work as analogous "aptamer beacons" that use the binding-induced conformational changes in aptamers for the optical detection of proteins and small molecules (Fig. 6D).[60] More recently still, this approach has been expanded to include polypeptides[54,58] (Fig. 6E) and even whole proteins[61] (Fig. 6F) as receptors.

Fig. 6 Optical sensors often employ colorimetric or fluorescent outputs. (A) Colorimetric sensors rely on either a shift in the absorbance wavelength of an optical reporter or an increase in its absorbance. (B) Fluorescent sensors typically employ binding-induced changes in luminescence generated via changes in either quenching or energy transfer. (C–F) We believe that the most promising fluorescent strategies for conducting continuous molecular monitoring use a binding-induced change in the conformation of the bioreceptor to change in the distance between a fluorescent label and a quencher or a fluorescent energy donor and acceptor. Examples include: (C) Molecular beacons, which are DNA molecules with fluorophore-and-quencher-modified, self-complementary ends that thus form a stem-loop. The binding of a target complementary to the loop sequence opens the stem, segregating the reporters and enhancing fluorescence. (D) Aptamer beacons are based on a similar concept, but employ the binding-induced folding of an aptamer (or, not shown, binding-induced dissociation of an aptamer-complement duplex) to generate the necessary conformational change. (E) The binding-induced folding approach has also expanded to polypeptides, (F) and even to genetically encoded sensors suitable for intracellular studies by employing fluorescent protein pairs as the reporters.

2.1.4 LoC using continuous optical sensing

Colorimetric and fluorescent sensors have been widely integrated into LoCs.[62,63] Originally, optical sensors were used to test new functionalities (e.g., mixing, separation, labeling)[64] of LoCs, rather than for molecular monitoring. More recently, however, we are observing an increasing number of LoCs using optical sensors to carry out such analysis. While the majority of this work still employs simple, buffered solutions in controlled environment, a nice example from the Soh group has demonstrated the ability to monitor clinically-relevant concentrations of glucose and insulin in blood withdrawn in real time directly from a living animal.[65] The device employed include three sections: labeling/mixing, sample treatment, and detection (Fig. 7). In the first, continuously-drawn blood is mixed with (i) microbeads modified with a fluorophore-modified aptamer whose intensity is quenched by a quencher-modified complementary strand, (ii) microbeads modified with anti-insulin antibodies and (iii) fluorophore-tagged anti-insulin antibodies. The binding of the glucose to the aptamer displaces the complementary strand recovering the fluorescence signal (the quencher is no longer in close proximity to the fluorophore). Instead, the presence of insulin induces the formation of an immune-sandwich, accumulating

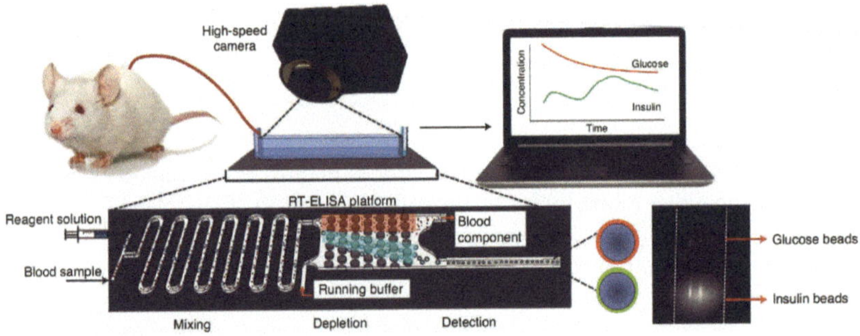

Fig. 7 Continuous molecular monitoring via fluorescent biosensors has been integrated into lab-on-a-chip architectures. Shown, for example, is a LoC capable to continuously measure insulin and glucose from the blood drawn continuously from an animal. The LoC includes three main functionalities. In the first, the blood is mixed with fluorophore-modified bioreceptors. In the second, blood cells and excess of labeled bioreceptor are removed from the flow stream. In the last, a high-speed camera records the resulting fluorescent signal. *Reproduced from Poudineh M, Maikawa CL, Ma EY, Pan J, Mamerow D, Hang Y, Baker SW, Beirami A, Yoshikawa A, Eisenstein M, Kim S, Vučković J, Appel EA, Soh HT. A fluorescence sandwich immunoassay for the real-time continuous detection of glucose and insulin in live animals.* Nat Biomed Eng. *2020;5:53–63. https://doi.org/10.1038/s41551-020-00661-1 with permission from Elsevier.*

the fluorophore-modified antibody on the beads. In the second section, the device removes interferents, such as red blood cells and any unbound receptor, allowing just the microbeads to reach the detection module. Finally, in the last section, a high-speed camera and two lasers (used to excite the fluorophores) are used to generate and measure a target-specific signal. Although, strictly speaking the system is not real-time (since from the extraction of the blood until the detection there is a 30 s lag) and its operative life is limited by the amount of reagent the reservoirs can hold (the reported measurements do not exceed 75 min), it represents the first example of how the automation and functionalities of LoCs can overcome important limitations for the use of optical sensors to support continuous monitoring.

2.2 Electrochemical sensors

Motivated by the difficult of deploying optical sensors in whole blood, much less in vivo, we and others have focused on electrochemical sensors supporting continuous, in-vivo molecular monitoring. These can be divided into two main subtypes: amperometric and voltametric techniques.

2.2.1 Amperometry

Amperometric methods such as amperometry,[66] chronoamperometry,[67] and double-step chronoamperometry[68] are based on the measurement of the current over time at a fixed potential (Fig. 8A).[69,70] Conceptually these techniques provide an excellent sampling frequency (down to sub-second resolution) when compared to voltammetry, since they do not require scanning the potential over broad windows.[71] Their sensitivity, however, tends to be poorer than that of voltammetry. Finally, unless the target itself exhibits useful redox properties (which would limit the generalizability of the technique), amperometric sensors require a redox reporter (catalytic reporters, such as nanoparticles or enzymes, that continuously produce electrons for amperometry, or molecular species that easily give up one or two electrons for chronoamperometry)[71,72] in order to generate a signal.

2.2.2 Voltammetry

In voltammetry the output signal is the variation in current seen upon changing the applied potential, with the exact details of method (e.g., cyclic voltammetry, stripping voltammetry, square wave voltammetry, differential pulse voltammetry) depending on the applied wave-form and the current sampling protocol (Fig. 8B).[73,74] Although both versatile and sensitive, voltametric methods require the use of a redox reporter.

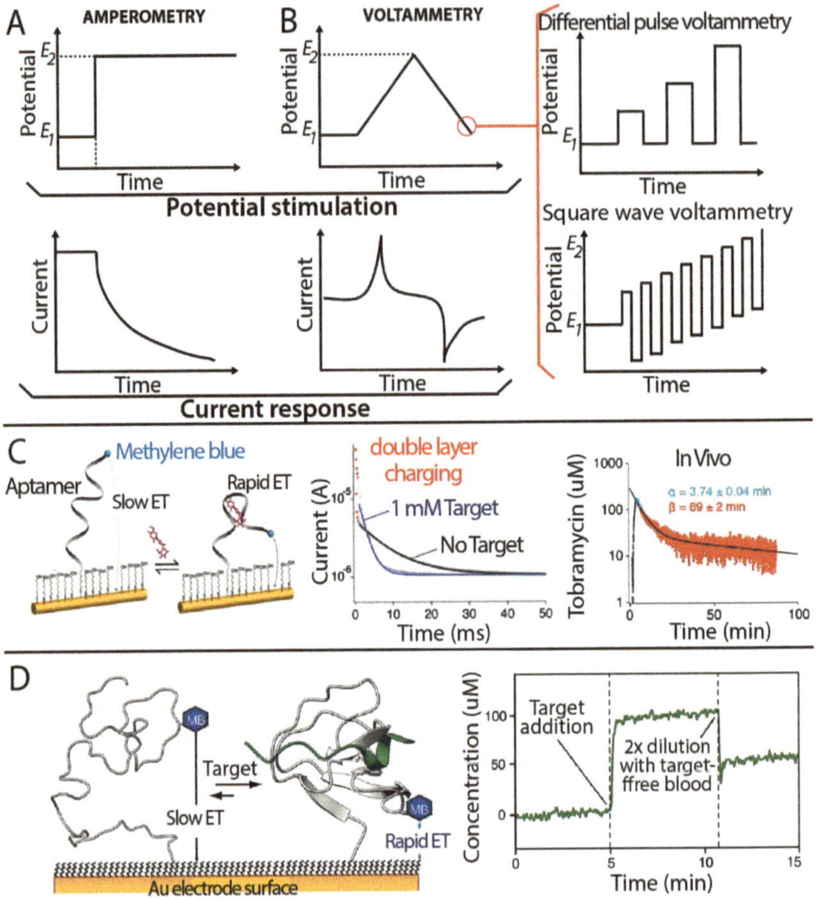

Fig. 8 Electrochemical biosensors have proven a promising means of performing continuous, real-time measurements in complex sample streams. (A) Amperometric methods (e.g., chronoamperometry is shown) measure changes in a current at a fixed potential. (B) Voltammetric methods, in contrast, measure variation in the current upon as the applied potential varies (e.g., differential pulse voltammetry and square wave voltammetry are shown). Examples of electrochemical biosensors include (C) electrochemical aptamer-based sensors (shown is an EAB sensor being interrogated via chronoamperometry) and (D) electrochemical protein-based sensors (shown here using square wave voltammetry as the interrogation method). Both methods are based on the binding-induced conformational change of a bioreceptor labeled with a redox reporter (in these cases, methylene blue) to generate a electrochemical output that varies rapidly and monotonically with target concentration.

2.2.3 Continuous sensing using electrochemical techniques

Both amperometric and voltametric techniques require a redox reporter[75,76] to generate a measurable signal, which, if employed as a label (i.e., to modify the target), would render continuous monitoring difficult (see Section 1.6). If the reporter is, instead, covalently attached to the transducer those techniques can provide excellent analytical performance to carry out continuous biosensing.[71] Harnessing this concept researchers succeeded in using electrochemical techniques to detect in real-time variations in concentrations of various types of target.

Probably the most promising approach in the category of generalizable electrochemical sensors for continuous analysis is the electrochemical aptamer-based (EAB) sensor platform.[77] EAB sensors have been shown able to monitor molecular analytes ranging from low-molecular-weight drugs and metabolites to full-length proteins directly in undiluted blood, urine, foodstuff.[47,78–83] Recently, they have even been shown to work in situ in the bodies of live animals.[78,84–88] By analogy to the similar optical biosensors described in Section 2.1.3, this sensing strategy relies on binding-induced conformational changes to drive signaling generation. Specifically, EAB sensors employ a double-modified aptamer: on one end it presents a functional group (e.g., thiol, carboxylic acid, amine) that supports attachment to an electrode surface, while on the other it presents a redox reporter (often methylene blue). Upon target binding, the aptamer changes its conformation, changing in turn the rate of electron transfer between the redox reporter and the electrode surface, and thus producing an easily measurable signal when the sensor is interrogated via either voltametric or chronoamperometric methods.[38,71] More recently we have had our first success in adapting the EAB sensing concept to a sensor employing FynSH3 as the bioreceptor (Fig. 8D),[89] in work that illustrates how this approach could be expanded to the use of protein-based receptors.

Binding-induced conformational changes are not the only electrochemical biosensors approaches that support continuous sensing capabilities. Another example is sensors that use nucleic acid or protein "scaffolds" anchored on the electrode surface to present or act as a bioreceptor.[90–94] The binding of a higher-molecular-weight target molecule to these constructs reduces electron transfer by steric hindrance, producing an easily measurable change in the electrochemical signal. An advantage of these is that their design is straightforward as it does not require the careful engineering of a bioreceptor structure to undergo a binding-induced conformational change. Their application is limited, however, to larger (in the order of tens of kilodaltons) targets, such as antibodies and other proteins.[90–93,95]

2.2.4 Electrochemical sensors in LoC applications

An increasing number of electrochemical biosensors have been integrated into LoCs for the detection of molecular analytes, even if most examples are of devices that perform single-point, rather than continuous, measurements.[96] This is reasonable, considering that electrochemical techniques require a redox reporter and that the label-free strategies mentioned in the previous section are relatively new (mostly developed in the last 5 years). However, there is a relevant example of the integration of an EAB sensors into a LoC, in which the authors monitored the drugs doxorubicin and kanamycin in a flowing stream of untreated blood taken directly from a living animal and directed into a LoC (Fig. 9).[97] Given that EAB sensors

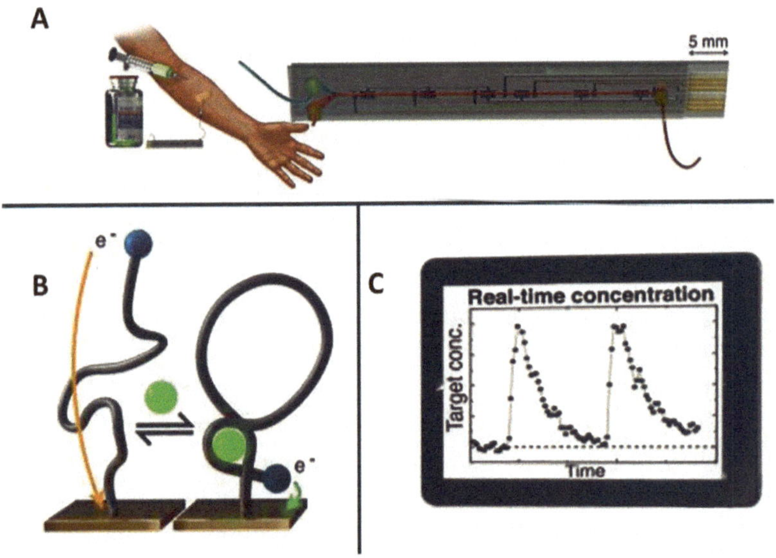

Fig. 9 Shown is the continuous monitoring of a molecular target using an EAB sensor integrated into a LoC. Specifically, in the work of Ferguson et al. (A) undiluted blood flows through a microfluidic channel containing a (B) drug-detecting EAB sensor. (C) The resulting output supplies a continuous estimation of the concentration of the drug in blood. *Reproduced from Ferguson BS, Hoggarth DA, Maliniak D, Ploense K, White RJ, Woodward N, Hsieh K, Bonham AJ, Eisenstein M, Kippin, TE, Plaxco, KW, Soh, HT. Real-time, aptamer-based tracking of circulating therapeutic agents in living animals.* Sci Transl Med. *2013;5:213ra165. https://doi.org/10.1126/scitranslmed.3007095 with permission from The American Association for the Advancement of Science.*

already provide all of the features required to carry out continuous monitoring (i.e., rapid, reversible binding, label-free), in this work the authors employed the microfluidic channel to guarantee both a constant contact of the sample on the electrode surface as well as a minimum sample volumes.

2.3 Electrical sensors

A different class of sensors that do not require a redox reporter to generate a measurable signal are the electrical sensors. Although there are a variety of electrical techniques (e.g., memristors,[98,99] chemiresistors[100] and diode biosensors[101]), the two most common types of electrical biosensors rely on the use of Electrical Impedance Spectroscopy (EIS) and Field Effect Transistors (FET).

2.3.1 Electrical impedance spectroscopy

The working principle of EIS consists in applying to a working electrode a small amplitude (~10 mV) sine wave potential that is overlaid to a direct current potential. The output signals are variations in the capacitance, inductance and resistance of the resulting alternating current over a wide frequency range (Fig. 10A).[102,103] EIS allows to carry out measurements without the need of any redox reporter and at a frequency that could support continuous molecular monitoring. Unfortunately, however, due to the false-positive signals that are associated with the non-specific adsorption

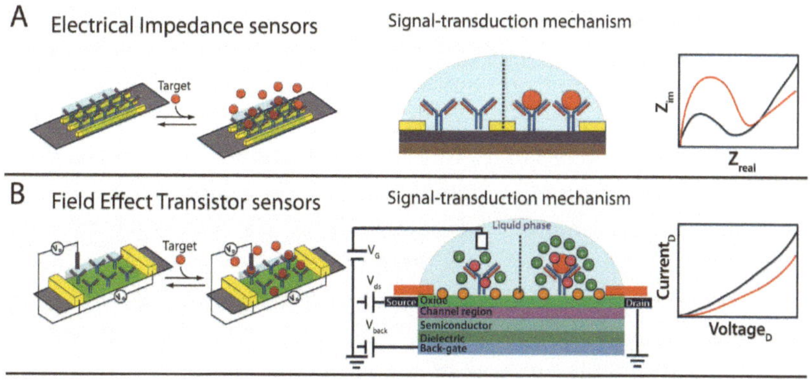

Fig. 10 The working principles of Electrical Impedance Spectroscopy (EIS) and Field Effect Transistor (FET) sensors. (A) EIS sensors use an applied sine-wave potential to monitor variations in the capacitance, inductance and resistance of the system induced by the recognition of the target by the bioreceptor. (B) FET sensors measure variations in the current passing through them when a charged target binds to their surface, altering the local electric field and, in turn, the conductance of a semiconductor.

of interferent on the sensor surface, EIS-based sensors fail when challenged in undiluted complex sample matrixes, limiting the application of this approach in real settings.

2.3.2 Field effect transistors

In FETs, the flow of current between a source and a drain is modulated by the application of a voltage to the gate. To convert FETs into biosensors, a bioreceptor is immobilized on or near the gate of a semiconductor connecting a source and drain. The binding of a charged target alters the local electric field, modulating the semiconductor's conductance (Fig. 10B). As is the case for EIS sensors, FET sensors are label-free and extremely sensitive. Unfortunately, while FET sensors often achieve exquisitely low limits of detection, they are also exquisitely sensitive to interference due to changes in solvent conditions (e.g., ionic strength, temperature) and the non-specific absorbance of interferents. Because of this, while they have been used to perform continuous molecular monitoring in the laboratory (e.g., single cell analysis or in cell culture[104]), their application to measurements in unprocessed clinical materials has proven difficult.[105]

2.3.3 Continuous sensing using electrical techniques

Neither EIS nor FET sensors require the modification of the receptor with the addition of a redox reporter or, for higher-molecular-weight targets, its re-engineering to undergo a binding-induced conformational change.[106] Conversely, however, a serious limitation of these transduction mechanisms is their susceptibility to false-positives arising due to the interference by materials typically found in clinical and foodstuff samples. And, although some progress has been made in solving this problem, it remains the major issue preventing their use in realistic clinical scenarios. For example, looking at EIS sensors for continuous monitoring, all of the examples we are aware of use highly diluted samples.[107] For this reason, EIS sensors have predominantly been used for either single-point measurements (achieving remarkable limits of detections, albeit, again only in dilute samples) or as tools to study the biophysical properties of single cells or cell layers (i.e., as a characterization technique rather than as a biosensor).[104]

The label-free, ultra-sensitive, continuous measurements produced by FETs make them seem an ideal sensing platform for molecular monitoring, but once again these attributes are overshadowed by intrinsic limitations. First, when deployed in realistically complex matrixes FET-based sensors are prone to false signals associated with non-specific adsorption on the sensor surface.[108] Second, they have been historically limited by the short Debye length (i.e., the maximum distance at which the potential applied on an electrode is observed in solution) seen at physiological salt concentrations.[109] For example, the Debye length seen in undiluted serum is ~0.7 nm, an order of magnitude smaller than the size of a typical antibody. For this reason, FET applications typically employ samples of extremely low ionic strength. Recently several research groups have proposed various solutions to these issues, ranging from improved blocking strategies (e.g., using ionic surfactants like polyethylene glycol)[110] to the use of novel nanomaterials[111–114] to minimize the effect of non-specific interactions and the use of smaller bioreceptors (i.e., aptamers and peptides)[109,115] to circumvent the Debye length problem. Despite these advances, however, we have been unable to find examples of continuous monitoring using FETs in untreated clinical samples, with the most relevant work carrying out measurement in undiluted, yet desalted, donkey serum.[116]

2.3.4 Combining electrical impedance spectroscopy with electrochemical, aptamer-based sensors

A more recent application of EIS to biosensing, which appears to circumvent the problems that have traditionally plagued the approach as a biosensor signal transduction mechanism, has been its use to interrogate EAB sensors continuously and in real time.[117] That is, to employ EIS to monitor binding-induced changes in the rate of electron transfer from the redox reporter on an EAB sensor's aptamer. In contrast to traditional EIS approaches, which, as we note above, are sensitive to non-specific adsorption, this approach has been shown to work for many hours in flowing, undiluted whole blood while collecting good-precision measurements at a time resolution of a few tens of milliseconds.

2.3.5 Lab on a chips and electrical measurements

The ultra-sensitivity and label-free capabilities of FET and EIS sensors together with their small size have been promoting their implementation into LoCs.[118–120] For example, the resulting devices have been employed to continuously monitor binding events in buffered solutions (for, e.g., nucleic acid hybridization, cell analysis),[43,121–123] or to automate incubation and washing steps in order to reduce either the ionic strength or the non-specific adsorption of interferent molecules (Fig. 11).[124,125] Currently, however, we could not find examples of LoCs that carried out continuous molecular monitoring in undiluted clinical matrixes using electrical

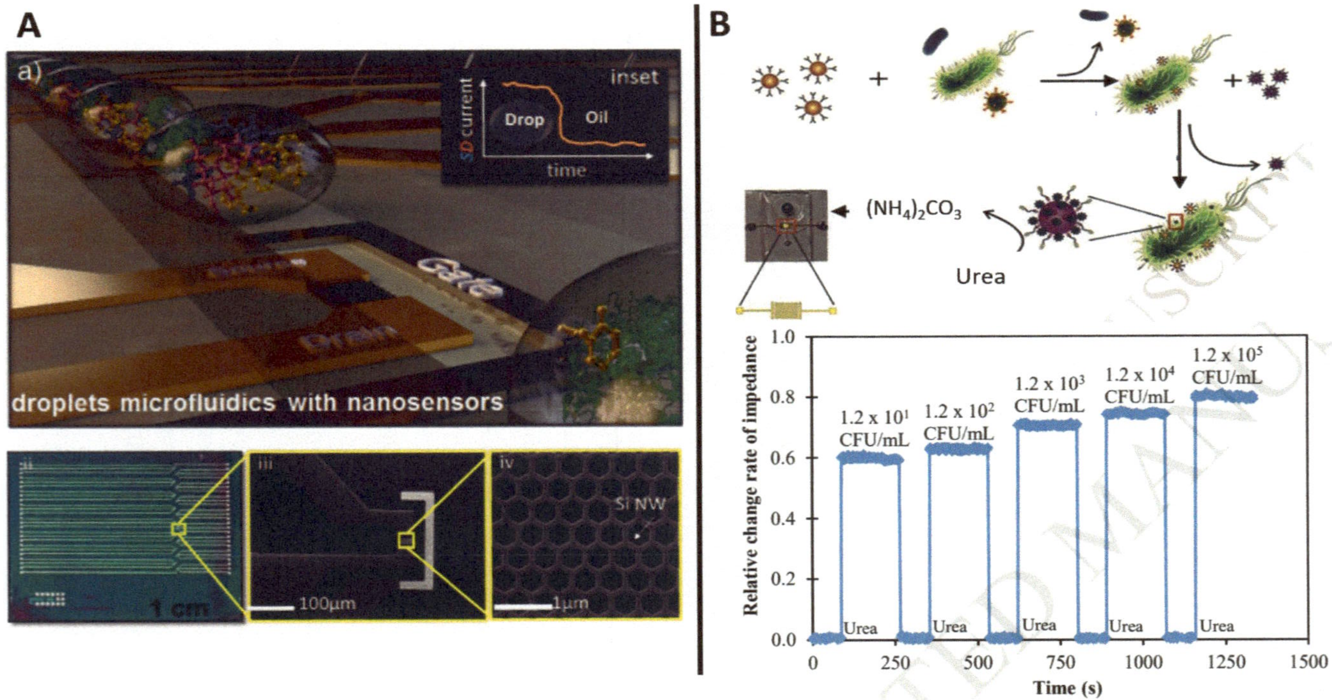

Fig. 11 The miniaturizability of electrical devices makes them ideal transducers to integrate into LoCs. Likewise, although the continuous molecular monitoring in unprocessed clinical samples has not been achieved using either FET or EIS, the possibility to automate the necessary sample processing via a LoC device has allowed their use for the continuous monitoring of chemical reactions. Shown, for example, are (A) a LoC device using FETs to continuously monitor β-galactosidase activity in individual emulsion-droplet micro-reactors. And (B) the use of an EIS-based LoC to measure *E. coli* counts in samples previously pre-treated via washing and labeling steps. *Panel (A) Reproduced from Belyaev D, Schütt J, Ibarlucea B, Rim T, Baraban L, Cuniberti G. Nanosensors-assisted quantitative analysis of biochemical processes in droplets.* Micromachines. *2020;11:138. https://doi.org/10.3390/mi11020138 with permission from MDPI. Panel (B) Reproduced from Yao L, Wang L, Huang F, Cai G, Xi X, Lin J. A microfluidic impedance biosensor based on immunomagnetic separation and urease catalysis for continuous-flow detection of E. coli O157:H7.* Sens Actuators B. *2018;259:1013–1021. https://doi.org/10.1016/j.snb.2017.12.110 with permission from Elsevier.*

methods. We thus believe that, without further advances (such as the use of deformed graphene channel[113,126] or the introduction of dielectric polymers on its surface[114]), the use of electrical sensors in LoCs will be limited to either single-point measurements of complex samples or the characterization of biological events in buffered solutions.

2.4 Mechanical biosensors

Sensing platforms relying on mass- or steric-bulk-driven detection have attracted the interest of biosensing community for more than three decades.[24,127–132] Because these are intrinsic properties of all molecular targets, mass- and bulk-based approaches are label-free and generalizable. Under the definition of mass-sensitive sensors falls several techniques, the most commonly employed being quartz crystal microbalances (QCM),[24,131] surface acoustic wave (SAW) sensors,[130,133] and microcantilevers (in liquid phase).[132,134] A limitation of these biosensors, however, is that, once again, non-specific absorption has limited their applicability to clinical measurements.[24] Specifically, while such platforms do support measurements in diluted biological fluids, such as diluted urine,[135,136] serum, or plasma,[24] they fail in undiluted clinical materials.

2.4.1 Quartz crystal microbalances

QCM is perhaps the most widely-used mass-based biosensor platform (Fig. 12A). This exploits the piezoelectric effect to create resonant acoustic waves (or oscillations) in a quartz crystal[24,131] (Fig. 12A, center) propagating perpendicularly with respect to the crystal surface.[137] Because the resonant frequency of the crystal depends sensitively on its mass, the resonant frequency can be used to track the binding of a target molecule to a surface-attached bioreceptor (Fig. 12A, right).[24,131] The resulting devices are sensitive, relatively low-cost and can be made quite small.[137]

2.4.2 Surface acoustic wave

As is true for QCM, SAW sensors rely on the generation of acoustic waves to monitor the absorption of mass on the sensing surface. In these devices, however, the waves travel parallel to the surface of the piezoelectric substrate rather than perpendicular[130,133,137,138] (Fig. 12B). Target binding to a surface-attached bioreceptor alters the speed, phase, and amplitude of this wave in a manner quantitatively related to the amount of bound target mass.[130,133,137] Due, in part, to their ease of fabrication (SAW devices are already mass produced as components of modern smartphones) and their

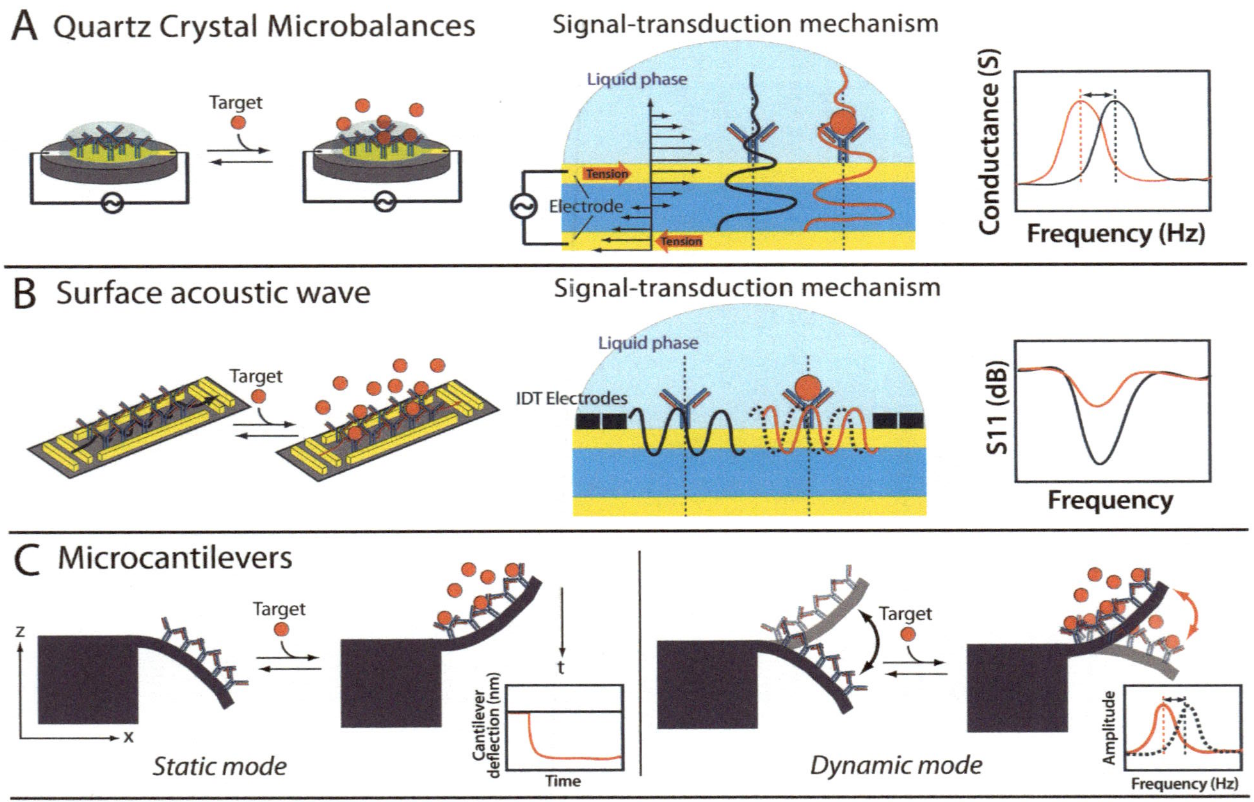

Fig. 12 A number of biosensor classes detect changes in mass or steric-bulk associated with target binding. (A) In QCM, acoustic waves are propagated perpendicularly to the surface of a piezoelectric quartz crystal. Any change in the surface mass of the crystal, such as that produced by the binding of a target molecule to a surface-bound bioreceptor, alters the resonant frequency of the crystal, generating a measurable signal. (B) SAW sensors uses a similar operation principle. But employ a wave traveling parallel to the surface of the piezoelectric material. (C) Microcantilevers have two operating modes: static and dynamic. The former measures the bending degree of the cantilevers upon a change in its the steric bulk of a bound target, while the latter measures the variation in the resonant frequency of the cantilevers associated with a change in mass.

high operating frequencies (which results in higher sensitivity compared to QCM), SAW-type biosensors have seen extensive recent exploration.[130,133,137] However, because the longitudinal component of acoustic waves are more sensitive by the contact with the liquid phase, this platform display an higher damping of acoustic energy (i.e., lower signal) than QCM.[137] This can be solved using alternative piezoelectric materials (e.g., $LiTaO_3$) and proper crystal cutting angles, which allow to reduce the loss of energy and achieve the required sensitivity.[137]

2.4.3 Microcantilevers

The development of micro- and nanofabrication approaches have allowed the miniaturization of mechanical transducers through the use of micro- and nanosized moving elements (usually cantilever-shaped) whose deformation and resonant vibrational frequency are affected by the binding of a sufficiently large or massive target to a surface-bound receptor.[132,134,139] In the case of steric-bulk detection, crowding of target molecules on the surface leads to steric stress, bending the cantilever, producing a static displacement (Fig. 12C, left). This, in turn, is detected by monitoring the angle at which a laser beam focused on the cantilever is reflected.[132,134,140] A limitation of this approach is that their correct function requires that only one cantilever surface presents the receptor; if both sides are functionalized, the stress is equivalent on both and bending will be observed. In contrast, in dynamic-mode detection the change in cantilever mass associated with target binding alters its resonant frequency (Fig. 12C, right), and thus differential modification of the two sides of the cantilever is not required.[134,141] This approach has proven harder to adopt for use in biosensors due to the strong vibrational dampening seen in water. An elegant solution to this has been to place the target-containing solution in an internal channel within the cantilever, which allows the vibrations of the cantilever to occur in air rather than emersed in liquid.[139,142]

2.4.4 Continuous sensing using mass- and bulk-based techniques

Mass- and bulk-based biosensors can potentially support the label-free, continuous measurements of any molecule of sufficient mass or steric bulk. Specifically, the physical change in the propagation of acoustic waves or in the nanomechanical properties of the sensors upon the binding occurs very rapidly. At the same time, the interface electronic system allows to track the resonance frequency variation continuously.[131,143] For example, using these sensors it has been possible to perform the continuous monitoring of specific proteins and nucleic acids[24,134,144] so as, for example, to characterize their binding kinetics.

2.4.5 *Mass- and bulk-based sensors in LoC applications*

Due to their ability to support high frequency, high-sensitivity measurements, significant efforts has been expended to integrate mass- and bulk-based biosensors into LoC systems.[24,62,145–147] The use of microfluidic devices has opened a new approach to solve this giving a further opportunity to make the overall analytical procedures automatized (Fig. 13A). Indeed, this allows to perform the measurements with a better control which improve the rapidity and robustness of the sensing systems (Fig. 13A).[24,62,148] Due to their small size, which allows them to be incorporated in microfluidic channels without the need to alter the channel geometry, microcantilever approaches have dominated this literature to date.[62,139] More recently, also SAW biosensors have also been integrated into LoC applications, providing an alternative route for the continuous detection of biomarkers.[62,145]

2.5 Refractometric sensors

Refractometric sensors detect target binding to a bioreceptor on the sensor surface via the change in the refractive index caused when the target displaces water.[149,150] Several types of sensors harness this mechanism, with the most common ones being employed as biosensors being those based on Surface Plasmon Resonance (SPR) and Interferometry.

2.5.1 *Surface plasmon resonance*

As the name indicates, SPR sensors measure variations of the refractive index on a metal surface that supports a surface plasmon, that is the resonant oscillation of conduction electrons (produced by incident light) at the interface between negative and positive permittivity materials (Fig. 14A).[151] The binding of a target molecule by a bioreceptor on this surface changes the refractive index at the interface, modifying the propagation of the surface plasmon. This can be monitored by measuring the resulting changes in the coupling angle, coupling wavelength, intensity, or phase. As for the electrical and mass-based techniques, SPR sensors do not require the labeling of the target, making them suitable for continuous measurements. And because they do not require modification of the receptor with a reporter, these approaches can be quite convenient. Because of this, SPR has achieved great commercial success; the Biacore instrument (now Cytiva), for example, is routinely used in research laboratories to measure protein-protein and drug-protein binding.

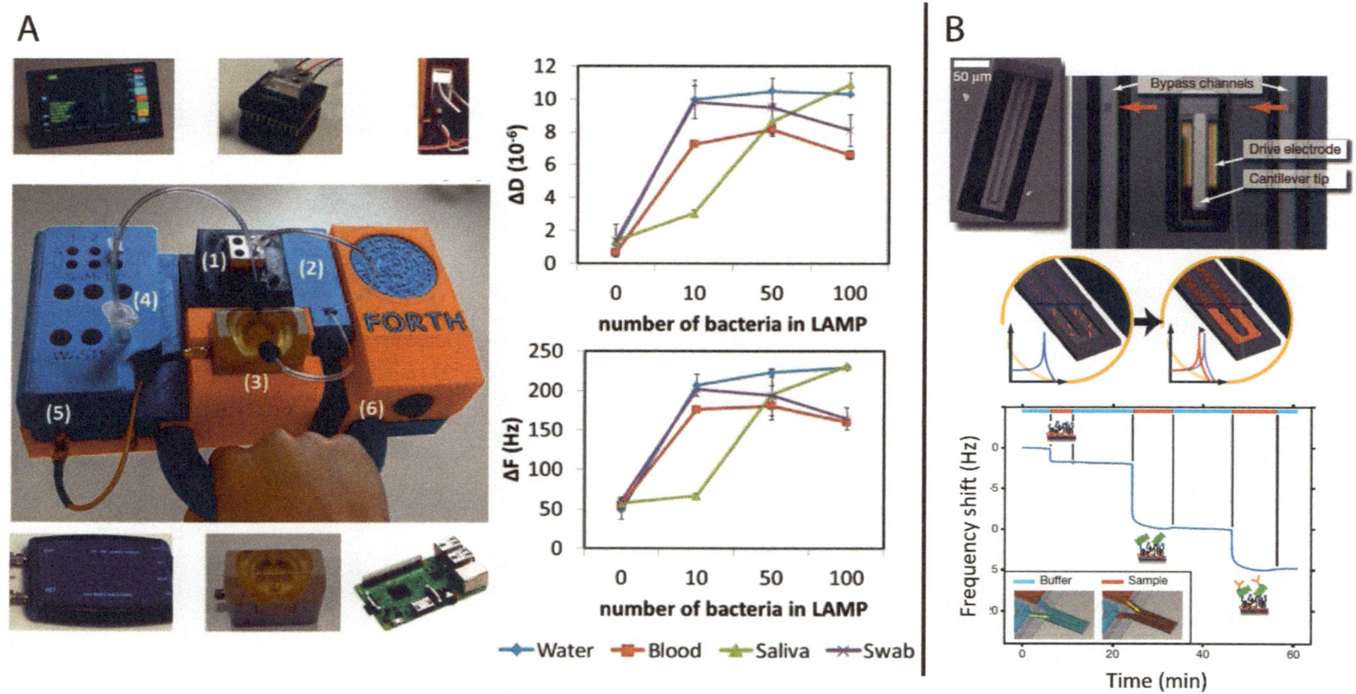

Fig. 13 Examples of LoC devices using mass- and bulk-detecting biosensors. (A) The integration of a QCM sensor with a portable-fluidic platform in support of point-of-care genetic testing. (B) Silicon cantilevers embedded in microchannels have achieved the continuous detection of proteins targets. *Panel (A) Reproduced from Papadakis G, Pantazis AK, Ntogka M, Parasyris K, Theodosi GI, Kaprou G, Gizeli E. 3D-printed point-of-care platform for genetic testing of infectious diseases directly in human samples using acoustic sensors and a smartphone.* ACS Sensors. *2019;4:1329–1336. https://doi.org/10.1021/acssensors.9b00264 with permission from American Chemical Society. Panel (C) Reproduced from Burg TP, Godin M, Knudsen SM, Shen W, Carlson G, Foster JS, Babcock K, Manalis SR. Weighing of biomolecules, single cells and single nanoparticles in fluid.* Nature. *2007;446:1066–1069. https://doi.org/10.1038/nature05741 with permission from Nature.*

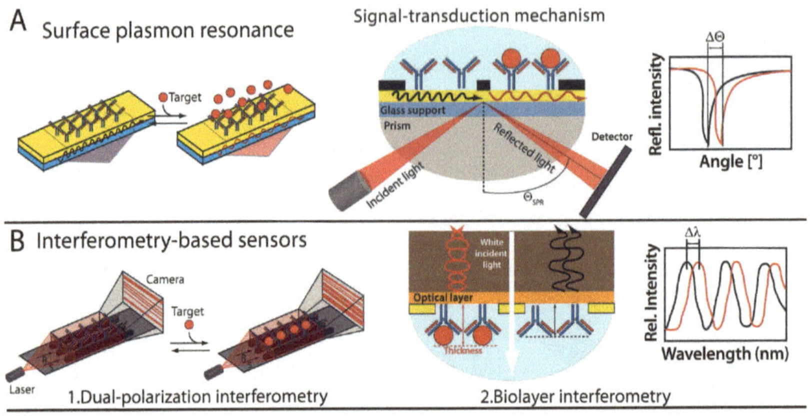

Fig. 14 Shown are the working principles of refractometric biosensors. (A) SPR sensors monitor changes in surface plasmons to measure variations in the refractive index that occur when the binding of a target to a surface-attached bioreceptor displaces water. (B) There are two most common interferometry-based techniques: DPI and BLI. The former uses an evanescent laser light to monitor both the refractive index and the thickness of the layer adsorbed to the sensor surface. The latter employs white light to measure only the optical thickness of the material adsorbed to the sensor surface, which in wavelength dependence of an interference pattern.

2.5.2 Interferometry

There are several sensing strategies relying on interferometry to generate a measurable signal, however, the two most common used in continuous biosensing are dual-polarization interferometry (DPI) and biolayer interferometry (BLI) (Fig. 14B).[152,153] The former uses the evanescent wave of a laser beam and two waveguides, one for sensing and one for referencing. By rotating the laser's polarization, it is possible to measure both the refractive index (that can be associated with the fold density of the probe) and the layer's thickness (that is related to the accumulation of the target protein on the sensor surface), obtaining a complete picture of the binding event (i.e., quantifying conformational changes, measuring reaction rates, affinities and thermodynamics). Instead, the latter measures the signal generated by a change in the wavelength of white light on the sensor (a fiber optic) surface in comparison to a reference layer. In DPI, in contrast, the accumulation of the target analyte in response to its recognition by the bioreceptor increases the optical thickness of the sensor surface, generating a measurable shift in wavelength. Conveniently, either DPI- nor BLI-based sensors require target labeling to function.

2.5.3 Continuous monitoring using refractometry techniques

Refractometric-based sensors are label-free and single-step, and thus could potentially support continuous molecular monitoring. However, as for electrical and mass-based sensors they can either carry out continuous molecular monitoring in buffered solutions or they can provide single-point measurements of biomarkers in complex media upon washing or dilution steps. A second limitation is that, because their signaling depends on the displacement of water (SPR) or variation of the thickness of the material adsorbed on the sensor surface (DPI, BLI), these approaches work better for the detection of higher molecular weight targets than for small molecules, such as drugs and metabolites. Consistent with this, SPR has been used for semi-continuous measurements.[154] The other approaches, however, have largely confined to characterize binding interactions between biomolecules.[155,156]

2.6 Refractometric sensors in LoC applications

Refractometric techniques have seen only quite limited integration into LoC devices, with most examples employing SPR (Fig. 15).[157–160 161,162]

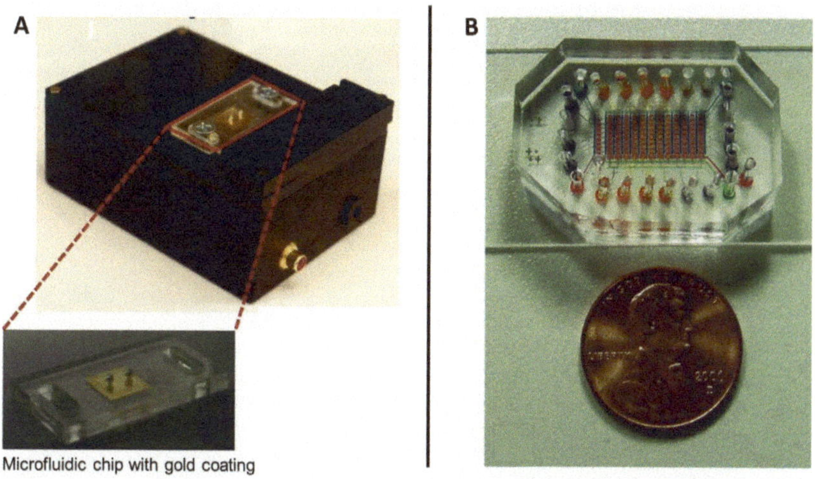

Fig. 15 While refractometric biosensors have been integrated into LoCs, their complexity has limited such applications. Most of the few reported examples employed SPR to automate multiplexed measurements. We have found no examples refractometric biosensors being used to perform continuous molecular monitoring in LoC applications. *Panel (A) Reproduced from Luo Y, Yu F, Zare RN. Microfluidic device for immunoassays based on surface plasmon resonance imaging.* Lab Chip. *2008;8:694–700. https://doi.org/10.1039/b800606g with permission from the Royal Society of Chemistry, Panel (B) Reproduced from Tokel O, Yildiz UH, Inci F, Durmus NG, Ekiz OO, Turker B, Cetin C, Rao S, Sridhar K, Natarajan N, Shafiee H, Dana A, Demirci U. Portable microfluidic integrated plasmonic platform for pathogen detection.* Sci Rep. *2015;5:1–9. https://doi.org/10.1038/srep09152 with permission from Nature.*

This should not come as a surprise if we consider the complex setup required to carry out these types of measurements and the difficulties involved in miniaturizing and making such setup portable. However, considering the excellent capabilities that these techniques have to characterize binding events, the development of a portable platform that can provide the required stability and ease of use, while at the same time limiting the non-specific adsorption of contaminants, would be a potential breakthrough in several sensing applications.

3. Conclusions

The integration of continuous molecular detection into LoC devices for the advancement of both is likely to be a growth area in the molecular sensing field. The use of continuous molecular sensing in LoCs will allow for the creation of sense-and-response functionalities. Likewise, the addition of LoC pre-treatment capabilities to continuous monitoring approaches could render some of them suitable for use with clinical sample materials.

References

1. Aronson JK, Ferner RE. Biomarkers—a general review. *Curr Protoc Pharmacol.* 2017;2017:9.23.1–9.23.17. https://doi.org/10.1002/cpph.19.
2. Boycott AE, Damant GCC, Haldane JS. The prevention of compressed-air illness. *J Hyg (Lond).* 1908;8:342–443. https://doi.org/10.1017/S0022172400003399.
3. Lippi G, Cervellin G. Canine olfactory detection of cancer versus laboratory testing: myth or opportunity. *Clin Chem Lab Med.* 2012;50:435–439. https://doi.org/10.1515/cclm.2011.672.
4. Parolo C, Merkoçi A. Paper-based nanobiosensors for diagnostics. *Chem Soc Rev.* 2013;42:450–457. https://doi.org/10.1039/C2CS35255A.
5. De La Escosura-Muñiz A, Parolo C, Merkoçi A. Immunosensing using nanoparticles. *Mater Today.* 2010;13:24–34. https://doi.org/10.1016/S1369-7021(10)70125-5.
6. Parolo C, Sena-Torralba A, Bergua JF, et al. Tutorial: design and fabrication of nanoparticle-based lateral-flow immunoassays. *Nat Protoc.* 2020;15:3788–3816. https://doi.org/10.1038/s41596-020-0357-x.
7. Lee H, Hong YJ, Baik S, Hyeon T, Kim DH. Enzyme-based glucose sensor: from invasive to wearable device. *Adv Healthc Mater.* 2018;7:1701150. https://doi.org/10.1002/adhm.201701150.
8. Bally L, Thabit H, Hovorka R. Glucose-responsive insulin delivery for type 1 diabetes: the artificial pancreas story. *Int J Pharm.* 2018;544:309–318. https://doi.org/10.1016/j.ijpharm.2017.12.022.
9. Boiroux D, Duun-Henriksen AK, Schmidt S, et al. Adaptive control in an artificial pancreas for people with type 1 diabetes. *Control Eng Pract.* 2017;58:332–342. https://doi.org/10.1016/j.conengprac.2016.01.003.
10. Resalat N, Hilts W, El Youssef J, Tyler N, Castle JR, Jacobs PG. Adaptive control of an artificial pancreas using model identification, adaptive postprandial insulin delivery, and heart rate and accelerometry as control inputs. *J Diabetes Sci Technol.* 2019;13:1044–1053. https://doi.org/10.1177/1932296819881467.

11. Chen C, Zhao XL, Li ZH, Zhu ZG, Qian SH, Flewitt AJ. Current and emerging technology for continuous glucose monitoring. *Sensors (Switzerland)*. 2017;17:182. https://doi.org/10.3390/s17010182.
12. Findlay JWA, Dillard RF. Appropriate calibration curve fitting in ligand binding assays. *AAPS J*. 2007;9:E260–E267. https://doi.org/10.1208/aapsj0902029.
13. Sassolas A, Blum LJ, Leca-Bouvier BD. Immobilization strategies to develop enzymatic biosensors. *Biotechnol Adv*. 2012;30:489–511. https://doi.org/10.1016/j.biotechadv.2011.09.003.
14. Nicholson LB. The immune system. *Essays Biochem*. 2016;60:275–301. https://doi.org/10.1042/EBC20160017.
15. Du Y, Dong S. Nucleic acid biosensors: recent advances and perspectives. *Anal Chem*. 2017;89:189–215. https://doi.org/10.1021/acs.analchem.6b04190.
16. Gold L, Tuerk C. Systematic evolution of ligands by exponential enrichment: RNA ligands to bacteriophage T4 DNA polymerase. *Science*. 1990;249:505–510. https://doi.org/10.1126/science.2200121.
17. Wu YX, Kwon YJ. Aptamers: the "evolution" of SELEX. *Methods*. 2016;106:21–28. https://doi.org/10.1016/j.ymeth.2016.04.020.
18. Moccia M, Caratelli V, Cinti S, et al. Paper-based electrochemical peptide nucleic acid (PNA) biosensor for detection of miRNA-492: a pancreatic ductal adenocarcinoma biomarker. *Biosens Bioelectron*. 2020;165, 112371. https://doi.org/10.1016/j.bios.2020.112371.
19. Gao S, Zheng X, Jiao B, Wang L. Post-SELEX optimization of aptamers. *Anal Bioanal Chem*. 2016;408:4567–4573. https://doi.org/10.1007/s00216-016-9556-2.
20. Chen A, Yang S. Replacing antibodies with aptamers in lateral flow immunoassay. *Biosens Bioelectron*. 2015;71:230–242. https://doi.org/10.1016/j.bios.2015.04.041.
21. Byun J. Recent progress and opportunities for nucleic acid aptamers. *Life*. 2021;11:1–18. https://doi.org/10.3390/life11030193.
22. Ruscito A, DeRosa MC. Small-molecule binding aptamers: selection strategies, characterization, and applications. *Front Chem*. 2016;4:14. https://doi.org/10.3389/fchem.2016.00014.
23. Lu Y, Liu J. Functional DNA nanotechnology: emerging applications of DNAzymes and aptamers. *Curr Opin Biotechnol*. 2006;17:580–588. https://doi.org/10.1016/j.copbio.2006.10.004.
24. Lim HJ, Saha T, Tey BT, Tan WS, Ooi CW. Quartz crystal microbalance-based biosensors as rapid diagnostic devices for infectious diseases. *Biosens Bioelectron*. 2020;168, 112513. https://doi.org/10.1016/j.bios.2020.112513.
25. Sadighbayan D, Hasanzadeh M, Ghafar-Zadeh E. Biosensing based on field-effect transistors (FET): recent progress and challenges. *TrAC, Trends Anal Chem*. 2020;133: 116067. https://doi.org/10.1016/j.trac.2020.116067.
26. Goeders KM, Colton JS, Bottomley LA. Microcantilevers: sensing chemical interactions via mechanical motion. *Chem Rev*. 2008;108:522–542. https://doi.org/10.1021/cr0681041.
27. Zeng S, Baillargeat D, Ho H-P, Yong K-T. Nanomaterials enhanced surface plasmon resonance for biological and chemical sensing applications. *Chem Soc Rev*. 2014;43:3426–3452. https://doi.org/10.1039/c3cs60479a.
28. Iglesias-Mayor A, Amor-Gutiérrez O, Costa-García A, de la Escosura-Muñiz A. Nanoparticles as emerging labels in electrochemical immunosensors. *Sensors (Switzerland)*. 2019;19:5137. https://doi.org/10.3390/s19235137.
29. Resch-Genger U, Grabolle M, Cavaliere-Jaricot S, Nitschke R, Nann T. Quantum dots versus organic dyes as fluorescent labels. *Nat Methods*. 2008;5:763–775. https://doi.org/10.1038/nmeth.1248.

30. Hvolbæk B, Janssens TVW, Clausen BS, Falsig H, Christensen CH, Nørskov JK. Catalytic activity of Au nanoparticles. *Nano Today*. 2007;2:14–18. https://doi.org/10.1016/S1748-0132(07)70113-5.
31. Hunt HK, Armani AM. Label-free biological and chemical sensors. *Nanoscale*. 2010;2:1544–1559. https://doi.org/10.1039/c0nr00201a.
32. De Rica R, Stevens MM. Plasmonic ELISA for the ultrasensitive detection of disease biomarkers with the naked eye. *Nat Nanotechnol*. 2012;7:821–824. https://doi.org/10.1038/nnano.2012.186.
33. Yang Y, Gao W. Wearable and flexible electronics for continuous molecular monitoring. *Chem Soc Rev*. 2019;48:1465–1491. https://doi.org/10.1039/c7cs00730b.
34. Parolo C, Idili A, Ortega G, et al. Real-time monitoring of a protein biomarker. *ACS Sensors*. 2020;5:1877–1881. https://doi.org/10.1021/acssensors.0c01085.
35. Landry JP, Ke Y, Yu G-L, Zhu XD. Measuring affinity constants of 1450 monoclonal antibodies to peptide targets with a microarray-based label-free assay platform. *J Immunol Methods*. 2014;417:86–96. https://doi.org/10.1016/j.jim.2014.12.011.
36. Prechl J. A generalized quantitative antibody homeostasis model: maintenance of global antibody equilibrium by effector functions. *Clin Transl Immunol*. 2017;6:e161. https://doi.org/10.1038/cti.2017.50.
37. Bao J, Krylova SM, Reinstein O, Johnson PE, Krylov SN. Label-free solution-based kinetic study of aptamer-small molecule interactions by kinetic capillary electrophoresis with UV detection revealing how kinetics control equilibrium. *Anal Chem*. 2011;83(22):8387–8390. https://doi.org/10.1021/ac2026699.
38. Arroyo-Currás N, Dauphin-Ducharme P, Scida K, Chávez JL. From the beaker to the body: translational challenges for electrochemical, aptamer-based sensors. *Anal Methods*. 2020;12:1288–1310. https://doi.org/10.1039/d0ay00026d.
39. Mishra RK, Vinu Mohan AM, Soto F, Chrostowski R, Wang J. A microneedle biosensor for minimally-invasive transdermal detection of nerve agents. *Analyst*. 2017;142:918–924. https://doi.org/10.1039/c6an02625g.
40. Wongkaew N, Simsek M, Griesche C, Baeumner AJ. Functional nanomaterials and nanostructures enhancing electrochemical biosensors and lab-on-a-chip performances: recent progress, applications, and future perspective. *Chem Rev*. 2019;119:120–194. https://doi.org/10.1021/acs.chemrev.8b00172.
41. Qiu Z, Shu J, He Y, et al. CdTe/CdSe quantum dot-based fluorescent aptasensor with hemin/G-quadruplex DNzyme for sensitive detection of lysozyme using rolling circle amplification and strand hybridization. *Biosens Bioelectron*. 2017;87:18–24. https://doi.org/10.1016/j.bios.2016.08.003.
42. Cinti S, Marrone R, Mazzaracchio V, Moscone D, Arduini F. Novel bio-lab-on-a-tip for electrochemical glucose sensing in commercial beverages. *Biosens Bioelectron*. 2020;165, 112334. https://doi.org/10.1016/j.bios.2020.112334.
43. Medina-Sánchez M, Ibarlucea B, Pérez N, et al. High-performance three-dimensional tubular nanomembrane sensor for DNA detection. *Nano Lett*. 2016;16:4288–4296. https://doi.org/10.1021/acs.nanolett.6b01337.
44. Smith EJ, Xi W, Makarov D, et al. Lab-in-a-tube: ultracompact components for on-chip capture and detection of individual micro-/nanoorganisms. In: *Lab Chip*. Royal Society of Chemistry; 2012:1917–1931. https://doi.org/10.1039/c2lc21175k.
45. Rodrigues RO, Sousa PC, Gaspar J, Bañobre-López M, Lima R, Minas G. Organ-on-a-chip: a preclinical microfluidic platform for the progress of nanomedicine. *Small*. 2020;16:2003517. https://doi.org/10.1002/smll.202003517.
46. Medina-Sánchez M, Miserere S, Morales-Narváez E, Merkoçi A. On-chip magneto-immunoassay for Alzheimer's biomarker electrochemical detection by using quantum dots as labels. *Biosens Bioelectron*. 2014;54:279–284. https://doi.org/10.1016/j.bios.2013.10.069.

47. Wu Y, Midinov B, White RJ. Electrochemical aptamer-based sensor for real-time monitoring of insulin. *ACS Sensors*. 2019;4:498–503. https://doi.org/10.1021/acssensors.8b01573.
48. Li Z, Askim JR, Suslick KS. The optoelectronic nose: colorimetric and fluorometric sensor arrays. *Chem Rev*. 2019;119:231–292. https://doi.org/10.1021/acs.chemrev.8b00226.
49. Liu L, Hao Y, Deng D, Xia N. Nanomaterials-based colorimetric immunoassays. *Nanomaterials*. 2019;9:316. https://doi.org/10.3390/nano9030316.
50. Wang H, Wu T, Li M, Tao Y. Recent advances in nanomaterials for colorimetric cancer detection. *J Mater Chem B*. 2021;**9:921–938**. https://doi.org/10.1039/d0tb02163f.
51. McCracken KE, Yoon JY. Recent approaches for optical smartphone sensing in resource-limited settings: a brief review. *Anal Methods*. 2016;8:6591–6601. https://doi.org/10.1039/c6ay01575a.
52. Miller BS, Parolo C, Turbé V, Keane CE, Gray ER, McKendry RA. Quantifying biomolecular binding constants using video paper analytical devices. *Chem A Eur J*. 2018;24:9783–9787. https://doi.org/10.1002/chem.201802394.
53. Lakowicz JR. *Principles of Fluorescence Spectroscopy*. Springer; 2006. https://doi.org/10.1007/978-0-387-46312-4.
54. Pazos E, Vázquez O, Mascareñas JL, Vázquez ME. Peptide-based fluorescent biosensors. *Chem Soc Rev*. 2009;38:3348–3359. https://doi.org/10.1039/b908546g.
55. Li B, Yu Q, Duan Y. Fluorescent labels in biosensors for pathogen detection. *Crit Rev Biotechnol*. 2015;35:82–93. https://doi.org/10.3109/07388551.2013.804487.
56. Kant K, Shahbazi MA, Dave VP, et al. Microfluidic devices for sample preparation and rapid detection of foodborne pathogens. *Biotechnol Adv*. 2018;36:1003–1024. https://doi.org/10.1016/j.biotechadv.2018.03.002.
57. Sin ML, Mach KE, Wong PK, Liao JC. Advances and challenges in biosensor-based diagnosis of infectious diseases. *Expert Rev Mol Diagn*. 2014;14:225–244. https://doi.org/10.1586/14737159.2014.888313.
58. Oh KJ, Cash KJ, Plaxco KW. Beyond molecular beacons: optical sensors based on the binding-induced folding of proteins and polypeptides. *Chem A Eur J*. 2009;15:2244–2251. https://doi.org/10.1002/chem.200701748.
59. Tyagi S, Kramer FR. Molecular beacon probes that fluoresce on hybridiztion. *Nat Publ Group*. 1996;14:303–308.
60. Moutsiopoulou A, Broyles D, Dikici E, Daunert S, Deo SK. Molecular aptamer beacons and their applications in sensing, imaging, and diagnostics. *Small*. 2019;15:1902248. https://doi.org/10.1002/smll.201902248.
61. Kohn JE, Plaxco KW. Engineering a signal transduction mechanism for protein-based biosensors. *Proc Natl Acad Sci U S A*. 2005;102:10841–10845. https://doi.org/10.1073/pnas.0503055102.
62. Lafleur JP, Jönsson A, Senkbeil S, Kutter JP. Recent advances in lab-on-a-chip for biosensing applications. *Biosens Bioelectron*. 2016;76:213–233. https://doi.org/10.1016/j.bios.2015.08.003.
63. Luka G, Ahmadi A, Najjaran H, et al. Microfluidics integrated biosensors: a leading technology towards lab-on-a-chip and sensing applications. *Sensors*. 2015;15:30011–30031. https://doi.org/10.3390/s151229783.
64. Montón H, Medina-Sánchez M, Soler JA, Chałupniak A, Nogués C, Merkoçi A. Rapid on-chip apoptosis assay on human carcinoma cells based on annexin-V/quantum dot probes. *Biosens Bioelectron*. 2017;94:408–414. https://doi.org/10.1016/j.bios.2017.03.034.
65. Poudineh M, Maikawa CL, Ma EY, et al. A fluorescence sandwich immunoassay for the real-time continuous detection of glucose and insulin in live animals. *Nat Biomed Eng*. 2020;5:53–63. https://doi.org/10.1038/s41551-020-00661-1.

66. Zosel J. Amperometry. In: *Encycl. Appl. Electrochem.* Springer New York; 2014:65–73. https://doi.org/10.1007/978-1-4419-6996-5_216.
67. Inzelt G. Chronoamperometry, chronocoulometry, and chronopotentiometry. In: *Encycl. Appl. Electrochem.* Springer New York; 2014:207–214. https://doi.org/10.1007/978-1-4419-6996-5_217.
68. Dhillon S, Kant R. Theory of double potential step chronoamperometry at rough electrodes: reversible redox reaction and ohmic effects. *Electrochim Acta.* 2014;129:245–258. https://doi.org/10.1016/j.electacta.2014.02.096.
69. Hayat A, Catanante G, Marty JL. Current trends in nanomaterial-based amperometric biosensors. *Sensors (Switzerland).* 2014;14:23439–23461. https://doi.org/10.3390/s141223439.
70. Wang J. Carbon-nanotube based electrochemical biosensors: a review. *Electroanalysis.* 2005;17:7–14. https://doi.org/10.1002/elan.200403113.
71. Arroyo-Currás N, Dauphin-Ducharme P, Ortega G, Ploense KL, Kippin TE, Plaxco KW. Subsecond-resolved molecular measurements in the living body using chronoamperometrically interrogated aptamer-based sensors. *ACS Sensors.* 2018;3:360–366. https://doi.org/10.1021/acssensors.7b00787.
72. Parolo C, De La Escosura-Muñiz A, Polo E, Grazú V, De La Fuente JM, Merkoçi A. Design, preparation, and evaluation of a fixed-orientation antibody/gold-nanoparticle conjugate as an immunosensing label. *ACS Appl Mater Interfaces.* 2013;5: 10753–10759. https://doi.org/10.1021/am4029153.
73. Sinha A, Dhanjai, Jain R, Zhao H, Karolia P, Jadon N. Voltammetric sensing based on the use of advanced carbonaceous nanomaterials: a review. *Microchim Acta.* 2018;185. https://doi.org/10.1007/s00604-017-2626-0.
74. Meirinho SG, Dias LG, Peres AM, Rodrigues LR. Voltammetric aptasensors for protein disease biomarkers detection: a review. *Biotechnol Adv.* 2016;34:941–953. https://doi.org/10.1016/j.biotechadv.2016.05.006.
75. Parolo C, Medina-Sánchez M, Montõn H, De La Escosura-Muñiz A, Merkoçi A. Paper-based electrodes for nanoparticles detection. *Part Part Syst Charact.* 2013;30:662–666. https://doi.org/10.1002/ppsc.
76. Wang J. Electrochemical glucose biosensors. *Chem Rev.* 2008;108:814–825. https://doi.org/10.1021/cr068123a.
77. Xiao Y, Lubin AA, Heeger AJ, Plaxco KW. Label-free electronic detection of thrombin in blood serum by using an aptamer-based sensor. *Angew Chem Int Ed.* 2005;44:5456–5459. https://doi.org/10.1002/anie.200500989.
78. Idili A, Arroyo-Currás N, Ploense KL, et al. Seconds-resolved pharmacokinetic measurements of the chemotherapeutic irinotecan *in situ* in the living body. *Chem Sci.* 2019;10:8164–8170. https://doi.org/10.1039/C9SC01495K.
79. Idili A, Parolo C, Ortega G, Plaxco KW. Calibration-free measurement of phenylalanine levels in the blood using an electrochemical aptamer-based sensor suitable for point-of-care applications. *ACS Sensors.* 2019;4:3227–3233. https://doi.org/10.1021/acssensors.9b01703.
80. Somerson J, Plaxco KW. Electrochemical aptamer-based sensors for rapid point-of-use monitoring of the mycotoxin ochratoxin a directly in a food stream. *Molecules.* 2018;23:912. https://doi.org/10.3390/molecules23040912.
81. Li H, Somerson J, Xia F, Plaxco KW. Electrochemical DNA-based sensors for molecular quality control: continuous, real-time melamine detection in flowing whole milk. *Anal Chem.* 2018;90:10641–10645. https://doi.org/10.1021/acs.analchem.8b01993.
82. gang Yu Z, Sutlief AL, Lai RY. Towards the development of a sensitive and selective electrochemical aptamer-based ampicillin sensor. *Sens Actuators B.* 2018;258:722–729. https://doi.org/10.1016/j.snb.2017.11.193.

83. Lai RY, Plaxco KW, Heeger AJ. Aptamer-based electrochemical detection of picomolar platelet-derived growth factor directly in blood serum. *Anal Chem.* 2007;79:229–233. https://doi.org/10.1021/ac061592s.
84. Arroyo-Currás N, Somerson J, Vieira PA, Ploense KL, Kippin TE, Plaxco KW. Real-time measurement of small molecules directly in awake, ambulatory animals. *Proc Natl Acad Sci.* 2017;114:645–650. https://doi.org/10.1073/pnas.1613458114.
85. Li H, Li S, Dai J, et al. High frequency, calibration-free molecular measurements: in situ in the living body. *Chem Sci.* 2019;10:10843–10848. https://doi.org/10.1039/c9sc04434e.
86. Ploense KL, Dauphin-Ducharme P, Arroyo-Curras N, et al. Real-time pharmacokinetic and pharmacodynamic measurements of drugs within the brains of freely behaving rats. *FASEB J.* 2020;34:1. https://doi.org/10.1096/fasebj.2020.34.s1.07174.
87. Dauphin-Ducharme P, Yang K, Arroyo-Currás N, et al. Electrochemical aptamer-based sensors for improved therapeutic drug monitoring and high-precision, feedback-controlled drug delivery. *ACS Sensors.* 2019;4:2832–2837. https://doi.org/10.1021/acssensors.9b01616.
88. Idili A, Gerson J, Kippin T, Plaxco KW. Second-resolved, in-situ measurements of plasma phenylalanine disposition kinetics in living rats. *Anal Chem.* 2021;93:4023–4032. https://doi.org/10.1021/acs.analchem.0c05024.
89. Kurnik M, Pang EZ, Plaxco KW. An electrochemical biosensor architecture based on protein folding supports direct real-time measurements in whole blood. *Angew Chem.* 2020;132:18600–18603. https://doi.org/10.1002/ange.202007256.
90. Parolo C, Greenwood AS, Ogden NE, et al. E-DNA scaffold sensors and the reagentless, single-step, measurement of HIV-diagnostic antibodies in human serum. *Microsyst Nanoeng.* 2020;6:13. https://doi.org/10.1038/s41378-019-0119-5.
91. Kang D, Parolo C, Sun S, Ogden NE, Dahlquist FW, Plaxco KW. Expanding the scope of protein-detecting electrochemical DNA "scaffold" sensors. *ACS Sensors.* 2018;3:1271–1275. https://doi.org/10.1021/acssensors.8b00311.
92. Yousefi H, Mahmud A, Chang D, et al. Detection of SARS-CoV-2 viral particles using direct, reagent-free electrochemical sensing. *J Am Chem Soc.* 2021;143:1722–1727. https://doi.org/10.1021/jacs.0c10810.
93. Puiu M, Idili A, Moscone D, Ricci F, Bala C. A modular electrochemical peptide-based sensor for antibody detection. *Chem Commun.* 2014;50:8962. https://doi.org/10.1039/c4cc02858a.
94. Cinti S, Cinotti G, Parolo C, et al. Experimental comparison in sensing breast cancer mutations by signal on and signal off paper-based electroanalytical strips. *Anal Chem.* 2020;92:1674–1679. https://doi.org/10.1021/acs.analchem.9b02560.
95. Das J, Gomis S, Chen JB, et al. Reagentless biomolecular analysis using a molecular pendulum. *Nat Chem.* 2021;13:428–434. https://doi.org/10.1038/s41557-021-00644-y.
96. Economou A, Kokkinos C, Prodromidis M. Flexible plastic, paper and textile lab-on-a chip platforms for electrochemical biosensing. *Lab Chip.* 2018;18:1812–1830. https://doi.org/10.1039/c8lc00025e.
97. Ferguson BS, Hoggarth DA, Maliniak D, et al. Real-time, aptamer-based tracking of circulating therapeutic agents in living animals. *Sci Transl Med.* 2013;5:213ra165. https://doi.org/10.1126/scitranslmed.3007095.
98. Carrara S, Sacchetto D, Doucey M-A, Baj-Rossi C, De Micheli G, Leblebici Y. Memristive-biosensors: a new detection method by using nanofabricated memristors. *Sens Actuators B.* 2012;171:449–457. https://doi.org/10.1016/j.snb.2012.04.089.
99. Ibarlucea B, Akbar TF, Kim K, et al. Ultrasensitive detection of Ebola matrix protein in a memristor mode. *Nano Res.* 2018;11:1057–1068. https://doi.org/10.1007/s12274-017-1720-2.

100. Fu Y, Romay V, Liu Y, et al. Chemiresistive biosensors based on carbon nanotubes for label-free detection of DNA sequences derived from avian influenza virus H5N1. *Sens Actuators B*. 2017;249:691–699. https://doi.org/10.1016/j.snb.2017.04.080.
101. Noroozi AA, Abdi Y. A graphene/Si Schottky diode for the highly sensitive detection of protein. *RSC Adv*. 2019;9:19613–19619. https://doi.org/10.1039/c9ra03765a.
102. Chang B-Y, Park S-M. Electrochemical impedance spectroscopy. *Annu Rev Anal Chem*. 2010;3:207–229. https://doi.org/10.1146/annurev.anchem.012809.102211.
103. Grossi M, Riccò B. Electrical impedance spectroscopy (EIS) for biological analysis and food characterization: a review. *J Sens Syst*. 2017;6:303–325. https://doi.org/10.5194/jsss-6-303-2017.
104. Bausch CS, Heyn C, Hansen W, et al. Ultra-fast cell counters based on microtubular waveguides. *Sci Rep*. 2017;7:41584. https://doi.org/10.1038/srep41584.
105. Macchia E, Manoli K, Holzer B, et al. Selective single-molecule analytical detection of C-reactive protein in saliva with an organic transistor. *Anal Bioanal Chem*. 2019;411:4899–4908. https://doi.org/10.1007/s00216-019-01778-2.
106. Nakatsuka N, Yang K-A, Abendroth JM, et al. Aptamer–field-effect transistors overcome Debye length limitations for small-molecule sensing. *Science*. 2018;362:319–324. https://doi.org/10.1126/science.aao6750.
107. Russell C, Ward AC, Vezza V, et al. Development of a needle shaped microelectrode for electrochemical detection of the sepsis biomarker interleukin-6 (IL-6) in real time. *Biosens Bioelectron*. 2019;126:806–814. https://doi.org/10.1016/j.bios.2018.11.053.
108. Picca RA, Manoli K, Macchia E, et al. Ultimately sensitive organic bioelectronic transistor sensors by materials and device structure design. *Adv Funct Mater*. 2020;30, 1904513. https://doi.org/10.1002/adfm.201904513.
109. Chu CH, Sarangadharan I, Regmi A, et al. Beyond the Debye length in high ionic strength solution: direct protein detection with field-effect transistors (FETs) in human serum. *Sci Rep*. 2017;7:1–15. https://doi.org/10.1038/s41598-017-05426-6.
110. Park S, Kim M, Kim D, Kang SH, Lee KH, Jeong Y. Interfacial charge regulation of protein blocking layers in transistor biosensor for direct measurement in serum. *Biosens Bioelectron*. 2020;147, 111737. https://doi.org/10.1016/j.bios.2019.111737.
111. Chang SM, Palanisamy S, Wu TH, et al. Utilization of silicon nanowire field-effect transistors for the detection of a cardiac biomarker, cardiac troponin I and their applications involving animal models. *Sci Rep*. 2020;10:1–11. https://doi.org/10.1038/s41598-020-78829-7.
112. Gao N, Zhou W, Jiang X, Hong G, Fu TM, Lieber CM. General strategy for biodetection in high ionic strength solutions using transistor-based nanoelectronic sensors. *Nano Lett*. 2015;15:2143–2148. https://doi.org/10.1021/acs.nanolett.5b00133.
113. Hwang MT, Heiranian M, Kim Y, et al. Ultrasensitive detection of nucleic acids using deformed graphene channel field effect biosensors. *Nat Commun*. 2020;11:1–11. https://doi.org/10.1038/s41467-020-15330-9.
114. Gao N, Gao T, Yang X, et al. Specific detection of biomolecules in physiological solutions using graphene transistor biosensors. *Proc Natl Acad Sci U S A*. 2016;113: 14633–14638. https://doi.org/10.1073/pnas.1625010114.
115. Filipiak MS, Rother M, Andoy NM, et al. Highly sensitive, selective and label-free protein detection in physiological solutions using carbon nanotube transistors with nanobody receptors. *Sens Actuators B*. 2018;255:1507–1516. https://doi.org/10.1016/j.snb.2017.08.164.
116. Zheng G, Patolsky F, Cui Y, Wang WU, Lieber CM. Multiplexed electrical detection of cancer markers with nanowire sensor arrays. *Nat Biotechnol*. 2005;23:1294–1301. https://doi.org/10.1038/nbt1138.

117. Downs AM, Gerson J, Ploense KL, Plaxco KW, Dauphin-Ducharme P. Subsecond-resolved molecular measurements using electrochemical phase interrogation of aptamer-based sensors. *Anal Chem.* 2020;92:14063–14068. https://doi.org/10.1021/acs.analchem.0c03109.
118. Medina-Sánchez M, Martínez-Domingo C, Ramon E, Merkoçi A. An inkjet-printed field-effect transistor for label-free biosensing. *Adv Funct Mater.* 2014;24:6291–6302. https://doi.org/10.1002/adfm.201401180.
119. Zhang Q, Subramanian V. Printable DNA sensor using organic transistors. In: *Transducers Eurosensors '07—4th Int. Conf. Solid-State Sensors, Actuators Microsystems*; 2007:1987–1990. https://doi.org/10.1109/SENSOR.2007.4300551.
120. Yao L, Wang L, Huang F, Cai G, Xi X, Lin J. A microfluidic impedance biosensor based on immunomagnetic separation and urease catalysis for continuous-flow detection of E. coli O157:H7. *Sens Actuators B.* 2018;259:1013–1021. https://doi.org/10.1016/j.snb.2017.12.110.
121. Khan NI, Mousazadehkasin M, Ghosh S, Tsavalas JG, Song E. An integrated microfluidic platform for selective and real-time detection of thrombin biomarkers using a graphene FET. *Analyst.* 2020;145:4494–4503. https://doi.org/10.1039/D0AN00251H.
122. Egunov AI, Dou Z, Karnaushenko DD, et al. Impedimetric microfluidic sensor-in-a-tube for label-free immune cell analysis. *Small.* 2021;17, 2002549. https://doi.org/10.1002/smll.202002549.
123. Belyaev D, Schütt J, Ibarlucea B, Rim T, Baraban L, Cuniberti G. Nanosensors-assisted quantitative analysis of biochemical processes in droplets. *Micromachines.* 2020;11:138. https://doi.org/10.3390/mi11020138.
124. Cheng HL, Fu CY, Kuo WC, et al. Detecting miRNA biomarkers from extracellular vesicles for cardiovascular disease with a microfluidic system. *Lab Chip.* 2018;18: 2917–2925. https://doi.org/10.1039/c8lc00386f.
125. Sinha A, Tai TY, Li KH, et al. An integrated microfluidic system with field-effect-transistor sensor arrays for detecting multiple cardiovascular biomarkers from clinical samples. *Biosens Bioelectron.* 2019;129:155–163. https://doi.org/10.1016/j.bios.2019.01.001.
126. Nguyen EP, De Carvalho Castro Silva C, Merkoçi A. Recent advancement in biomedical applications on the surface of two-dimensional materials: from biosensing to tissue engineering. *Nanoscale.* 2020;12:19043–19067. https://doi.org/10.1039/d0nr05287f.
127. Ilic B, Czaplewski D, Craighead HG, Neuzil P, Campagnolo C, Batt C. Mechanical resonant immunospecific biological detector. *Appl Phys Lett.* 2000;77:450–452. https://doi.org/10.1063/1.127006.
128. Shiokawa S, Moriizumi T. Design of saw sensor in liquid. *Jpn J Appl Phys.* 1988;27:142–144. https://doi.org/10.7567/JJAPS.27S1.142.
129. Martin SJ, Granstaff VE, Frye GC. Characterization of a quartz crystal microbalance with simultaneous mass and liquid loading. *Anal Chem.* 1991;63:2272–2281. https://doi.org/10.1021/ac00020a015.
130. Länge K, Rapp BE, Rapp M. Surface acoustic wave biosensors: a review. *Anal Bioanal Chem.* 2008;391:1509–1519. https://doi.org/10.1007/s00216-008-1911-5.
131. Ferreira GNM, da-Silva AC, Tomé B. Acoustic wave biosensors: physical models and biological applications of quartz crystal microbalance. *Trends Biotechnol.* 2009;27: 689–697. https://doi.org/10.1016/j.tibtech.2009.09.003.
132. Tamayo J, Kosaka PM, Ruz JJ, Paulo ÁS, Calleja M. Biosensors based on nanomechanical systems. *Chem Soc Rev.* 2013;42:1287–1311. https://doi.org/10.1039/c2cs35293a.

133. Länge K, Gruhl FJ, Rapp M. Surface acoustic wave (SAW) biosensors: coupling of sensing layers and measurement. *Methods Mol Biol.* 2013;949:491–505. https://doi.org/10.1007/978-1-62703-134-9_31.
134. Alvarez M, Lechuga LM. Microcantilever-based platforms as biosensing tools. *Analyst.* 2010;135:827. https://doi.org/10.1039/b908503n.
135. Abadian PN, Buch PJ, Goluch ED, Li J, Zhang Z. Real-time monitoring of urinary encrustation using a quartz crystal microbalance. *Anal Chem.* 2018;90:1531–1535. https://doi.org/10.1021/acs.analchem.7b04047.
136. Buchapudi KR, Huang X, Yang X, Ji HF, Thundat T. Microcantilever biosensors for chemicals and bioorganisms. *Analyst.* 2011;136:1539–1556. https://doi.org/10.1039/c0an01007c.
137. Mujahid A, Afzal A, Dickert FL. An overview of high frequency acoustic sensors—QCMs, SAWs and FBARs—chemical and biochemical applications. *Sensors (Switzerland).* 2019;19:4395. https://doi.org/10.3390/s19204395.
138. Gray ER, Turbé V, Lawson VE, et al. Ultra-rapid, sensitive and specific digital diagnosis of HIV with a dual-channel SAW biosensor in a pilot clinical study. *Npj Digit Med.* 2018;1:35. https://doi.org/10.1038/s41746-018-0041-5.
139. Burg TP, Godin M, Knudsen SM, et al. Weighing of biomolecules, single cells and single nanoparticles in fluid. *Nature.* 2007;446:1066–1069. https://doi.org/10.1038/nature05741.
140. Savran CA, Knudsen SM, Ellington AD, Manalis SR. Micromechanical detection of proteins using aptamer-based receptor molecules. *Anal Chem.* 2004;76:3194–3198. https://doi.org/10.1021/ac049859f.
141. Johnson BN, Mutharasan R. Biosensing using dynamic-mode cantilever sensors: a review. *Biosens Bioelectron.* 2012;32:1–18. https://doi.org/10.1016/j.bios.2011.10.054.
142. Burg TP, Manalis SR. Suspended microchannel resonators for biomolecular detection. *Appl Phys Lett.* 2003;83:2698–2700. https://doi.org/10.1063/1.1611625.
143. Alassi A, Benammar M, Brett D. Quartz crystal microbalance electronic interfacing systems: a review. *Sensors (Switzerland).* 2017;17:2799. https://doi.org/10.3390/s17122799.
144. Kim HH, Jeon HJ, Cho HK, Cheong JH, Moon HS, Go JS. Highly sensitive microcantilever biosensors with enhanced sensitivity for detection of human papilloma virus infection. *Sens Actuators B.* 2015;221:1372–1383. https://doi.org/10.1016/j.snb.2015.08.014.
145. Agostini M, Greco G, Cecchini M. Full-SAW microfluidics-based lab-on-a-chip for biosensing. *IEEE Access.* 2019;7:70901–70909. https://doi.org/10.1109/ACCESS.2019.2919000.
146. Thies JW, Thürmann B, Vierheller A, Dietzel A. Particle-based microfluidic quartz crystal microbalance (QCM) biosensing utilizing mass amplification and magnetic bead convection. *Micromachines.* 2018;9:194. https://doi.org/10.3390/mi9040194.
147. Thies JW, Kuhn P, Thürmann B, Dübel S, Dietzel A. Microfluidic quartz-crystal-microbalance (QCM) sensors with specialized immunoassays for extended measurement range and improved reusability. *Microelectron Eng.* 2017;179:25–30. https://doi.org/10.1016/j.mee.2017.04.023.
148. Papadakis G, Pantazis AK, Ntogka M, et al. 3D-printed point-of-care platform for genetic testing of infectious diseases directly in human samples using acoustic sensors and a smartphone. *ACS Sensors.* 2019;4:1329–1336. https://doi.org/10.1021/acssensors.9b00264.
149. Gauglitz G. Critical assessment of relevant methods in the field of biosensors with direct optical detection based on fibers and waveguides using plasmonic, resonance, and interference effects. *Anal Bioanal Chem.* 2020;412:3317–3349. https://doi.org/10.1007/s00216-020-02581-0.

150. Peltomaa R, Glahn-Martínez B, Benito-Peña E, Moreno-Bondi MC. Optical biosensors for label-free detection of small molecules. *Sensors (Basel)*. 2018;18:4126. https://doi.org/10.3390/s18124126.
151. Homola J. Surface plasmon resonance sensors for detection of chemical and biological species. *Chem Rev*. 2008;108:462–493. https://doi.org/10.1021/cr068107d.
152. Mechaly A, Cohen H, Cohen O, Mazor O. A biolayer interferometry-based assay for rapid and highly sensitive detection of biowarfare agents. *Anal Biochem*. 2016;506:22–27. https://doi.org/10.1016/j.ab.2016.04.018.
153. Escorihuela J, González-Martínez MÁ, López-Paz JL, Puchades R, Maquieira Á, Gimenez-Romero D. Dual-polarization interferometry: a novel technique to light up the nanomolecular world. *Chem Rev*. 2015;115:265–294. https://doi.org/10.1021/cr5002063.
154. Kim DH, Cho IH, Park JN, Paek SH, Cho HM, Paek SH. Semi-continuous, real-time monitoring of protein biomarker using a recyclable surface plasmon resonance sensor. *Biosens Bioelectron*. 2017;88:232–239. https://doi.org/10.1016/j.bios.2016.08.035.
155. Wang Y, Xue Y, Wang S, Huang J, Yang X. Real-time analysis of specific binding between apolipoprotein E isoforms and amyloid β-peptide by dual polarization interferometry. *Anal Chem*. 2021;93:1472–1479. https://doi.org/10.1021/acs.analchem.0c03542.
156. Xue Y, Wang Y, Wang S, Yan M, Huang J, Yang X. Label-free and regenerable aptasensor for real-time detection of cadmium(II) by dual polarization interferometry. *Anal Chem*. 2020;92:10007–10015. https://doi.org/10.1021/acs.analchem.0c01710.
157. Estevez MC, Alvarez M, Lechuga LM. Integrated optical devices for lab-on-a-chip biosensing applications. *Laser Photonics Rev*. 2012;6:463–487. https://doi.org/10.1002/lpor.201100025.
158. Wang DS, Fan SK. Microfluidic surface plasmon resonance sensors: from principles to point-of-care applications. *Sensors (Switzerland)*. 2016;16:1175. https://doi.org/10.3390/s16081175.
159. Lee SH, Lindquist NC, Wittenberg NJ, Jordan LR, Oh SH. Real-time full-spectral imaging and affinity measurements from 50 microfluidic channels using nanohole surface plasmon resonance. *Lab Chip*. 2012;12:3882–3890. https://doi.org/10.1039/c2lc40455a.
160. Qu JH, Dillen A, Saeys W, Lammertyn J, Spasic D. Advancements in SPR biosensing technology: an overview of recent trends in smart layers design, multiplexing concepts, continuous monitoring and in vivo sensing. *Anal Chim Acta*. 2020;1104:10–27. https://doi.org/10.1016/j.aca.2019.12.067.
161. Luo Y, Yu F, Zare RN. Microfluidic device for immunoassays based on surface plasmon resonance imaging. *Lab Chip*. 2008;8:694–700. https://doi.org/10.1039/b800606g.
162. Tokel O, Yildiz UH, Inci F, et al. Portable microfluidic integrated plasmonic platform for pathogen detection. *Sci Rep*. 2015;5:1–9. https://doi.org/10.1038/srep09152.

CHAPTER TWELVE

Microfluidics device for drug discovery, screening and delivery

Rupesh Maurya[a], Nisarg Gohil[a], Gargi Bhattacharjee[a], Khalid J. Alzahrani[b], Suresh Ramakrishna[c,d], and Vijai Singh[a,*]

[a]Department of Biosciences, School of Science, Indrashil University, Rajpur, Mehsana, Gujarat, India
[b]Department of Clinical Laboratories Sciences, College of Applied Medical Sciences, Taif University, Taif, Saudi Arabia
[c]Graduate School of Biomedical Science and Engineering, Hanyang University, Seoul, South Korea
[d]College of Medicine, Hanyang University, Seoul, South Korea
*Corresponding author: e-mail addresses: vijaisingh15@gmail.com; vijai.singh@indrashiluniversity.edu.in

Contents

Abstract

Microfluidics and lab-on-chip are two progressive technologies widely used for drug discovery, screening and delivery. It has been designed in a way to act as a platform for sample preparations, culturing, incubation and screening through multi-channels. These devices require a small amount of reagent in about micro- to nanolitre volume. Microfluidics has the capacity to perform operations in a programmable manner and is easy to fine tune the size, shape and composition of drugs by changing flow rate and precise manipulations. Microfluidics platform comes with the advantage of mixing fluid in droplet reactors. Microfluidics is used in the field of chemistry, biomedical, biology and nanotechnology due to its high-throughput performance in various assays. It is potent enough to be used in microreactors for synthesis of particles and encapsulation of many biological entities for biological and drug delivery applications. Microfluidics therefore has the scope to be uplifted from basic to advanced diagnostic and therapeutic applications.

Progress in Molecular Biology and Translational Science, Volume 187
ISSN 1877-1173
https://doi.org/10.1016/bs.pmbts.2021.07.028
Copyright © 2022 Elsevier Inc. All rights reserved.

1. Introduction

In recent times, climate changes, global warming and depleting resources of petrochemicals have significantly raised a number of concerns for sustainability toward fulfilling global demand of chemicals, fuels, medicine, drugs, food and other such commodities.[1,2] This is one of the challenging issues for scientists looking for ways to tackle it by bringing innovation in research, developing technologies and making proper use of its applications. At the same time, with increasing number of health issues, finding its treatment options is on high priority which can be resolved by developing innovative and novel technologies. Novel and potent drugs and medicines are needed to treat diseases in order to save human and animal lives. Drugs can be chemically synthesized but many of the chemicals are too complex and involve long-steps of synthesis. It can be time-consuming, labor-intensive and expensive and also non-sustainable due to mainly using raw materials of petrochemical origin. Natural metabolites or chemicals contain a number of bioactive molecules that can be produced by plants, bacteria, marine organisms and others.[3,4]

Modern drug discovery demands identification and testing of drugs. It requires optimization to increase affinity, selectivity and potency, bioavailability, and stability.[5] A highly potent drug and metabolite need to be identified from a library for its use in the treatment of diseases. Therefore, screening of drugs of high potency is one of the challenging issues due to lack of methods or assays to analyze. It can be identified by using a biosensor that allows its identification and later on its extraction and purification for further applications.[6] In order to discover and screen drugs, microfluidics and lab-on-chip technologies play a vital role to test and validate the results rapidly in a cost-effective manner. Recent advances in microfluidics and lab-on-chip have led to the discovery and high-throughput screening of drugs, bioactive molecules, metabolites and several others of this sort. It requires small volume to test and screen potent molecules. Microfluidics can be integrated with automation, thus enabling systematic drug development, screening and systematic testing.[7,8]

Droplet microfluidics is another platform used for synthesis of complex drugs with uniform size, desired features and monodisperse distribution. It can be used for fabrication of drug delivery systems including microcapsules, microparticles, microgels, emulsions and more.[9,10] Microfluidics is used for drug delivery by sustainable release of drug molecules to target locations such as cells or tissues by fine-tuning of flow-rate, size and shape.[11–13] In this chapter, we highlight microfluidics technology for drug discovery, screening and delivery.

2. Microfluidics for drug discovery and screening

Microfluidics offers a powerful platform for efficient production of functional materials, nanoparticle-based drug delivery vehicles and high-throughput drug screening. Vladisavljević et al.[14] suggested that microfluidics should be applied in several areas such as biomedicine, chemicals and nutraceuticals. It has been used for drug development of precise size, shape, dispersion and composition by fine-tuning the flow-rate and pressure. It can be used to scale-up droplets and flow-rate for drug discovery. Pang et al.[15] indicated that multichannel microfluidics platforms have the ability to rapidly screen and identify most potent drug molecules from a library. It is easy to test physico-chemical profiles, chiral separation, and enzymes/protein and metabolites analysis. Dittrich and Manz[16] recommended microfluidics platform for bioassay, separation and chemical synthesis of important metabolites. Since reduction in volume up to microliter scale does not change the nature of reactions, it can be used for drug discovery towards generating better treatment.

Currently, microfluidics is being used for drug development. The properties of drug, i.e., its composition, shape, size and pharmacokinetics can all be modified by fine-tuning the flow-rate and geometries of the chip. Cell or tissue culture can be performed using microfluidics chips for *in vitro* drug screening. This way it is very easy to perform studies in a high-throughput manner with high-rate of reproducibility at low-cost.[17] Fig. 1 depicts a microfluidics technology for manipulation of genes or production of drugs. It comprises of several steps including continuous flow, mixing and manipulation of droplet by changing flow-rate and its scale-up.[17] Damiati et al.[11] reported about microfluidics devices that enable production of stable, uniform and monodispersed particles with high encapsulated efficiency. As compared to 2D cell culture and animal models, 3D microfluidics platforms can easily mimic *in vivo* cells in an inexpensive and hassle-free manner. This allows multiplexing and high-throughput screening at cellular, organ and organism-level.

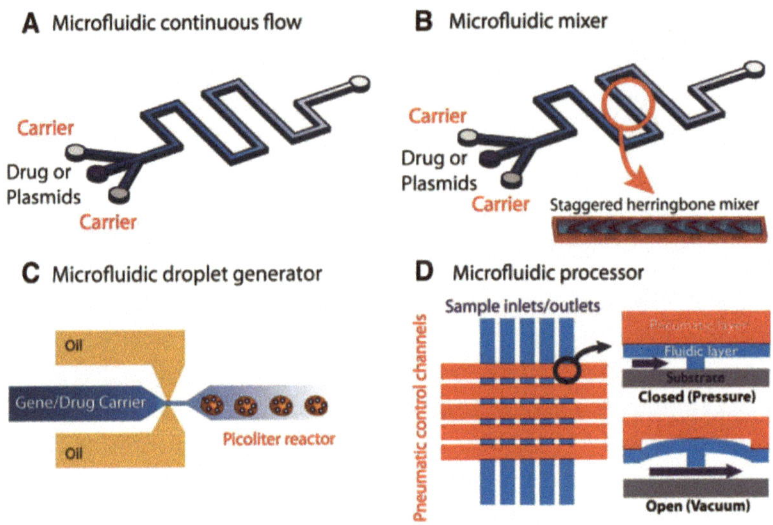

Fig. 1 A microfluidics technology for production of gene or drug. General design classes of microfluidic platforms for gene/drug carrier production. (A) A continuous-flow microfluidic fabricator for diffusion of drug and carrier. (B) Passive micro-mixer integrated with synthesizer for rapid mixing of carrier and drugs. (C) Microdroplet for production of precise control of composition and size of carrier. (D) Pneumatic microfluidics processor for production of large scale drug carrier library. *Reproduced with permission from Tsui JH, Lee W, Pun SH, Kim J, Kim DH. Microfluidics-assisted* in vitro *drug screening and carrier production.* Adv Drug Deliv Rev. *2013;65:1575–1588, (C) Elsevier.*

3. Microfluidics for drug delivery

Kuramoto et al.[18] developed microfluidics devices for rapid fabrication of small quantities of several non-viral vectors for gene therapy applications. It is permitted to manufacture multifunctional envelope-type nanodevices (MENDs) which are made up of DNA-core and lipid bilayer membrane shell. Chip design of MEND was kept simple, easy to use, rapid and cost-effective. The size of MEND depends upon the concentration and flow-rate of reaction. This study provides insight for gene therapy with required concentration of drug. Similarly, Kitazoe et al.[12] used MENDs which are currently very promising drug delivery systems. It was integrated with short mixing that allows the development and maintenance of efficient liposome-based non-viral vectors that have efficient penetration in cells. This helps to obtain higher gene transfection efficiency.

Koh et al.[19] developed non-viral vectors for gene therapy using polyethylenimine (PEI) and plasmid DNA (pDNA) complexes (PEI/pDNA). Large size distribution, poor gene transfection and low cytotoxicity were common observations in conventional methods, where vortexing was used for mixing and preparing complexes. In that study, microfluidics was used to improve the size, cytotoxicity and transfection efficiency for gene therapy. As a result, they obtained complexes that are smaller and uniform with higher cell viability and improved gene expression. In addition to that, Sant et al.[20] developed microfluidics devices for precise control of size and shape of a single drug delivery vehicle. It was developed for biomimetic gastrointestinal (GI) tract models along with integration of physiological cell types, brush-like structure and flow pattern that can be used to evaluate drug delivery efficiency. Similarly, Ghosh et al.[21] have designed extended-release GI luminal delivery systems for better administration and therapeutic use of drugs. In this system, the drug is released at prolong time period, that in turn enhances the efficacy of drug delivery. Depicted in Fig. 2 are different organ-on-a-chip for *in vitro* drug screening including liver, kidney, lung and gut that are grown on chip for modeling, mimicking and analyzing its behavior. *In vitro* cell culture is rapid and cheaper than *in vivo* models. To test drugs, mimicking human organs and tissues play a vital role. Nowadays, it is grown in lab-on-chip devices for more precise *in vitro* drug testing and screening applications.[17]

Hood et al.[22] designed microfluidics devices mediating the formation of liposomes, wherein the devices are coupled with sample preparation and drug loading chamber in a single step, while maintaining continuous flow for nanoscale vesicle synthesis with stably loaded drugs. In these devices, the large transmembrane pH and ion gradients allows introduction of drugs into the liposome. This allows the formation of drug-laden liposomes (drug: lipid molar ratio (3:1)) within 3 min. This work laid the foundation for formulation of liposomal drugs with high concentration of drug with minimal reagents waste that can be used in future for point-of-care drug delivery. Riahi et al.[13] reported about the key role of microfluidics devices in development of effective drug delivery systems. It has the capability for precise handling and small liquid transport that could emerge as an advanced drug delivery method. It has been designed and fabricated to deliver drugs directly into the target tissue.

Huang et al.[23] reported that small interference RNAs (siRNAs) for cancer therapy possesses a challenge for mediating targeted drug delivery especially in *in vivo* delivery. Microfluidics hydrodynamic focusing devices were

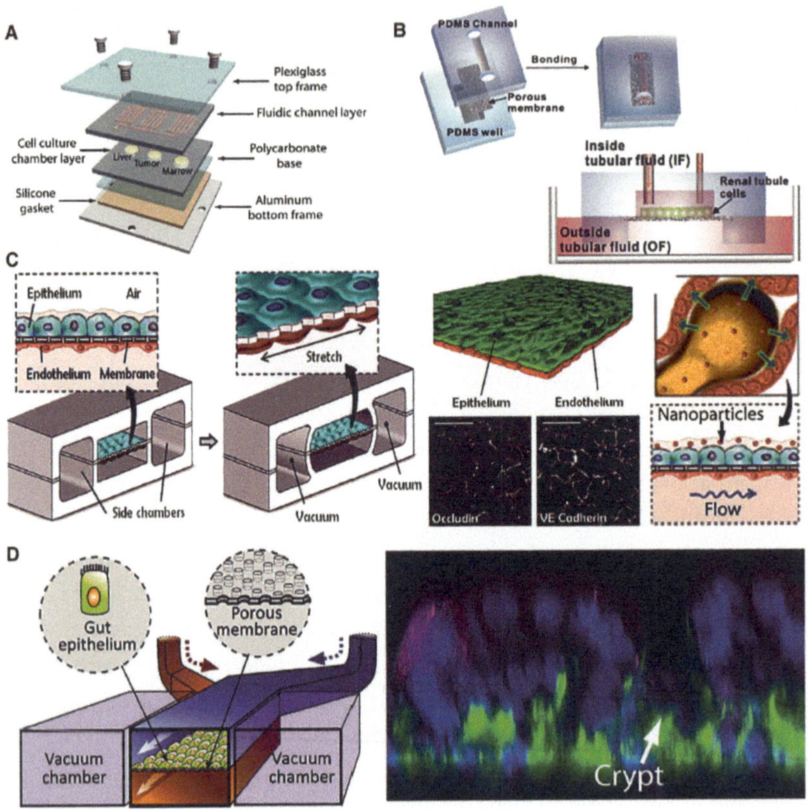

Fig. 2 *In vitro* drug screening using organ-on-a-chip. (A) Liver-on-a-chip pharmacokinetic modeling. (B) Kidney-on-a-chip for using fluid flow in renal tubules that induces cells growth. (C) Lung-on-a-chip for mimicking and form alveolar capillary barrier. (D) Gut-on-a-chip which forms villi-like structure that support microflora growth and also showed confocal cross section of tissue grown on lab-on-a-chip. *Reproduced with permission from Tsui JH, Lee W, Pun SH, Kim J, Kim DH. Microfluidics-assisted* in vitro *drug screening and carrier production.* Adv Drug Deliv Rev. *2013;65:1575–1588, (C) Elsevier.*

designed for polymer-lipid hybrid nanoparticles (P/LNPs) loaded with VEGF siRNA. Microfluidics helps to produce smaller-sized molecules and enables its distribution with low vehicle cytotoxicity and strong *in vitro* tumor cell inhibition. P/LNPs induce downregulation of VEGF mRNA and protein expression that leads to tumor inhibition. Therefore, P/LNPs produced by microfluidics is a safe and effective therapeutic siRNA delivery for *in vivo* and *in vitro* cancer treatment. Recently, Jung et al.[24] designed a chip for self-assembled drug (doxorubicin) encapsulated polymeric nanoparticles that enables higher uniformity for enhanced anticancer inhibition *in vivo* and *in vitro*. It has shown enhanced cellular uptake and higher anticancer activity with pH dependent drug release *in vitro* and *in vivo*, thus improving tumor suppression.

4. Droplet microfluidics platforms

Droplet microfluidics is related with droplets of fluidic sample in very small quantities. It allows control over dispensing and subsequent manipulation of droplets on a micro-fabricated device. Droplet microfluidics plays a significant role in performing high-throughput and sensitive screening. Most important part of this very technology is that, droplet microfluidics requires lesser volume of samples which ultimately leads to reduction in cost; also large number of samples can be screened in a day.[25,26] Several analytical techniques have been employed for detection, such as, bright-field microscopy to study dynamics of droplet generation,[27] fluorescence microscopy for 3D imaging and quantitative analysis,[28] Raman spectroscopy for monitoring droplet polymerization in devices,[29] chip lipidomics,[30] laser induced fluorescence, capillary electrophoresis, mass spectrometry, nuclear magnetic resonance spectroscopy, absorption detection, chemiluminescence and many such significant tools.[31]

Biochemical analysis is very crucial when it comes to disease diagnosis at an early stage. Analytes often exist in very low amounts or concentration and hence it is difficult to analyze the same in a short time. Droplet microfluidics has the ability of rapid and high-throughput detection. Therefore, it helps in detection of complex samples. Droplet microfluidics has been tested for quantitative and sensitive detection of rare mutation.[32] β-Lactamase is naturally expressed by *Mycobacterium tuberculosis* which serves as a marker for diagnosis of tuberculosis. But while achieving in picoliter droplet dilution, it may happen that the drops contain one or no cell. Encapsulation of a single cell in drops supports the precise detection of the cell.[33] Droplet microfluidics has also been used for detection of protein-protein interaction. It has been found that only four droplets are sufficient for quantitative detection of avidin and streptavidin in microdroplets.[34] Chen et al.[35] reported quantification of angiogenin in cow's milk in a very less amount of sample (1 nL) through this approach. It is also significant for genetic detection,[36] quantitative detection of protein expression,[37] analysis of protein-protein interaction,[38] detection of DNA, protein and ions.[35]

5. Droplet microfluidics for drug discovery, screening and delivery

As discussed above droplet microfluidics is a rapidly growing area and considered as a powerful platform for precise synthesis of novel functional drug delivery vehicles and drugs for therapeutic applications. Brouzes et al.[39] developed droplet microfluidics for high-throughput screening of single mammalian cells. Integration of microfluidics allowed encapsulation of single cells into solution-independent aqueous microdroplets (1 pL–10 nL). The arrangement was dispersed with immiscible carrier oil for manipulation of the reactor, thus enabling high-throughput cytotoxicity screening. It has been used to test high viability of encapsulated human monocytic U937 cells over 4 h. Wong et al.[40] designed droplet microfluidics for drug screening pertaining to cancer cells. A single cell was dispersed in droplets and monitored for 24 h following drug treatment, and viable cells were later examined. This system ensured a rapid and low-cost drug screening of primary cancers in 24 h and may help to design personalized cancer therapies in future.

This technology has the potential to support sustainable release of drugs to target locations using improved drug delivery vehicles. It was designed to maintain interaction between the carriers and drugs, and ultimately regulate the rate at which drugs are released. Droplet microfluidics shows great potential toward synthesis of complex drug delivery systems with uniform size, desired features and monodisperse size distribution.[9] Beneyton et al.[41] used droplet microfluidics to screen microorganisms in a high-throughput manner. The highly monodispersed volume (which is in picoliter), makes it easy to inject, fuse, split, incubate and sorted based on fluorescent intensities. A single cell of bacteria or yeast can be compartmentalized in droplet (10 nL) and easily screened for enzymes in a high-throughput manner even in very less volume. In order to pursue goals of better treatment outcomes, there are urgent requirements to be fulfilled for development of new drugs and strategies with enhanced bioavailability and specificity.

Beneyton et al.[41] reported filamentous fungi to be a good source of industrial enzymes. These fungi have the ability to produce large quantities of enzymes but suffers from low output and expensive extraction cost. In one of the works, nanolitre droplet microfluidics system was designed for high-throughput screening of enzymes from filamentous fungi library. Microfluidics allowed (i) compartmentalization of single spore in 10 nL droplet, (ii) germination and growth of mycelium, and (iii) screening of enzymes using high-throughput technology (Fig. 3).

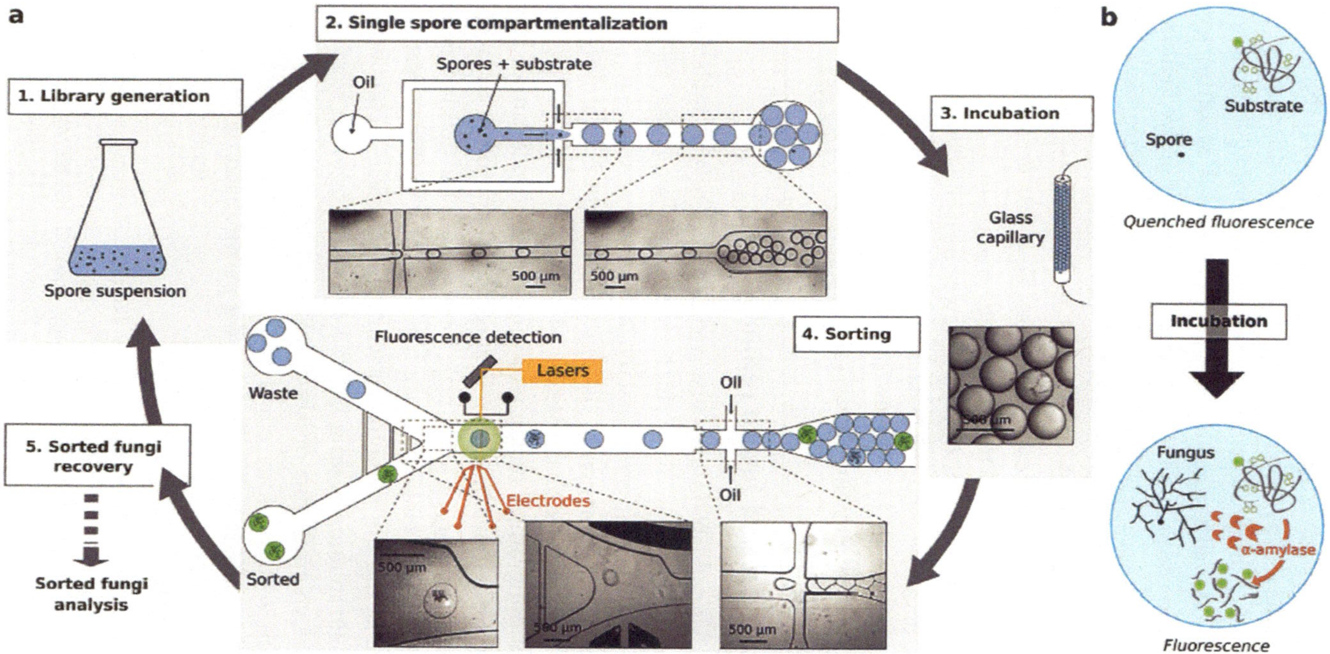

Fig. 3 Droplet microfluidics for screening. (A) Complete scheme of the system. (1) Production of a whole-genome fungal mutated library. (2) Spores were compartmentalized in 10 nanolitre droplets with a fluorogenic enzyme substrate. It was present in suitable spore to droplet ratio. (3) The emulsion was incubated for 24 h at 30 °C. Therefore, fungi can grow and produce enzyme and digest substrate inside droplet. (4) Droplet was reloaded and sorted based on fluorescent intensity. (5) Fungi recovered from sorted droplet and characterized. (B) α-Amylase fluorogenic assay where substrate was hydrolyzed and induced fluorescence. *Adapted from Beneyton T, Wijaya IP, Postros P, Najah M, Leblond P, Couvent A, et al. High-throughput screening of filamentous fungi using nanoliter-range droplet-based microfluidics.* Sci Rep. *2016;6:27223.*

Recently, de Carvalho et al.[42] developed droplet microfluidics to encapsulate nanoparticles that are coated with bovine serum albumin (NPs-BSA) to induce hydrophilic and hydrophobic nanocarriers. A hybrid microgel was synthesized with aid of droplet microfluidics using silk fibroin (SF) and chondroitin sulfate (CS) as well as alginate (ALG). The release of nanoparticles in microgel depends on composition and properties of gel. It showed encapsulated efficiency (EE) from 83% to 98% maintaining a better sustainable release that can be used for drug delivery for treating chronic diseases. Similarly, Fang et al.[43] have developed a digital droplet infusion system which is low-cost, high-precision and uses continuous flow that is converted into droplet aliquots. It has been used to achieve high-resolution of digital droplet transfer volume (57 nL) which is three-fold lower than the clinical standard infusion pumps. It can be adjusted in ranges from 0.1 to 10 mL h^{-1}. This device has the potential to be used for the next-generation drug delivery system with high-precision at low-cost.

6. Conclusion and future remarks

Microfluidics and lab-on-chip play a vital role in several applications including diagnosis, drug discovery, high-throughput screening, cell and organ growth-on-chip, *in vitro* toxicity and drug testing and many more. A pressing need has arisen to discover, screen and deliver drugs for precise treatment in translational science. Microfluidics can be integrated with automation using simple design which may also include sample preparation, identification, screening and validation to discover drugs and screen from library. Several organ-on-chip have been grown for mimicking and modeling diseases for developing future medicine and improve treatment and devise cure. It has the potential to work on interface technology to determine the right solution of drug and make use of it for better treatment. In the near future, microfluidics or lab-on-chip can be designed with automation where an artificial body could be used for testing of large scale libraries and identify the most potent drug for personalized treatment.

Acknowledgments

N.G. acknowledges the Indian Council of Medical Research, Government of India for financial assistance as Senior Research Fellowship (File No. 5/3/8/63/ITR-F/2020). The financial support from Gujarat State Biotechnology Mission (GSBTM) (Project ID: 5LY45F), Gujarat, India to G.B. and V.S. is duly acknowledged. The support from the National Research Foundation of Korea (NRF) (2021M3A9H3015389) to S.R. is also duly acknowledged. R.M., N.G., G.B. and V.S. thank Indrashil University, Rajpur, Mehsana, India for providing the infrastructure facility to carry out this study.

References

1. Neff RA, Parker CL, Kirschenmann FL, Tinch J, Lawrence RS. Peak oil, food systems, and public health. *Am J Public Health*. 2011;101:1587–1597.
2. Schwartz BS, Parker CL, Hess J, Frumkin H. Public health and medicine in an age of energy scarcity: the case of petroleum. *Am J Public Health*. 2011;101:1560–1607.
3. Dias DA, Urban S, Roessner U. A historical overview of natural products in drug discovery. *Metabolites*. 2012;2:303–336.
4. Mohs RC, Greig NH. Drug discovery and development: role of basic biological research. *Alzheimers Dement*. 2017;3:651–657.
5. Helleboid S, Haug C, Lamottke K, et al. The identification of naturally occurring neoruscogenin as a bioavailable, potent, and high-affinity agonist of the nuclear receptor RORα (NR1F1). *J Biomol Screen*. 2014;19:399–406.
6. Dincer C, Bruch R, Costa-Rama E, et al. Disposable sensors in diagnostics, food, and environmental monitoring. *Adv Mater*. 2019;31:1806739.
7. Neuži P, Giselbrecht S, Länge K, Huang TJ, Manz A. Revisiting lab-on-a-chip technology for drug discovery. *Nat Rev Drug Discov*. 2012;11:620–632.
8. Sun J, Warden AR, Ding X. Recent advances in microfluidics for drug screening. *Biomicrofluidics*. 2019;13:061503.
9. Chou WL, Lee PY, Yang CL, Huang WY, Lin YS. Recent advances in applications of droplet microfluidics. *Micromachines*. 2015;6:1249–1271.
10. Suea-Ngam A, Howes PD, Srisa-Art M, deMello AJ. Droplet microfluidics: From proof-of-concept to real-world utility? *Chem Commun*. 2019;55:9895–9903.
11. Damiati S, Kompella UB, Damiati SA, Kodzius R. Microfluidic devices for drug delivery systems and drug screening. *Genes (Basel)*. 2018;9:103.
12. Kitazoe K, Park YS, Kaji N, et al. Fabrication of functionalized double-lamellar multifunctional envelope-type nanodevices using a microfluidic chip with a chaotic mixer array. *PLoS One*. 2012;7:39057.
13. Riahi R, Tamayol A, Shaegh SAM, Ghaemmaghami A, Dokmeci MR, Khademshosseini A. Microfluidics for advanced drug delivery systems. *Curr Opin Chem Eng*. 2015;7:101–112.
14. Vladisavljević GT, Khalid N, Neves MA, et al. Industrial lab-on-a-chip: Design, applications and scale-up for drug discovery and delivery. *Adv Drug Deliv Rev*. 2013;65:1626–1663.
15. Pang HM, Kenseth J, Coldiron S. High-throughput multiplexed capillary electrophoresis in drug discovery. *Drug Discov Today*. 2004;9:1072–1080.
16. Dittrich PS, Manz A. Lab-on-a-chip: microfluidics in drug discovery. *Nat Rev Drug Discov*. 2006;5:210–218.
17. Tsui JH, Lee W, Pun SH, Kim J, Kim DH. Microfluidics-assisted *in vitro* drug screening and carrier production. *Adv Drug Deliv Rev*. 2013;65:1575–1588.
18. Kuramoto H, Park YS, Kaji N, et al. On-chip fabrication of mutifunctional envelope-type nanodevices for gene delivery. *Anal Bioanal Chem*. 2008;391:2729–2733.
19. Koh CG, Kang X, Xie Y, et al. Delivery of polyethylenimine/DNA complexes assembled in a microfluidics device. *Mol Pharm*. 2009;6:1333–1342.
20. Sant S, Tao SL, Fisher OZ, Xu Q, Peppas NA, Khademhosseini A. Microfabrication technologies for oral drug delivery. *Adv Drug Deliv Rev*. 2012;64:496–507.
21. Ghosh A, Li L, Xu L, et al. Gastrointestinal-resident, shape-changing microdevices extend drug release *in vivo*. *Sci Adv*. 2020;6:4133.
22. Hood RR, Vreeland WN, DeVoe DL. Microfluidic remote loading for rapid single-step liposomal drug preparation. *Lab Chip*. 2014;14:3359–3367.
23. Huang X, Lee RJ, Qi Y, et al. Microfluidic hydrodynamic focusing synthesis of polymer-lipid nanoparticles for siRNA delivery. *Oncotarget*. 2017;8:96826–96836.
24. Jung S, Lee J, Lim J, et al. Polymeric nanoparticles controlled by on-chip self-assembly enhance cancer treatment effectiveness. *Adv Healthc Mater*. 2020;9:2001633.

25. Guo MT, Rotem A, Heyman JA, Weitz DA. Droplet microfluidics for high-throughput biological assays. *Lab Chip*. 2012;12:2146–2155.
26. Teh SY, Lin R, Hung LH, Lee AP. Droplet microfluidics. *Lab Chip*. 2008;8:198–220.
27. Pirbodaghi T, Vigolo D, Akbari S, deMello A. Investigating the fluid dynamics of rapid processes within microfluidic devices using bright-field microscopy. *Lab Chip*. 2015; 15:2140–2144.
28. Jiang H, Zhu T, Zhang H, et al. Droplet-based light-sheet fluorescence microscopy for high-throughput sample preparation, 3-D imaging and quantitative analysis on a chip. *Lab Chip*. 2017;17:2193–2197.
29. Barnes SE, Cygan ZT, Yates JK, Beers KL, Amis EJ. Raman spectroscopic monitoring of droplet polymerization in a microfluidic device. *Analyst*. 2006;131:1027–1033.
30. Kim HS, Waqued SC, Nodurft DT, Devarenne TP, Yakovlev VV, Han A. Raman spectroscopy compatible PDMS droplet microfluidic culture and analysis platform towards on-chip lipidomics. *Analyst*. 2017;142:1054–1060.
31. Zhu Y, Fang Q. Analytical detection techniques for droplet microfluidics—a review. *Anal Chim Acta*. 2013;787:24–35.
32. Pekin D, Skhiri Y, Baret JC, et al. Quantitative and sensitive detection of rare mutations using droplet-based microfluidics. *Lab Chip*. 2011;11:2156–2166.
33. Lyu F, Xu M, Cheng Y, Xie J, Rao J, Tang SK. Quantitative detection of cells expressing BlaC using droplet-based microfluidics for use in the diagnosis of tuberculosis. *Biomicrofluidics*. 2015;9:044120.
34. Benz C, Retzbach H, Nagl S, Belder D. Protein-protein interaction analysis in single microfluidic droplets using FRET and fluorescence lifetime detection. *Lab Chip*. 2013; 13:2808–2814.
35. Chen J, Liu Y, Ye T, Xiang X, Ji X, He Z. A novel droplet dosing strategy-based versatile microscale biosensor for detection of DNA, protein and ion. *Sensors Actuators B Chem*. 2015;215:206–214.
36. Ranc TD, Zec HC, Puleo C, Lee AP, Wang TH. Droplet microfluidics for amplification-free genetic detection of single cells. *Lab Chip*. 2012;12:3341–3347.
37. Huebner A, Srisa-Art M, Holt D, et al. Quantitative detection of protein expression in single cells using droplet microfluidics. *Chem Commun*. 2007;(12):1218–1220.
38. Srisa-Art M, Kang DK, Hong J, et al. Analysis of protein–protein interactions by using droplet-based microfluidics. *Chembiochem*. 2009;10:1605–1611.
39. Brouzes E, Medkova M, Savenelli N, et al. Droplet microfluidic technology for single-cell high-throughput screening. *Proc Natl Acad Sci*. 2009;106:14195–14200.
40. Wong AH, Li H, Jia Y, et al. Drug screening of cancer cell lines and human primary tumors using droplet microfluidics. *Sci Rep*. 2017;7:9109.
41. Beneyton T, Wijaya IP, Postros P, et al. High-throughput screening of filamentous fungi using nanoliter-range droplet-based microfluidics. *Sci Rep*. 2016;6:27223.
42. de Carvalho BG, Taketa TB, Garcia BBM, Han SW, de la Torre LG. Hybrid microgels produced via droplet microfluidics for sustainable delivery of hydrophobic and hydrophilic model nanocarriers. *Mater Sci Eng C*. 2021;118:111467.
43. Fang Z, Li AI, Liu H, Pan T. Digital droplet infusion. *Lab Chip*. 2021;21:502–512.

Index

Note: Page numbers followed by "*f*" indicate figures and "*t*" indicate tables.

V

Z

FSC
www.fsc.org
MIX
Papier | Fördert
gute Waldnutzung
FSC® C083411

Kolibri 360 GmbH
Ferdinand-Jühlke-Straße 7
99095 Erfurt, Deutschland
produktsicherheit@kolibri360.de